PENGUIN BOOKS

BLUEPRINTS

Maitland A. Edey began his career as a writer and editor at *Life* magazine, went on to become editor-in-chief of Time-Life Books, and has devoted recent years to writing. He is married to Helen Edey, a retired physician, and lives in New York City and on Martha's Vineyard.

Donald C. Johanson is the paleoanthropologist who discovered "Lucy," and is Director of the Institute of Human Origins in Berkeley, California. The Institute is assisting the Royal College of Surgeons in raising funds to endow Charles Darwin's home, Down House, in Downe, Kent, England.

BLUEPRINTS

SOLVING THE MYSTERY OF EVOLUTION

Maitland A. Edey
Donald C. Johanson

PENGUIN BOOKS

PENGUIN BOOKS
Published by the Penguin Group
Viking Penguin, a division of Penguin Books USA Inc.,
375 Hudson Street, New York, New York 10014, U.S.A.
Penguin Books Ltd, 27 Wrights Lane, London W8 5TZ, England
Penguin Books Australia Ltd, Ringwood, Victoria, Australia
Penguin Books Canada Ltd, 10 Alcorn Avenue, Suite 300,
Toronto, Ontario, Canada M4V 3B2
Penguin Books (N.Z.) Ltd, 182–190 Wairau Road,
Auckland 10, New Zealand

Penguin Books Ltd, Registered Offices:
Harmondsworth, Middlesex, England

First published in the United States of America
by Little, Brown and Company, 1989
Reprinted by arrangement with Little, Brown and Company
Published in Penguin Books 1990

3 5 7 9 10 8 6 4

LIBRARY OF CONGRESS CATALOGING IN PUBLICATION DATA
Edey, Maitland Armstrong, 1910–
Blueprints: solving the mystery of evolution/Maitland A. Edey,
Donald C. Johanson.
p. cm.
Includes bibliographical references.
ISBN 0 14 01.3265 1
1. Evolution. I. Johanson, Donald C. II. Title.
QH366.2.E26 1990
575—dc20 89–71054

Printed in the United States of America
Designed by Jeanne Abboud

To Don Cutler
without whose encouragement
and to William Phillips
without whose patience
this book would not have been published

Contents

Acknowledgments

We wish to thank Tim Berra of Ohio State University, Mansfield, for a thorough reading of our entire manuscript and for the many useful suggestions he made for improving it; Vincent Sarich of the University of California, Berkeley, who also read the entire manuscript and in particular gave us valuable guidance on how primate relationships are traced through analysis of blood proteins; Carl R. Woese of the University of Illinois, Urbana, for his remarkable patience in steering us through the complexities of his specialty, bacterial evolution; Concetta Clarke of the Hampton Library, Bridgehampton, New York, for her ingenuity and persistence in running down hard-to-find books and magazines, also for help in locating picture sources and with our bibliography; and finally our editor, Debbie Roth, whose sensitive and perceptive touch has marked every page of our manuscript.

BLUEPRINTS

Introduction

What is the most profound question that human beings can ask about themselves? It has to be: Where do we come from? That leads, of course, to: Where does all life come from? Don Johanson and I have both been interested in that question for many years. I, as a journalist, have written about the origins of humankind. Don, as a collector of their fossils, has helped develop our knowledge of our extinct ancestors. His most important contribution to date has been his discovery in Ethiopia of the partial skeleton of an erect-walking creature that was more than three million years old and scarcely more than three and one-half feet tall. It was not an ape, but it was not yet a human either. And it was a female. Popularly known as "Lucy," it has become the most famous fossil of a human ancestor ever found.

Those ancestors—near-humans, pre-humans, almost-humans—along with ourselves and some cousins that never made it and became extinct—are collectively known as hominids. They have two things in common. They all walked on their hind legs like human beings, and they all had their origins in Africa. After Don had found Lucy and some others of her same type, he and I wrote a book: *Lucy: The Beginnings of Humankind.* It traced the story of hominid evolution. Now we were meeting to discuss what to do next.

"Do you realize," said Don, "that nearly half the people in the United States don't believe in evolution?"

"I know it's a big figure," I said.

"We consider ourselves a literate society. We think we're educated. We believe in science. We base our culture on technology. Technology *is* science; it's the practical application of it, to learn about the world and put that knowledge to work."

"So, what's new about that?" I asked.

"Nothing. What bothers me is that some people try to make science selective. You can't do that. You can't accept one part of science because it brings you good things like electricity and penicillin and throw away another part because it brings you some ideas you don't like about the origin of life."

"You're talking about Creationism."

"I am. How can so many people in this country, supposedly brought up to think scientifically, to do experiments in school and college, to admire and reward its doctors, its astronomers and biologists — how can they honestly believe in the sudden creation of life on earth? How can they believe that it all happened in a flash less than ten thousand years ago?"

"Because it's in the Bible. A lot of people take the Bible literally."

"Yes, but — "

"There are people in this country, people who have gone through high school and college, who think the earth is flat."

"Come on."

"They do. There's an active Flat Earth Society here in the United States. They can find statements in the Bible to back them up."

"That's not the way they should read the Bible," said Don. "By the way, which Bible are they reading? There's the King James Version; there's the Revised Standard Version; there's the Douay Bible. There are a lot of Bibles and they don't all say exactly the same thing. Monks have been saving bits of the Bible for nearly two thousand years, scratching down copies of it, making mistakes. The Bible isn't literally sacrosanct, for goodness' sake."

"Some people think it is. They think you shouldn't monkey with the hand of God."

"But it's not God's hand. It's the fumbling hands of a lot of scribes. It's our best effort to state the word of God. And as our knowledge and scholarship grow, our interpretations of the Bible should grow accordingly. We should be humble enough to realize that we don't get everything right the first time."

"We should," I said, "but in fact it sometimes works in reverse.

Creationists not only believe firmly in the Adam-and-Eve version of life's start, but they actively try to suppress any competing scientific version of it. Try finding a good course on evolution in a Bible-Belt high school. Look at some of the biology books being handed out there. Evolution isn't even mentioned."

"Well — " Don drew a deep breath. "Somebody should do something about that."

—And we were off.

We had told Lucy's tale by piecing together clues from a number of scientific disciplines: geology, botany, anatomy, climatology, molecular physics, and others. The cement that held those clues together was, of course, fossils. The story of human origins that is emerging today from those fossils, after a century of false starts, inspired hunches, hopeful expeditions, marvelous finds, spectacular surprises that led to hasty rewriting of the script, is a tribute to the scientific method working at the very edge of its power.

Apes into humans! Our first book looked at the bare bones of that story, the WHAT of human evolution, but it said little about a larger and more interesting question: HOW or WHY it happened.

"We should get into that," I said.

"You want to do a cram course on evolution?"

"Why not? It's the most startling idea the human mind has had to grapple with, ever. What would you compare it to?"

"Well, you could start with Copernicus. He got rid of the idea that the earth was at the center of the universe, that everything revolved around it."

"Okay, give Copernicus a B-plus."

"Then Isaac Newton should get an A-plus. He figured out how everything is held together, why the earth revolves around the sun. That was a pretty big idea. Or you could jump to Einstein and relativity, and all the people who came after him and are still wondering how the universe itself got started. That's an even bigger idea."

Don was right. Those are enormous ideas. But they all come out of the human brain, a brain that somehow had to be organized from the chaos of matter and energy that makes up the rest of the universe. That story — how life itself emerged from the random whizzing of particles, and how we know as much about that process as we do today — is without doubt the greatest intellectual detective story of all time.

For practical purposes, that story starts about two hundred years ago, with the beginnings of modern science. It builds on the discoveries and insights of scores of people. It has been fiercely argued all along the way. Today its general outlines are accepted by all respectable scientists. They still fight over the details, but the grand scheme seems firmly locked in place—for everyone except the Creationists, who, in the face of a Himalaya of evidence, still deny the possibility of evolution.

That denial is not only a pity but also a danger, for our knowledge has progressed to a point where we can begin thinking about directing our own evolution through genetic engineering. Starting about now, we will increasingly be in a position to decide what we ourselves will be like. Armed with that power, shouldn't we understand the process? Shouldn't we know something about its history, its potential? Don and I think so. For if we don't understand evolution how can we possibly deal with it? That is why Don and I decided to write this book. If we can open up just a few minds we will feel rewarded.

Intimations: Many Questions But No Good Answers

On the mountains of truth you can never climb in vain: either you will reach a point higher up today, or you will be training your powers so that you will be able to climb higher tomorrow.

—NIETZSCHE

Many scientific theories have, for long periods of time, stood the test of experience, until they had to be discarded owing to man's decision, not merely to make other experiments, but to have different experiences.

—ERICH HELLER

Astrology fosters astronomy. Mankind plays *his way up.*

—G. C. LICHTENBERG

1

Six Who Helped Lay the Groundwork

It did not all begin with Darwin, as many people think. He had some help.

"Who was it," Don asked, "who said: 'I stand on the shoulders of giants?' Was it Isaac Newton?"

"I don't know," I said.*

"I think it was Newton. Anyway, it's a great line. Nothing comes out of nowhere in science. Even Darwin didn't just jump out of the woodwork."

"He came pretty close," I said.

"Not that close. People were beginning to wonder more and more about evolution and the origin of life in the century before Darwin. He was standing on a lot of shoulders."

"Then we should examine those shoulders," I said.

That led to a series of arguments between us as to whose work might have been essential for Darwin to have done what he did. In the course of them we were struck by how close some of those earlier men had come to hitting the evolutionary bull's-eye. They

* I subsequently learned that Bernard of Chartres said (in about 1100): "We, like dwarfs on the shoulders of giants, can see more and farther, not because we are keener and taller, but because of the greatness by which we are carried and exalted." Newton may have drawn his inspiration from Bernard.

observed certain things. They had brilliant ideas about them, but none succeeded in pulling all those ideas together to form a scientifically acceptable explanation of how evolution worked. Darwin did. And he did need some shoulders — in four areas that had constrained the earlier speculators.

Four big problems: the first was the problem of the age of the earth. Evolution takes time, and according to the Bible there wasn't any. Just before 1650 a learned Irish archbishop, James Ussher, decided to calculate the age of the earth from evidence in the Bible. Working his way back through all the "begats," he concluded that Adam and Eve had been created by God in 4,004 B.C. Other scholars confirmed his finding. They even improved on it, pinpointing the actual moment of human creation at 9:00 A.M. on Sunday, October 23. This date thereafter assumed a heavy weight of authority. It became a baseline for measuring earthly time, a point at which scholarship and science met and blended with Church doctrine. Later geological evidence would have a hard time dislodging it.

The second problem had to do with the possibility of change among living forms. Because of a general acceptance of the biblical account of Creation, almost everybody took it for granted that species were "fixed." They had been created by God in their own shapes and could not change. That idea would turn out to be even harder to dislodge.

The third problem was actually basic to the other two. While they both turned on specific issues brought up by contradictions between the Bible and scientific observation, this third one hinged on the nearly universal acceptance of *all* Scripture as sacrosanct. Furthermore, not only was the Bible not questioned, ***it should not be!*** To do so would cause the questioner severe social risk, and even possible physical harm. That put a real constraint on scientific inquiry, although there were risk takers here and there. Darwin's own grandfather, for one, speculated about evolution in print, but his speculations were cautious and veiled. Few others dared go even that far.

Among the more daring souls was Voltaire. By holding up all kinds of conventional ideas to ridicule, he challenged smug minds to look at themselves. That made it pretty hot for him at home, and he spent considerable periods of his life outside France. At about the time he died the French Revolution broke out. It was brutally egalitarian, ruthlessly anticlerical. Its effect on scientific thought elsewhere was predictable. Many people regarded the excesses of the revolution as the result of too much dangerous freethinking in France and blamed men like Voltaire. There was a backlash against science, as a dubious logical sequence emerged, running something like this: if you are against the Revolution and its mindless atrocities, then you must support the things that the Revolution is trying to destroy. Since it is trying to destroy faith in the Bible, that faith must be supported. Anything that undermines it should be regarded with hostility.

Since the work of scientists had already begun raising serious questions about some of the Bible's most fundamental statements, many men — often without the slightest suspicion that their scientific premises were being warped by preconceived ideas — were still trying to practice science from a platform of biblical belief. Most of Darwin's immediate predecessors, and some of his learned contemporaries, found themselves in a predicament.

The fourth problem constraining early speculations about evolution was much simpler and would be more quickly solved. It sprang from the confused and imprecise knowledge that naturalists had about living organisms. It would be disposed of simply by collecting and classifying specimens. The collector did not have to bother himself with speculations about how those specimens came into being. Increased knowledge was enough of a goal in itself, and it gave rise to a healthy collecting industry in the eighteenth and nineteenth centuries. Darwin himself would be caught up in it as a young man.

Having identified those four problems as the greatest roadblocks to the construction of a workable theory of evolution, Don and I then sat down to hammer out a list of names of the men we thought had done the most to clear those roadblocks away. Most historians of science, we believe, will agree with our choice of the following men:

Linnaeus	1707–1778
Buffon	1707–1788
Hutton	1728–1799
Lamarck	1744–1829
Malthus	1766–1834
Cuvier	1769–1832

With the exception of Malthus, whose ideas about population growth echo ominously today on an ever more crowded planet, these are not exactly household names. But it is through the lenses of these men's minds that the noble modern structures of geology, biology, and evolution should be viewed. We can do no better, in attempting to acquaint ourselves with the climate of thought into which Darwin was born, than to note just what each of these spectacularly gifted men did. Chronologically it is proper to start with the first two; it is proper ideologically also, for they could be called the organizers of what would follow.

Although born in the same year, the first two could not have had more different starts. One was a Swede of modest origins. The other was a titled Frenchman who, through his aristocratic connections would become head of the Royal Botanical Garden in Paris; his full name was Georges Louis Leclerc, Comte de Buffon. The Swede's true name was Carl Linné, better known to us as Carolus Linnaeus, a Latinization that he bestowed on himself. It is difficult now to comprehend the enormous renown that both enjoyed during their lifetimes. They had much in common. Both had powerful intellects and immense industry. Both got started early on careers in natural history. Both specialized in botany. Both rode the same surging wave of curiosity that was sweeping over Europe, an avid desire for explanations and descriptions of the natural world. Both wrote books of epic influence. Both had sad ends.

CAROLUS LINNAEUS: "I WILL SORT AND NAME MANY THINGS."

LINNAEUS (1707–1778) overcame his rather small beginnings. He managed to get himself to Holland where he studied botany as a young man. Back in Sweden, he began organizing and naming specimens of plants. Organization was desperately needed. The outthrust of colonization and exploration that all the European countries were engaged in, and the impact of exotic animal and plant specimens that were consequently pouring into port from ship after ship were not only sending naturalists into paroxysms of excitement but were also drowning them. The Western world was caught up in a passion for naming things. Every man was putting his own label on his own specimens according to his own ideas of how they should be categorized. Nobody was listening to anybody else. The relationships of entire groups of organisms were in chaos.

Tarpon were different from toads—it didn't take a very sharp taxonomist to figure that out. But what about the many different kinds of toads? Into what sort of useful arrangement could they be put? What about worms; were all worms just different examples of the same basic worm model? Looking at worms, a man could be engulfed by them unless he understood something that is well understood today: worms are not all basically alike. There are fundamental differences among many of them, as great as the differences between a chicken and a rhinoceros. Until matters like that were clarified, the natural sciences could not be pursued effectively.

Stop a moment to consider what the lack of a well-organized catalog means in trying to understand the nature of living forms and their relationships. Every high school student today who enters a biology class faces a crystal-clear order in which all living things fit. The paramecium to be seen through a microscope, the frog to be dissected, the mouse to be bred—all have been assigned logical places on the multibranched tree of life. Two hundred years ago there was no such clarity. The water was still muddy when Darwin was a boy. Growing up, he was surprised to be told by an eminent Dr. Holland, who would later become physician to Queen Victoria, that whales were cold-blooded. Darwin knew better, but like a well-bred young man he held his tongue.

It was gross misconceptions like Dr. Holland's, and a thousand others, that Linnaeus decided to devote his life to correcting. To do it he needed an orderly system of labeling, and he invented one. Named after him and in universal use today, it is the Linnaean binomial system of nomenclature. He made mistakes, of course, many of them; they are still being corrected. But the system itself endures. It is so simple and so fundamental to all of natural history that most beginning students in biology don't even stop to consider where it came from, any more than they do the filter papers they use or the agar cultures on which they grow bacteria.

What Linnaeus did was to give two Latin names (hence binomial) to every living organism. The first was its *genus* name, the second its *species* name. A genus is a group, large or small, of similar creatures that obviously are very closely related, but contains within the group individual types that are consistently different from one another. Those different types are species. Linnaeus then organized his genera into larger units that were alike in some more general respects, then into still larger ones, and so on.

A good example of Linnaean classification could start with one of the most beautiful of American songbirds, the scarlet tanager, *Piranga olivacea.* Those two names label a particular kind of bird, just as my two names label me: Edey (the small group of my closest relatives), Maitland (my own specific name). Similarly *olivacea* describes a single kind of tanager that falls within a closely related group of five others that make up the genus *Piranga:* the tanagers of North America. The beauty of the Linnaean binomial system is that it permits those five very closely related birds to be related more loosely to a larger group comprising all the tanagers of the world. They — dozens of them — make up the next larger Linnaean group, a *family.*

The world's tanagers resemble each other more closely than they do other families: chickadees, thrushes, or blackbirds, for example. To get all of them together, a still larger group, an *order,* must be recognized; it includes all smallish so-called perching birds. But the order of perching birds falls far short of taking care of all birds. There are eagles, storks, ostriches, hummingbirds, ducks — each as different from the others as they are from the perching birds. Together they comprise a *class.* Into that class falls everything on earth that has feathers and lays eggs.

The Linnaean system, as used today, does not stop there. It puts birds into a still larger group, a *phylum,* noting that they have one important attribute in common with certain other creatures that are not birds: a backbone. All these backboned creatures fall into the phylum of Chordata, and within it is found a subphylum of vertebrates which includes four other classes of backboned animals: fishes, amphibians, reptiles, and mammals. The scarlet tanager, in short, shares with shark, bullfrog, rattlesnake, and human a common backbone with a brain at one end.

Since there are creatures without backbones — insects and octopuses, for example, and others without brains either — clams, jellyfish, and amoebas, it is clear that other phyla must be created to accommodate them. Science today recognizes twenty-six such phyla. Together these phyla make up the *kingdom* of multicelled animals. A similar kingdom, proceeding from species through genus, family, order, class, and phylum, was established for the world's plants.

The task that Linnaeus set himself was the organization within his binomial system of this avalanche of living forms, concentrating on his speciality: plants. In his day there were far fewer species known, but to balance that, the known ones were not nearly as well understood as they now are. The characteristics that locate them securely today—often something as obscure as a small but telltale skeletal peculiarity in a bird, or the reproductive apparatus of a flower (rather than the flower's apparent relationship to another flower because it crudely resembles it in color and shape)—have been slowly but steadily worked out. The work continues on seven continents (for there are now known to be insects and lichens inhabiting the continent of Antarctica). Thousands of men and women have devoted their lives to it. According to the entomologist E. O. Wilson, 1.7 million species of living organisms have been described; there may actually be as many as fifty million insect species alone.

Linnaeus, who put this colossal effort in motion, appeared suddenly on the scientific stage when he was only twenty-eight, upon the publication of the first edition of his book on classification, *Systema Naturae.* It would go through many editions during his lifetime as new species and new insights required. Although the first edition of his work was only 142 pages long, the sixteenth consisted of three volumes and ran to 2,300 pages as Linnaeus was flooded with material from all over the world. It became the ambition of collectors everywhere to nibble off a crumb of fame by being mentioned by Linnaeus. His influence was enormous and persisted well after his death. It was the echo of it that propelled men like Darwin, and a contemporary, Alfred Russel Wallace, to go on collecting expeditions to remote places.

Two ironies mark the life and work of Linnaeus. The first is that the master namer would suffer a stroke which so damaged his brain that at the end of his life he could no longer remember his own name, let alone the thousands he had bestowed on other organisms.

The second is a more subtle one. Without the orderly arrangement of types that Linnaeus pioneered, later insights into evolution could not have been made. Linnaeus himself started his career with a stout belief in the fixity of species. But the longer he worked, and the more variation he observed among the specimens he was trying to classify, the more he began to doubt that. Later editions of his work omitted earlier statements about fixity. Now for the irony: while he may have begun to suspect that some vague process like evolution was at work deep in the recesses of nature, his *names,* those labels that were beginning to produce a miraculous order in the great mess of living things, had the opposite effect. They put a heavy stamp of approval on the concept of the fixity of forms. Label something, stick a pin through it, and it ceases to wriggle about on the shelf or in the mind. It becomes static, embedded, as in amber. That appearance of fixity would persist into the next century and make the acceptance of evolution more difficult.

GEORGES LOUIS LECLERC, COMTE DE BUFFON: "I WILL DESCRIBE AND EXPLAIN EVERYTHING."

GREAT as the ambition of Linnaeus was, that of the Comte de Buffon (1707–1788) was even greater. He set out to describe the entire world, its origins, and everything in it, ending up with a forty-four-volume encyclopedia of natural history, *Histoire Naturelle, Générale et Particulaire,* that was translated into other languages as fast as the volumes came out. It was the most wide-ranging and influential scientific work of its century, and by far the most popular, for it combined elegantly written descriptions and life histories of enormous numbers of plants and animals with discourses about astronomy, the age of the earth, and the processes of life.

As others were beginning to do, Buffon perceived that species were not fixed. He noted the success of stockmen and fruit growers in changing and improving their breeds by selecting the best ones for further propagation. He observed that pigeon fanciers were producing types that not only bred true but were different from anything found in nature. He summarized all this with a statement that deserves to be underlined, proposing that:

> *Each family, among the animals as well as among the plants, comes from the same origin, and even that all animals are come from but one animal, which, in the succession of eras . . . has produced all the races of animals that now exist.*

Reading that, written nearly a century before Darwin, makes one shiver. What a mind Buffon had, and what sparking leaps it took. He noted—as would Malthus later—that life multiplied faster than food, and that that meant a struggle for survival. He also noted that there were differences among individuals in a species. Darwin would put those ideas together to find a mechanism for evolution. Buffon never quite did, although he danced all about the subject in the most tantalizing way. His growing conviction that species were not fixed led him to the bold conjecture (quoted previously) that they were probably related in the remote past. Seeing the need for a remote past, if such relationships were to be traced through slow change, he provided one. He calculated that the earth, hot at birth, had become sufficiently cool to support life at least 70,000 years ago, then went on to propose that in another 70,000 years it would have become so cold that life would vanish. Such speculations about changes in temperature carried him a step further, to a suggested solution to the puzzle of fossils: there were indeed extinct types, and they had become extinct because of the cooling of the earth.

As one who sought to put the universe in order, Buffon, of necessity, had to become a classifier himself. He rejected Linnaeus's system, even going so far as to insist that the binomial Latin names of Linnaeus be put on the underside of the labels of his specimens in the Royal Botanical Garden, where he would not see them. His own system of animal classification was preposterous. He graded animals according to how useful they were to humans, and started with the horse, "the noblest conquest man has ever made." Foolish as that now seems, it does make clear what a strong hold the idea of the centrality of humans in the cosmos had on humans themselves—before Darwin.

We see in Buffon an extraordinarily restless imagination trying to grapple with too many huge ideas at once. They jostled each other in a deep green sea of information, which became deeper the more Buffon learned. Way down there they were dimly seen, their edges unclear, their shapes shifting as they drifted through his mind just beyond the penetration of any illuminating shaft of a unifying theory. There they swam, fathoms below the clear upper currents of conventional thought: the world was young, species were fixed. Buffon believed the opposite: the world was old, species changed.

The tragedy of Buffon was that he did not hold tightly to this view. Professors at the Sorbonne, affronted by his heretical statement about species quoted on page 14, examined Buffon's *History* more carefully and came up with a fourteen-point indictment against him.

He immediately recanted: "I declare that I had no intention of contradicting the Scriptures, that I believe most firmly in everything they say about the Creation, both as to order of time and matter of fact. I abandon everything in my book respecting the formation of the earth, and in general all that may contradict the account of Moses."

In evaluating Buffon, one must realize that he was a courtier, in a world where the court was everything. Men spent their lives scheming to get close to the throne or to its chief ministers. Only in that way could patronage and privilege be obtained. Buffon managed it by catching the attention of the king's mistress, Madame de Pompadour. She diverted herself and her royal patron by dabbling in the universal craze for knowledge about the natural world. Her salon became peppered with men of science and learning, to whom fashionable men and women could listen, and with whom they could stretch their fashionable minds with fashionable talk about fashionable subjects. It was there that Buffon shone, with his soaring speculations about the earth's origins and his absorbing discourses about the life histories of animals and plants. He was an elegant young man, an exquisite dresser and an impressive speaker.

As long as Pompadour was up, so was he. He got his appointment as director of the Royal Botanical Garden through her influence. When she was down—and she fell catastrophically at one point in her career when the king grew tired of her—Buffon was without a patron. That explains his inelegant haste to scramble back to respectability upon the indictment from the Sorbonne. Later, with Pompadour restored to favor, Buffon's career resumed its majestic course. He was elected to the Academy of Sciences, France's highest intellectual honor (Pompadour again). He married an heiress and became very rich. Volumes of his *History* continued to pour out—using a near-assembly-line system, with assistants now working for him.

As Buffon grew older, his approaching end grew sadly clearer in his work. He continued his mountainous efforts but their quality deteriorated. He had never been a truly scrupulous scientist but rather a polymath: a collector, organizer, writer, publisher, self-promoter, fountain of ideas, and public great man. Today he would have been a hero on television. Flaws in his work were revealed as tougher scientific minds picked on him. Eventually he became a sort of national monument, a mausoleum, but one without its explanatory sarcophagus: a magnificent facade, empty inside. He suffered a crushing blow when his adored son, whom he called Buffonet, failed in an important appointment, making it clear that the father's influence at court was gone. Worse, Buffonet's pretty young wife was debauched by the Duc d'Orléans, and thereafter by many others. She became a public scandal. But up to the moment of his death Buffon himself remained one of the revered men of France. Kings called on him. A statue to him was erected in the Royal Botanical Garden. Medals were plastered on him like scales on a fish. Rousseau, on a visit, kissed the floor of his house. Twenty thousand people came to his funeral.

Then it all unraveled. The Revolution erupted within a year of Buffon's death. His statue was thrown down by the revolutionary mob. He was scarcely cold in the ground when his remains were dug up and thrown away, in the words of Will Durant, "by revolutionists who could not forgive him for having been a nobleman, and his son was guillotined."

That son would have the last word. His murderers are forgotten, but he is remembered today for what he said as the blade was about to drop: "Citizens, my name is Buffon." Nothing else. To many Frenchmen that meant something. Buffon's legacy is his *History,* which opened the minds of thousands of people to the possibility of looking at the world and its creatures in new bold ways. That he, a careerist, retreated expediently from his boldest propositions in no way detracts from his having thought of them, or from the influence they had. Conventional thought would not be the same again.

Conventional thought, both lay and scientific, from the sixteenth century on into the nineteenth, had as its center of biological doctrine a concept known as the *Great Chain of Being.* It was science's way of accommodating the known world with the Bible and was regarded as the organizing principle for all life. It was totally static, having at its core a divine act (or acts) of universal Creation. Living things were linked according to an immutable master plan, in a logical progression of forms from the simplest single-celled creatures to increasingly complex multicelled ones and culminating in the noblest of all: human beings. There was no need for time in this scenario, nor any possibility of change. The ape was the next highest thing to humans, and their closeness could be viewed without alarm because there would be no suggestion of a relationship other than that arbitrarily stamped on both by the Creator. That one could be descended from the other, or that they might have a common ancestor, was unthinkable. How could they? They were as they were and always had been.

The protean speculations of Buffon notwithstanding, most natural historians conducted their studies in the shadow of the Great Chain of Being. They read Buffon, of course, and were variously stimulated or shocked, but they were under no compulsion to take his ideas seriously because he did not go about nailing them down with argument and example, as one would a theorem in geometry. They were tossed into his *Natural History* here and there as inspired speculations. Buffon, in fact, banged and chopped in too many places. It would take a much more concentrated kind of picking to cause serious cracks in the facade of the Great Chain of Being. Those would come first from geology, from a Western world increasingly filling with alert-eyed men looking at the earth's crust and trying to make some sense of it.

The majority of those who were doing the looking were not scientists in the presently accepted meaning of the word. Outside of such disciplines as astronomy, mathematics, chemistry, and optics—many whose practitioners enjoyed the patronage of kings, who wore exceptional men as badges of their own brilliance—science scarcely existed as an acceptable way of passing one's time in the seventeenth and eighteenth centuries. It was certainly not something by which an ordinary man could make a living or toward which a hopeful father would aim a son. Consequently those who were interested in what would become science—particularly the natural sciences—tended to be gifted amateurs. They were mostly men of means and education, brushed by an increasing awareness of the complexity of a world just being brushed in its turn by exploration and technology. Much of true science lay just ahead; these men would create it. They were largely self-taught, propelled by curiosity. More than one English gentleman or country parson, with an orderly life and time on his hands, filled it up with botany or collecting. Many of these men traded specimens and information. They were great letter writers.

Such a one was the Reverend John Ray (1627–1705). He became

a fine botanist and even anticipated the work of Linnaeus by attempting a crude classification system of his own. He is notable today for his efforts to square biblical doctrine with alarming nuggets of evidence that did not fit that doctrine. One was the impression of a fossil fern in an extremely old rock. Either the fern itself was older than the accepted date of Creation or its intricate pattern had been created entirely by chance by geological forces. He struggled to accept the latter view, but common sense eventually persuaded him it could not be. Later some stumps from a long-gone forest buried beneath the silt of a Belgian estuary and now raised again—how many centuries or hundreds of centuries later—brought him up short. Suddenly the enduring surface of the earth became erosive and insubstantial. Mountains rose and fell. Time stretched. Ray reassured himself by noting: "Whatever may be said for ye antiquity of the Earth itself and bodies lodged in it, ye race of mankind is new."

Another man was nudged by time in quite a different way. Edward Lhwyd lived in a rugged part of Wales whose valley floors were covered with large boulders. He paid no attention to them; they had always been there. But one day another huge stone came crashing down from the mountainside. Lhwyd inquired around and learned that this was the first event of its kind to have occurred within the memory of anyone alive in those valleys. If it took fifty or a hundred years for one stone to fall, how long would it take for thousands? And there were thousands; he could see them.

Individual oddities like these cropped up increasingly to plague the minds of clear-thinking observers. Gradually the conviction grew that the world might indeed be old. One way to explain such oddities, and still have them conform to biblical doctrine, was to assume a series of great catastrophic events, each of which could have caused worldwide rumpling and devastation of the earth's surface, the most recent one being Noah's Flood. This theory of catastrophism became widely popular in the eighteenth century and was used to interpret many anomalies. Under it, for example, the Flood, in one wrenching, sucking torrent, could have accounted for all of Lhwyd's boulders, strewing them like seeds across the valley floors to sit there until Lhwyd came by to count them. The one that fell could have been the *only* one to fall in a thousand years. That kind of reasoning did not satisfy the gimlet minds of a few, but it satisfied most.

Catastrophism had its subdivisions and its dissenters because it had difficulty in accounting for all geological conditions. There was the problem of sedimentation, of staggering amounts of material obviously laid down over staggering amounts of time by water action — material of all different kinds: mud, clay, gravel, sand, more mud. A theory grew that all this sedimentation had taken place *before* the emergence of dry land, when the earth was entirely covered by a sheet of water. The continents had appeared later when the water drained down into deep subterranean caverns. This theory was known as Neptunism, getting its name from the Roman god of the sea. It gave a plausible explanation for sedimentation but failed to show how or why the water went, or where the caverns were. Like the Flood itself, Neptunism receded and catastrophism grew in strength.

JAMES HUTTON: "THE WORLD IS OLDER THAN YOU THINK."

CATASTROPHIST ideas did not, however, satisfy James Hutton (1728–1799), the third of the six heroes of this chapter. Hutton was a brooding Scot with a philosophical turn of mind. Attracted to mathematics and logic, and dazzled by the lucidity of Newtonian law, he saw no reason why its kind of rigor should be limited to the movement of celestial bodies. He decided that the earth could be examined in the same way. He became the world's first systematic geologist and provided science with the first scheme for earth processes that sensibly explained all of its phenomena.

His own ruthlessly reasonable mind made Hutton impatient with anything that was not logically provable. He turned his back on catastrophist or Neptunist doctrines because the results they were supposed to explain were based on assumptions that could not be verified. In their place he put an entirely new principle that came to have the name of uniformitarianism: a long word for a very simple idea.

Uniformitarianism says one thing: that what is now taking place on the earth's surface is no different from what has always taken place. The processes are alike, that is, uniform. The reason we are not apt to notice them is that they are slow and our lives are short. However, if we go looking for them we find that they do occur. Shorelines change; what was a seaport in the time of the Pharaohs is now twenty miles inland. Rocks split from the sides of mountains as a result of rain and frost. They tumble into valleys, breaking and crumbling as they go. They become gravel, then sand. We see it happening everywhere. Rivers become brown in spring floods, carrying unmeasurable tons of material downstream and ultimately into the sea. Every pebble that falls from Everest is a tiny episode in the life of that mightiest of all mountains, fated some day to dwindle and vanish.

What Hutton saw all around him was a universal falling, a sliding and flattening, but slowly—oh, so slowly. As Loren Eiseley put it: "So preternaturally acute was his sense of time that he could foretell in a running stream the final doom of a continent."

But if everything falls and is ground small, there must be some counteracting force, an uplift to balance all that falling. Hutton found it in the heat of the earth, in the heavings and bucklings of its crust caused by that heat, in the spewing of volcanoes. He noted that the material of the earth's crust consisted of two kinds of rock: sedimentary, the kind that was washed *down* as mud and sand and eventually compacted into rock again by pressure; and igneous, the kind that was forced *up,* often in molten form, from the earth's interior.

Having found two balancing forces, Hutton could then sit back and icily view a world "with no vestige of a beginning, no prospect of an end." Just a ceaseless churning, the same processes working over and over to eternity. Hutton was quick to emphasize that uniformity of process was no guarantee of uniformity of result. Different conditions, different combinations of conditions, would produce different results. But the processes themselves were unwavering.

Hutton's bleak view, reposing on a bed of limitless time, was so far removed from the cozier constructs of all but a few of his most attentive readers that he died without really having dented catastrophism. But for those who did attend, he had stretched the age of the earth in a way that made them blink. Because of the slowness of the geological events he was observing, the world simply had to be old. Uniformitarianism would be picked up by a much better known geologist, Charles Lyell. Lyell's books would accompany Darwin on the voyage of the *Beagle.* Without those books, it is conceivable that the theory of evolution could have eluded him, for Darwin would need time, and Hutton had supplied that.

JEAN BAPTISTE LAMARCK: "SPECIES CHANGE. THEY DO IT BY STRIVING."

WHETHER Hutton's speculations about endless geological process ever crossed the Channel to lodge themselves in the brain of a young French soldier is not known. But it is certain that Buffon's less bold, but still imperious, stretch of biblical time did enter that brain.

The soldier was Jean Baptiste Pierre Antoine de Monet, Chevalier de Lamarck (1744 – 1829). He was the son of a minor baron, one of a long line of hardy patriotic warriors living in a threadbare chateau in Picardy. Lamarck's home has been described by Donald Culross Peattie as the kind of place "where the servants have not been paid for an age, but there is always money for a new saddle or a new blade." If one's blood was blue, one could always try to improve one's chances by hanging out at court, but this does not seem to have occurred to Lamarck's family. Often the most patriotic are the least politically adept and the least rewarded; the Lamarcks were always hard pressed. When his father died, Jean, age seventeen and the youngest of eleven children, went off to fight the Germans. He distinguished himself in battle, drifted to Paris, lived for a while on a tiny war-disability pension, got interested in natural history, and wrote a book on botany—as who didn't at the time in France.

Lamarck's book, *Flore Francaise,* was a great success. It was a kind of semiamateur field guide to French plants and made no attempt to follow Linnaean classification. That, of course, was no drawback in the eye of that Linnaeus-scorner Buffon, who hired Lamarck as a tutor to his son Buffonet. Lamarck grew up scientifically under the wing of Buffon and eventually outstripped him. He must surely have concluded that Buffon's lordly pronouncemental approach to the universe left something to be desired, for he followed a quite different course himself.

Like all the emerging natural scientists of the time, he realized the importance of classification. He understood the absolute necessity of a system like Linnaeus's, and set out to reorganize one largely overlooked group, the marine invertebrates. That was typical of

Lamarck. He was an industrious, modest man, not cut to the courtier's cloth at all. He escaped the Terror, presumably through sheer unobtrusiveness. His selection of marine invertebrates was in character — they were almost universally regarded as the least interesting of God's creatures. He also began beating the drum, softly but persistently, for a systematic organization of things in museums, which heretofore had been horrible jumbles of specimens. He was the first to propose sorting specimens out and displaying them according to class and order so that visitors could get an inkling of the relationships between living forms. When a new National Museum of Natural History was formed by the revolutionary government in 1793, Lamarck was made head of the marine invertebrate section.

It is not likely that any of the new scientists roaring up the back of the Revolution, pushing to get to the top of the new anthill, could have recognized in marine invertebrates — those flabby, static, supremely dull, and almost universally ignored creatures — an opportunity to make a brilliant contribution to science. Lamarck made several. His first accomplishment was to sort them out sensibly. He started from scratch, tossing out the nearly indecipherable mess left by Linnaeus, whose interest in and comprehension of worms, clams, and starfish had been negligible. Lamarck created entire new groupings of them, arranging them in relationships that are still largely respected today. He is not remembered now as a classifier, but in his day he was one of the best.

A great bonanza for him was the richness of marine invertebrate fossils. Embedded in the rocks were far more of them than all the other kinds of fossils put together. The reason, of course, was that most marine invertebrates had hard shells, had existed for enormous amounts of time on the ocean bottom, and had accumulated there in fossil form by the billions. Some sediments, under the microscope, revealed themselves as consisting almost entirely of the remains of innumerable tiny marine organisms.

As he became more familiar with that wealth of fossil life, Lamarck realized that it was possible to make geological inferences about the rocks they were found in from the nature of the fossils themselves. In time he could tell an ancient shoreline, with its shallow-water types, from a deep sea bottom. That technique is commonplace today, but was unknown before Lamarck. Almost by accident he became a competent geologist. As it had for others, time grew in him. He threw aside Buffon's once-bold estimates as too constricting and began thinking, like Hutton, of measureless expanses during which there would be opportunity for the kind of change he was beginning to perceive. For he had noticed that existing species seemed to be linked to similar, possibly ancestral, ones lurking in those ancient rocks. "Time," he said, "is insignificant and never difficult for nature. It is always at her disposal and represents an unlimited power with which she accomplishes her greatest and smallest tasks."

Gradually Lamarck came to envision an improvement of the Great Chain of Being. Instead of being static — created all at once in all its variety, and changeless thereafter — it became a moving staircase on which there was a constant upward progression of forms. At the bottom were the simplest ones continually coming into being from the primordial ooze (in a way he did not attempt to explain, leaving that to the Creator). At the top were those that had been in existence longer and had had more time to approach perfection.

Why did they change? What was the motor that drove them toward perfection?

Lamarck, in accepting the ideas of Buffon and others that the world was old, deduced that conditions on its surface would necessarily have undergone considerable changes over great periods of time. Therefore all species, if they were to survive, would have to adapt to those changes. According to him they did it by learning and by striving, by constantly trying to adjust, and, in the process, altering their shape and their behavior. The classic example, usually cited to illustrate this idea of Lamarck's, is the giraffe, which, by repeatedly stretching its neck to get at the tender leaves at the tops of acacia trees, actually manages to develop a longer neck during its lifetime. Another example would be a weight lifter who, by constant exertion, develops extra-large arm and back muscles.

Lamarck did not stop there. He went on to a second deduction: that the longer neck of the original giraffe was passed on to its descendants. In other words, whatever useful changes in the length of its neck the giraffe managed to acquire during its own life would show up in a slightly longer-necked offspring which, during *its* life would also be a neck stretcher and bequeath a still longer neck, and so on, resulting finally in the otherwise inexplicable creature that now roams the African veldt. This idea, in itself, was far from new, but by tying it to a general principle of evolution, Lamarck gave it a new spin.

Lamarck's brilliant two-point proposal — the first carefully reasoned theory of evolution ever to be propounded — is now known as the theory of the inheritance of acquired characteristics. It drew a mixed reaction. To some it came as a revelation because it appeared to solve a number of problems for which there was no other answer. It took care of the problem of extinct fossil forms: they had failed to adjust fast enough or successfully enough and had perished. It took care of the coexistence of apes and humans: the apes were still stuck in the forest and were so busy trying to cope with the challenges of living there that they hadn't the time to progress further. It even addressed itself to a still more perplexing matter: that sameness and change seemed to exist together in all creatures. Dogs always produced dogs, cats always produced cats. And yet both, when carefully bred, could produce stranger and stranger kinds of dogs and cats. This, according to Lamarckian principle, could also happen in a state of nature. The whippet, the terrier, and the deerhound could each explain itself by constant striving to excel in a particular life: as a courser honing itself in generation after generation of effort to run down rabbits, as a scurrier to catch rats, as a powerful bringer-down of large game. And, by putting Lamarck in reverse, the disappearance of organs could be accounted for by failure to use them. An example would be the blind fish found in dark caves.

There has been speculation about whether Darwin's grandfather, Erasmus Darwin, who hit on the idea of acquired characteristics at about the same time as Lamarck, borrowed from Lamarck or whether Lamarck borrowed from him. The best evidence seems to be that the idea was in the air because of the growing preoccupation with the riddle of life, and that both arrived at it independently. Whatever the case, Erasmus Darwin's proposal was no more than an audacious idea cautiously concealed in a long poem, "The Temple of Nature," which he wrote in 1802:

Hence without parents, by spontaneous birth,
Rise the first specks of animated earth. . . .
Organic life beneath the shoreless waves
Was born and nursed in ocean's pearly caves;
First, forms minute, unseen by spheric glass,
Move in the mud or pierce the watery mass;
These, as successive generations bloom,
New powers acquire, and large limbs assume;

Whence countless groups of vegetation spring,
And breathing realms of fin and feet and wing.

The message was clear for anybody who bothered to pry it out of the poetry: life originated in the deep ocean slime and then, by degrees of increasing complexity, proceeded to fill the earth with all its forms. Other verses of the poem suggested that apes, by constant use, produced hands with opposable thumbs and that some of them, by the same kind of aimed intent, developed brains and an upright stance and became men. In short, if one needed the neck of a giraffe for reaching, one could gradually achieve it over many generations of trying.

Concerned about a possible backlash from the Church, Erasmus Darwin never went beyond burying his ideas in verse. He made no effort to give evidence for them. Not being a scientist, perhaps he was unable to. But Lamarck *was* a scientist. He did publish and he did document, and his name has come down to us as the promulgator of the idea of the inheritance of acquired characteristics.

If Lamarck had been right, he would have been as famous today as Darwin is. Unfortunately, he wasn't. Natural science was already beginning to move out of the pleasant realm of collecting, sorting, and speculating and was going on to the more austere country of testing the speculations. And it was easy to put Lamarck to the test: cut the tails from rats, breed them, and then cut the tails from the offspring. This was done for scores of generations. But the rats obdurately kept appearing with normal-sized tails, as did several breeds of dogs which had had their tails bobbed for centuries and their ears trimmed so that they would stand up smartly. Like the rats, they persisted in appearing with long tails and droopy ears. Closer to home was a three-thousand-year experiment in the circumcision of Jews. Shakespeare — though he was talking about something quite different than circumcision — unwittingly delivered a comical blow to Lamarck when he wrote: "There's a divinity that shapes our ends, rough-hew them how we will."

GEORGES LÉOPOLD CHRÉTIEN FRÉDÉRIC DAGOBERT, BARON CUVIER: "FOSSILS ARE REAL."

RIGHT or wrong, Lamarck's ideas failed to provoke anything like the storm that Darwin's would half a century later. For one reason, all of

France was gripped by the struggles of Napoleon to conquer Europe. For another, there was Lamarck's own failure to promote himself as skillfully as his early patron Buffon had done, or as skillfully as a slightly younger man, Georges Cuvier (1769–1832), would do. Finally, much of Lamarck's evidence was derived from the squashy little marine creatures in which he had immersed himself for many years and in which nobody took much interest.

Nobody, that is, except Cuvier. He had made some studies of jellyfish and was invited to continue his work at the Museum of Natural History in Paris. There Cuvier quickly branched out, embarking on a lifework that would establish him as one of the great anatomists of his century and would credit him with the invention of two new sciences. One was comparative anatomy. Whereas Lamarck had been interested primarily in sorting out and classifying whole organisms, Cuvier focused on the different parts of the organisms themselves. He *compared* them feature by feature, organ by organ, bone by bone. He established that every part had to function as a harmonious piece in the overall machine. All meshed for a single purpose: to allow the animal to live the life it led. He refined his ideas by proposing a theory, the "correlation of parts," which recognized that essential meshing. A certain kind of toe went with a certain kind of foot, and they in turn had something to say about the nature of the limbs and backbone — and so on. Cuvier went further: he said that the way an animal behaved governed its overall shape and the shape of every bone in its body.

That emphasis was exactly backward, said another French anatomist, Etienne Geoffroy Saint-Hilaire. He insisted that the shape governed the way the animal behaved. This was awkward, for Saint-Hilaire was the senior man; he had invited Cuvier to Paris to work with him. That did not deter Cuvier for a moment. A man of devouring ambition, a master diplomat, one of the handsomest men of his day, and iron-willed, Cuvier was ruthless in ridding himself of opposition. He demolished Saint-Hilaire in a famous confrontation over classification. As to their basic difference, neither man was entirely right — or wrong. Form and function were not as separable as they thought but are now understood to be interrelated. But Cuvier was also on safer ground ideologically; fixity of species was the prudent view. His comparative studies of living forms had made him a believer in it.

That put Cuvier in opposition to Lamarck. Here again he came out ahead. He was a far flashier and more forceful man, and he had far flashier fossil specimens than Lamarck's dreary marine worms. Cuvier's were the bones of huge mastodons and other even stranger creatures that had vanished from the face of the earth. He became extraordinarily skillful at interpreting them and fascinated both science and the public with his ability to reconstruct an entire animal from a few of its bones. In doing so he invented another science: paleontology.

That monsters from the past should have existed at all and could be reassembled by Cuvier, almost as if by magic, was a sensation. He established beyond further doubt that strange animals had inhabited the planet in the distant past and no longer did. One might think that Cuvier had himself constructed an argument for evolution, but his detailed anatomical comparisons had convinced him of the opposite. How could species be anything but fixed, since all their parts were shaped to make them fit only for the one thing they did? He explained their disappearance by falling back on catastrophism; the extinct types were the unlucky ones, swept away in one or another of a series of cataclysms. Others had survived in remote places and had migrated back; they were the ones that now inhabited the earth.

Considering his great skills as an anatomist and paleontologist, it is curious that Cuvier failed to acknowledge the evolutionary evidence that was staring him in the face. It is equally curious that he made another important discovery without apparently noticing its evolutionary significance either. He observed that the older a fossil was, the less closely it resembled living forms. He ignored the implications of this and turned his energies to the demolition of Lamarck and his totally unacceptable idea that species did change.

In this effort Cuvier was largely successful. By that time the careers of the two men had diverged dramatically. Lamarck, quietly pottering along, had drifted into obscurity and then into near oblivion. He became blind, had to give up his post at the museum, abandon teaching, and finally live on the charity of an impoverished daughter until his death in 1829. Cuvier, by contrast, soared upward. He became universally recognized as the greatest anatomist of his time. He entered politics and was elected to the Council of State. It was with a seigneurial flourish that he condescended to deliver a eulogy to his former colleague, blind, poor, forgotten, and now dead. But he need not have savaged him the way he did.

Some years before, Lamarck had taken a potshot at catastrophism, and that apparently still rankled. Cuvier, referring to a whole series of imagined Floods—which he needed to explain the disappearance of so many fossil species—had previously written: "Life in those times was often disturbed by these frightful events. Numberless living things were victims of such catastrophes; some, inhabitants of dry land, were engulfed in deluges; others, living in the heart of the seas, were left stranded when the ocean floor was suddenly raised up again; and whole races were destroyed forever, leaving only a few relics which the naturalist can scarcely recognize."

Lamarck's observation must have stung him: "A catastrophe that regulates nothing, that mixes up and scatters everything, is a very convenient way of solving the problem for naturalists who wish to explain everything but do not take the trouble to observe and investigate what actually happens in nature." To Cuvier that must have sounded as if he were being told that he didn't know anything about geology and hadn't even bothered to go out into the field to educate himself.

Whatever the case, when Cuvier stepped up to deliver his eulogy he began by paying tribute to scientists who adhere strictly to truth. He then quickly excluded Lamarck from that company, suggesting that he belonged with another group entirely, one of dreamers who have built "vast edifices on totally imaginary bases, like those enchanted palaces of our old romances, that can be made to vanish into thin air by exploding the magical idea on which their very existence depends." Having buried Lamarck as a competent scientist, he then tossed a last shovelful of dirt on his evolutionary ideas by remarking that if their baseless assumptions were accepted, then "only time and circumstance are needed for the . . . polyp to end up by transforming itself, gradually and indifferently, into a frog, a swan, or an elephant."

Lamarck's idea exactly. Why did Cuvier tread on it so viciously? At this remove we can only speculate. The simplest answer, and the one we have already suggested, is that Cuvier was a vain and vengeful man who, truly believing in catastrophism, could not resist a final spearing of a man who had opposed it. More interesting is the possibility that he may have begun to suspect that catastrophism was wrong. It is hard to believe that a scientist as experienced and intelligent as he was, particularly one working as brilliantly as he did in the field of fossils, would not have. But blind obduracy is not unknown in science. A person becomes wedded to a set of ideas and builds a career and a reputation on them. In time they become so much a part of the self that attacks on them are not tolerated. In the case of a very powerful scientist — and Cuvier was one — the sense of being able to "control" conflicting evidence begins to color judgment. When the evidence becomes intractable the scientist moves to destroy it. That is why the suspicion remains that Cuvier may have sensed more truth than he ever admitted in Lamarck's dismissal of catastrophism and his espousal of a theory of evolution. Otherwise, why the reaction? It is inexplicable, and so it seemed later to the French Academy of Science. It refused to print Cuvier's eulogy in the form in which he had presented it.

A cholera epidemic, Paris's first, swept the capital in 1832. It carried off Cuvier, and catastrophism with him. He was the last scientist of the first rank to support a concept that could no longer stand up to the snowballing evidence of a more sophisticated geology.

Lamarck fared better. As more and more evidence began to accumulate that species might not be fixed — evidence from fruit growers, stock breeders, and pigeon fanciers all over Europe — his idea became more plausible. Forget about the rats' tails; that was something done *to* them, not something they did themselves. There sprang up a debate as to whether the controlled breeding that took place on farms and in aviaries could be duplicated in nature. In other words, could there be *some* kind of selection among wild animals, spurred by their need and striving? Those who favored fixity said no. Those who favored flexibility said why not? As that argument rumbled along, the spark of Lamarckism continued to glow. It would be extinguished only by a better idea proposed by Darwin.

THOMAS MALTHUS:
"LIFE IS A STRUGGLE. ONLY THE FITTEST SURVIVE."

THE last of the six eighteenth-century men who would affect the evolutionary ideas of the nineteenth was not a natural scientist like the other five, but an early example of what would now be labeled a social scientist. He was Thomas Malthus (1766–1834), an English clergyman with a powerful concern for the downtrodden. Not that other clergymen were not concerned, but their reactions were different from his. They were parochial. They went about doing good works, preaching uplifting sermons, attending to the poor, interesting the local squire in rescuing the destitute. This, if Malthus had been living today, would have been described by him as a Band-Aid approach: patching sores. He wanted to get at the source of the sores. Like many of his countrymen he was appalled by the ferocity of the Revolution that had taken place across the Channel: as most of Europe saw it, the direct result of an incredibly listless and self-indulgent aristocracy paying no attention whatsoever to the masses. Economic reform, the rights of man — those explosive new social philosophies — had been born into the French inequity. Although they failed to divert its fatal drift, they did bring into being for the first time in Europe some careful thinking about the total human condition. It was against this background that Malthus published his *Essay on the Principles of Population.*

Trained in mathematics, Malthus noted a sinister relationship between the amount of food available anywhere and the number of mouths crowding the table. Animals, he pointed out, had a fearsome fecundity. No matter how fast the food supply went up, the supply of eaters went up faster. He reduced this to a formula: populations tend to increase geometrically; food supplies increase arithmetically. The result: an enormous and constant oversupply of eaters; a steady threat of starvation; populations ultimately controlled by starvation. That meant, according to Malthus, that there would be a continuous struggle among the eaters for the available food. Only the strongest in that struggle would survive.

Malthus's principle, so easily understood, so widely observed, caught on instantly. Life *was* a struggle. The weakest puppy in the litter was shoved from the nipple. The smallest hatchling in the nest was trampled by its fellows. Among caterpillars that stripped a bush bare in a day, the last to get there would shrivel. The feckless farmer would lose his property and be cast adrift. Even among the weeds jostling in the garden, the strongest choked the weakest.

As a prophet of doom for the human condition, Malthus fell into disrepute among sociologists and demographers in the twentieth century when it was observed that food supplies were growing more rapidly than populations in the developed countries. But he has been revived again in recent decades. Neo-Malthusian doctrines abound today on a planet whose population has roared past five billion and will hit six billion by the year 2000.

This explosion of people has been made possible by the opening up of new food sources, by intensive crop yields from better and better plant strains, by the heavy use of chemical fertilizers, by the mechanization of farming, by the tilling of increasingly large amounts of marginal land, and by health programs that cut down on infant mortality. To critics of Malthus, the people explosion is a natural result of the technological strides of the twentieth century. It is not something to worry about because further improvements in technology, *more* food, and deliberate self-regulation of their numbers by people practicing birth control will prevent the planet from ever becoming intolerably crowded.

Nonsense, say the neo-Malthusians. Malthus's model may have been a crude one but it is basically correct. In the most crowded parts of the world, where the standard of living is lowest, are found the highest birth rates—and mass-scale starvation. Marginal farmland destroys itself rapidly through erosion and dust bowl blowaways. Tropical forests cannot be cleared for agriculture because of the thin soil and heavy rainfall. Indeed the arable land of the planet is decreasing at the rate of thousands of acres a day—at just the time that the advanced countries are realizing that there is not an endless supply of petroleum, the stuff of which artificial fertilizers are made and on which they depend for high crop yields. When the oil runs out, what will those six billion people do? More to the point, what are nearly half of the present five billion doing right now? They are already dangerously underfed.

This is not the place to argue Malthus out into the twenty-first century. The subject is immensely complex. But it does remind us of an elementary biological truth that Malthus put his finger on: eaters tend to overproduce themselves unless regulated in some way. His way—famine—now seems to be a rather simplified partial answer to a question that behavioral and ecological studies have revealed to be far more intricate.

What came out of Malthus, and went marching forward to where it would be picked up by Darwin, was the sense of struggle: the survival of the fittest. That concept made a deep impression. For one reason, as noted above, it was obvious to all. For another, it suited the social attitudes of the upper classes in Europe. This was particularly true in England, and most particularly true in the England of the nineteenth century. For it was then that England was most energetic in carving out a world empire from "lesser breeds" of "inferior" culture and "inferior" skin color. To the British upper classes who were directing that huge colonial effort, Malthus obviously had it right. They were the fittest, and they were the survivors. They had proved their fitness by surviving, and their survival was explained by their inherited fitness. This is a neat circular argument that actually proves nothing. But it would be seized on by followers of Darwin who wished to pervert his theories to their own ends. This perversion would come to be known as social Darwinism. However much damage social Darwinism may have done, it is still to Malthus's enduring credit that he was able to broaden the shoulders — already made broad by Linnaeus, Lamarck, and those others — that Darwin would stand on.

When we pause to consider the immense amount of knowledge we have today about biology and the earth sciences, when we survey the panoply of instruments and laboratory techniques available to us for testing our ideas about the relationships of creatures, their true structure, and their tiniest functions, we become almost overwhelmed by the complexity of science. By contrast, the accomplishments of the eighteenth century suddenly seem rather simple, almost pathetically primitive.

They were anything but. When we are unsure of basic principles, we must make sure that we have hold of some rather elementary facts if we are ever to understand the principles. We have to make an orderly start somewhere. It has been said that the first step in science is to know one thing from another. Is it important for an understanding of the universe to know a daisy from a buttercup? Linnaeus thought so. And so it has proved to be. Out of his efforts came the present orderly arrangement of all living things, a profound accomplishment.

Equally profound was the wrenchingly difficult abandonment of the belief that the world was only a few thousand years old. That short, neat box of time gave way grudgingly to the inexorable findings of geology. The biblical "days" of Creation became figurative; they stretched. Three Frenchmen, Buffon, Lamarck, and Cuvier, all understood that. And if one of them, Cuvier, became entangled in the net of catastrophism, it was at least a catastrophism that looked back beyond Noah's Flood to another age deeper in time, and to another deeper yet, and yet another. It was not for him to count them all.

Time, in short, had at last been unshackled, intuitively by those three and in a logically demonstrable way by the Scot Hutton. We talk today about "gospel" truth as the truest kind of truth, the kind we swear by, the kind we know in our bones to be true, the last truth of all that we are willing to give up. It was that kind of truth, a universal belief shored up by all tradition, all prejudice, all faith, that had to be unraveled and restated. By the early nineteenth century it had been.

As to the fixity of species, that was still largely unresolved. Buffon didn't believe in it, did, didn't. Lamarck definitely didn't, although his explanation was not entirely satisfactory. What is important is that he turned the idea that species were not fixed into something worthy of respectable discussion. He had expressed a thought that had lodged in other minds and would have to wrestle there with another truer-than-truth belief: that living things had not changed materially since their creation. Lamarck had succeeded in bending that belief but did not break it.

Cuvier, as far as we can tell, did believe strongly in the fixity of species, although he was the one who, more than anybody, should not have. In his work with fossils he saw that ancient forms were different from living forms and that the older the fossils were, the greater those differences became. Furthermore he was able to demonstrate that some species had indeed become extinct. By doing so he shattered another truer-than-true belief: that extinction could not be. How did he miss the bigger truth? We will never know.

Did Cuvier read Malthus? Again we do not know, nor do we know if he would have changed his mind had he done so. However, it is fair to say that, given all the scientific advances that were made during the late eighteenth and early nineteenth centuries, and given the debate that they engendered even when the ideas behind them were far from established, there was a key still needed to unlock the riddle of species change. That key, although he could not have imagined it, was in the hand of the English clergyman, Thomas Malthus.

A Theory at Last: Charles Darwin and the Origin of Species

That man can interrogate as well as observe nature was a lesson slowly learned in his evolution.

—SIR WILLIAM OSLER

Nature has neither kernel nor shell; she is everything at once.

—GOETHE

How do we distinguish the oak from the beech, the horse from the ox, but by the bounding outline? . . . Leave out this line. . . . all is chaos again.

—WILLIAM BLAKE

2

The Voyage of the *Beagle*. First Suspicions about Change. Years of Lonely Labor.

The scientific and intellectual climate in England during the fifty years between Charles Darwin's birth and the publication of the *Origin of Species* is absolutely fascinating. That period was one of excruciating mental turbulence to all conscientious thinkers, as old convictions were besieged by new ideas, and two conflicting mind-sets found themselves on a collision course. The old conviction — belief in the Bible as *the* source of information about the age of the earth and the stability of species — was taken so much for granted by nearly everyone that it was not even a cause for discussion in any but narrow scientific circles. It was the familiar, comfortable, apparently impregnable intellectual fortress on which the Crown, the Church, the stability of Empire, the dependability of society, and the respectability of individuals and their thinking depended.

The other conviction — growing out of a mass of geological data — was making the first conviction increasingly indigestible to scientists. By 1809, the year in which Darwin was born, most geologists had accepted the idea of an old earth, although they continued to argue about catastrophism and Neptunism, about the number of catastrophes, their causes, and their duration. Most of the arguing took place out of the public eye. It is putting it too strongly to say that there was a deliberate conspiracy of silence among scientists on matters of biblical deviationism. Rather, it was a matter of prudence: we have important work to do; let's get on with it and not involve ourselves in public disputes that will only make life awkward for us. As a result, the new geological knowledge — by no means secure yet because of ongoing internal disagreement — was developed among specialists, disseminated in papers read before learned societies, discussed in correspondence and with students. All this went largely unnoticed by the public, which could not have cared less whether one dull rock specimen was older than another even duller one and certainly would have been put to sleep by the arguments pro and con, which were duller yet.

If geology was dull, and its implications ignored, questions about the role of the Deity as Creator were not. They were perceived as threatening to established order, and scientists who entertained them tended to keep their thoughts to themselves. Those scientists were not jumping at shadows; they could see heretical men being punished. When Darwin was two years old, the poet Shelley was expelled from Oxford for his pamphlet *The Necessity of Atheism.* Not one natural scientist stood up to support him, although there must have been quite a few who secretly did — if atheism meant not believing in the literal account of the Creation. The ideas of Lamarck had crossed the Channel by that time, and the seed of evolutionary thought, already discreetly planted by Darwin's grandfather Erasmus, was now taking firmer root in minds responsive to the more carefully thought-out argument that Lamarck gave to it. But it did not burst into flower; it still lacked a compelling rationale. Furthermore, unlike geology, which would gain acceptance through the passage of time and through being uncontroversial, the mutability of species was explosive in its implication. It cast doubt on the all-pervading belief that God had created humans.

One of the basic concepts of modern evolution theory is that it is not purposive. It does not aim itself in any direction but rather is a process of adapting to circumstance, of seizing opportunities offered by changes in the environment. In a word, evolution is unpredictable. A scientist looking at a fossil fish — armed with the theory but with no knowledge beyond fishes — could in no way predict a future mouse, raccoon, or kangaroo. A relationship can be worked out by hindsight but not by foresight.

Similarly, the career of Charles Darwin can be reviewed in retrospect so that it hangs together logically. Even so, the idea that a particular little boy, the rather ordinary son of an English country doctor, should have been the one who would revolutionize the concept that had been held of human origins, and the origins of all life, still seems — after all the study, all the examinations of his career, his thoughts, his notes, his letters — wildly unlikely.

Darwin, it is true, grew up with two assets not available to most boys. First, his family was wealthy, which meant that he could, as an adult, do more or less what he wanted. Second, he had a truly extraordinary grandfather. Erasmus Darwin was an English doctor whose reputation was such that he was asked by King George III to come to London and take up the post of Royal Physician. Erasmus Darwin declined. That would have meant giving up many other interests — and he had many and was passionately involved in them. He was a poet, a mathematician, a philosopher, a strongly liberal pursuer of human rights. Above all, he was interested in the natural sciences. He was by nature a freethinker, a foe of ecclesiastical dogma. He corresponded with many other men of humanistic and scientific bent, among them the geologist Hutton. Hutton's revelations about the age of the earth may well have stimulated his razor-sharp mind, already curious about the nature of life, to the Lamarckian speculations (quoted in the previous chapter) that he published in poetical form shortly after 1800.

Erasmus Darwin was an imposing man in every respect. He was tall and heavily built and in later life grew so fat that a semicircle had to be cut out of his dining table so that he could get his belly up to it. He was forceful in character, devastating in argument, and compelled his son Robert to follow him into medicine. The family was inordinately proud of the achievements of Erasmus Darwin and kept his memory green. There is no question that this huge man had an effect on the thinking of his grandson, even though he died seven years before Charles was born.

Charles's father, Robert Darwin, at 300 pounds, was also an imposing man. He lacked the brilliance and intellectual curiosity of Erasmus, but he, too, became an extremely successful physician. He practiced in the town of Shrewsbury in Shropshire, where he raised a family of six children, the next to the youngest being Charles. His wife, a sister of Josiah Wedgwood, proprietor of the famous pottery works, died when Charles was only eight. As a result, Charles was largely brought up by his older sisters. His father comes down to us as a stern, autocratic man whom Charles saw little of. Biographers have emphasized the coldness of this relationship in their efforts to explain certain aspects of Charles's character, and particularly the attacks of chronic illness that plagued him throughout his adult life. But blaming it all on the father is unfair. Robert Darwin was a typical English father of his time and class, the no-nonsense head of a large household who expected to get his way and got it. He may have intimidated young Charles, but no more than other fathers of the day did their own sons. Charles himself seems to have felt nothing but admiration and affection for his father. He also was strongly attached to his uncle Jos Wedgwood and to his Wedgwood cousins, a lively and stimulating group. There were close bonds between the two families. Charles spent much of his youth visiting the Wedgwood home at Maer, only twenty miles from Shrewsbury, and eventually married one of the Wedgwood girls: his first cousin Emma.

Neither the Darwins nor the Wedgwoods were quite landed "gentry" in the peculiarly narrow British nineteenth-century sense of the word. True, they owned land and were actually richer than many of their neighbors. But the Wedgwood fortune was new money, only one generation old. Furthermore it came from trade, something the gentry looked down on. Uncle Jos had, in fact, at one time moved to another part of England in an effort to distance himself from the potteries and break into society. In this he failed, so he returned to Maer. The Darwins, by contrast, were not so blatantly and regrettably commercial; they had a professional background and their position was somewhat more secure. They made it more so by gradually entrenching themselves through the passage of time, through the growing reputation of the doctor, and through his astuteness as a manager of money. He made sizable loans to the local gentry at good interest rates; his neighbors became his debtors and his friends.

On this increasingly solid base young Charles grew up as a country gentleman. He learned to ride and handle a gun as a boy. As a teenager he was already a first-class shot who could scarcely wait each year for the bird season to open; he lived for hunting. He loved the country, liked to take long walks, and from childhood enjoyed collecting things like shells, moths, and butterflies. He became passionately interested in beetles.

Small objects sometimes cast long shadows. Darwin's interest in beetles, through an unlikely chain of circumstances, can almost be said to have shaped his career. It affected his choice of friends and his attachment to certain teachers. One of the latter was instrumental in his being appointed naturalist aboard a survey vessel, the *Beagle,* on a voyage around the world. That trip supplied Darwin with much of the raw material—in the form of observations and specimens—that he would later shape into a theory of evolution. Beetles, in short, as much as any other thing, tilted Darwin in the direction he ultimately would go.

That he would go anywhere seemed at first highly unlikely. Put into a "good" school, Shrewsbury, he drifted through it, bored stiff by the endless lessons in Latin and Greek, and was sent off to Edinburgh University. The expectation was that he would follow in the footsteps of his older brother, Erasmus, who was already there as a full-fledged medical student, their father having planned medical careers for both of them just as his father had planned for him. In the Darwin family of two boys and four girls Erasmus was the brilliant older son, Charles the dull young plodder. Hopes for a glittering career rode on Erasmus, but he disappointed his father. He wandered away from medicine as he did from everything difficult, settling eventually into the stressless life of a fashionable bachelor dilettante in London. He frittered his time away as a friend and entertainer of abler people and died having accomplished nothing. It must have affected him strangely to see his less favored, less talented younger brother looming larger and larger on the landscape until, by the time of Erasmus's death, Charles was possibly the most famous natural scientist in the world. He was certainly the most notorious.

None of this could be seen in the student Charles. He was a lanky young man with a round face, a snub nose, and sandy reddish hair, interested in guns and dogs. He had an easy, amiable disposition which concealed two strong traits. One was a sense of extreme modesty bordering on a feeling of inferiority, of not "measuring up," that a strict father and a tendency to judge himself harshly had produced. His family situation didn't help. He had always played second fiddle to a smarter older brother. He was bossed and criticized by affectionate but exacting older sisters. He had accomplished little at school. Then—worst of all, even though he was only the poor second choice—he, like Erasmus, dropped out of medicine. In Erasmus's case it had been laziness. In Charles's it was an honest lack of interest in the subject, combined with a horror of surgery. While at Edinburgh he had witnessed two operations. In the days before anesthetics, these were gruesome experiences even for hardened stomachs. To Charles, whose stomach was queasy in the extreme and who throughout life was unusually sensitive to pain or suffering, watching a patient strapped to a table, writhing and shrieking as the knife went in and the blood spread, was more than he could endure. He withdrew from Edinburgh.

A frustrated and angry father wondered what to do with him next. He decided to send him to Cambridge, hoping that Charles would do what so many other nice men of means but no talent did: marry well and become a clergyman. Charles docilely agreed to that.

His docility, his unwillingness to engage in disputes, his desire to avoid confrontations, a sort of humble passivity that he carried with him through his life concealed the second important trait: a patient, unaggressive, but in the end enduring determination to do what he wanted. When he was not sure about that—and after leaving Edinburgh he certainly was not—he contented himself with trying to please others. All he knew then was that he liked the out-of-doors, riding, hunting, and collecting and that he was instinctively drawn to men of similar interests. Since the world was full of shell-collecting and insect-collecting students who would never go further, there was obviously something more in the seriousness and determination of Charles that, despite his modesty, recommended him to his teachers. During one summer break at Cambridge he went on a beetle-collecting trip in Wales with the entomologist F. W. Hope. Two years later he was taken on a geology trip by the Cambridge geologist Adam Sedgwick. But the man most important to him at the university was John S. Henslow, a botanist. Henslow liked him, and Charles admired Henslow extravagantly. He sought him out so assiduously that he became known to his undergraduate contemporaries as "the man who walks with Henslow. . . ."

Cambridge, in the eyes of Darwin's father, was the right place for the failed young medical student. It had a long reputation as a religious institution and had eased many young men of Darwin's amiable temperament and aimless bent into the Church. But it was changing. Shortly before Darwin arrived, Sedgwick and Henslow had organized a philosophical society whose design was to promote interest in science and natural history. It was the first organization of its kind at Cambridge. Henslow then started a natural history museum, the first in England.

Darwin's exposure to men like these intensified his interest in the natural world. It gave him a chance to consider the respectability of collecting beetles and flowers as something more than a pastime. He graduated from Cambridge, having lived a very pleasant life but having come no closer to the Church than before. He was not irreligious. On the contrary, he was a devout young man, but religion simply did not interest him. The natural sciences did, more and more. There is further proof that his zeal or the orderliness of his mind must have caught Henslow's eye. In the summer of 1831, just after Darwin's graduation, Henslow wrote his young student a letter saying that he had recommended him for the post of naturalist aboard H.M.S. *Beagle.* This was the watershed moment in Darwin's life. Should he take the post, give up his plans for the ministry? Was he qualified? Would he be accepted? Most important, would his father let him go?

The last question was settled first. Dr. Darwin took an extremely dim view of the proposal. Two years away from home? Wasting time chasing around the world to no sensible end other than collecting specimens? Was the ship seaworthy? Why had a couple of others turned the offer down? What would become of the second career that he had so carefully selected for a boy who seemed unable to settle into anything practical? Would he *ever* become a clergyman? Even if he did, what would his prospects be after taking off on such a harebrained adventure?

Charles assured his father that the last thing he would ever do was disobey him. But with the humble persistence that would propel him through life, he did manage to get his father to agree that his objections might be overcome by the opinion of someone he respected, and that he would not then stand in Charles's way. Charles immediately went to Maer and laid the matter before his uncle Jos Wedgwood. They returned to Shrewsbury together. Wedgwood made a strong case for going. Dr. Darwin, who had the highest regard for his brother-in-law, relented.

Charles then had to present himself in London for an interview with Captain Robert Fitzroy, commander of the expedition. Henslow in his letter had explained that the scientific requirements would not be too severe and that Charles, with the knowledge that he already had of geology, botany, and taxidermy, could probably handle them. He wrote: "Captain Fitz-Roy wants a man (I understand) more as a companion than a mere collector, and would not take anyone, however good a naturalist, who was not recommended to him likewise as a *gentleman.*"

Charles had no trouble in identifying himself as a gentleman, but in his self-depreciating way he wondered if he would be a splendid enough one for the captain, who was dazzlingly splendid. He was descended from the first Duke of Grafton, who had been made a duke by King Charles II because he was the king's bastard son by the beautiful and notorious Lady Barbara Villiers. Fitzroy was proud of his royal blood, somewhat stained as it was. He was a slender, autocratic young man with aquiline features, hardened opinions, and an explosive temper. He had entered the Royal Navy at the age of twelve, risen rapidly, and was now readying a small vessel, the *Beagle,* for a two-year Admiralty mapping survey of the coast of South America.

It was this finely tuned thoroughbred, oversensitive to any slight, quick to redress it, a commander who ran a tight ship and did not hesitate to put most of his crew in chains after a drunken Christmas revelry ashore, whom the young gentleman with modest mien and country manners had to face. Surprisingly, the two hit it off instantly. Darwin got the job (with no pay) and went home bubbling with admiration for Fitzroy. Some months later the *Beagle* sailed, having been almost completely rebuilt to Fitzroy's exacting specifications. Despite his youth—he was only twenty-six—and his autocratic ways, he was an excellent commander and a scrupulous mapper. He was determined to produce the best charts ever made of the South American coast and had equipped the *Beagle* with superior navigational and surveying equipment. To make absolutely sure of his longitudinal calculations, which depended on accurate time, he brought twenty-six chronometers aboard, six of them his own. He was the kind of man who, when crossed, felt that his honor or his reputation had been questioned. During the voyage he bought or chartered smaller local vessels to speed his work, paying for them out of his own pocket and later getting into rancorous dispute with the Admiralty over repayment. Refused reimbursement for one, he fell into such a black depression that he threatened to resign his command and was only persuaded to keep it by the arguments and strong support of his brother officers.

Fitzroy was also a strong fundamentalist, whose desire to have a naturalist on board sprang from his belief that the work such a man could do in collecting specimens from across the globe would confirm his own belief in divine Creation. What an irony that a man of this stamp should have provided Darwin with the opportunity to destroy that belief. How odd, also, that two such different temperaments could be locked into cramped quarters in a bouncy, uncomfortable little vessel for years — and emerge still on good terms with each other.

Much of the success of that relationship was due to Charles's amiability and to his caution in discussing what he suspected would be touchy matters. He had no trouble with his captain's theological ideas because he was a good Christian himself. Their only falling-out came in an argument over slavery (Charles was against it) during which Fitzroy became so incensed that he ordered Darwin, with whom he had shared his meals, to mess with the lesser officers. A few hours later he was all repentance for his outburst, begged Charles's pardon, and reinstalled him.

Fitzroy's mood swings were wild, exacerbated by the strains of command, tempered only by his grim determination to control them. All too often he failed. With Darwin he alternated between an old-school affability and a vaulted privacy. He never entirely revealed himself. Darwin was astonished to learn, shortly after the voyage was over, that Fitzroy had been engaged to be married all that time. He had never even hinted at the fact or mentioned his fiancée's name. His image of himself as an infallible leader and remote aristocrat had to be preserved at all costs, and the price was high. He drove himself throughout his life, rose to be an admiral, and served a term as Governor of New Zealand. Over the years his fundamentalism became rock hard. After the publication of the *Origin* he transformed Darwin into the blackest sort of enemy and ranted against evolution. He ended his life a suicide. How far the shared young adventure of these two was from the uncharted and bitter future: the beached, half-crazy, enraged old admiral railing at the once-respected friend who had become an anti-Christ, the Fiend Incarnate.

The trip itself was a revelation to Darwin. Although he was wretchedly seasick throughout the voyage and subsisted on biscuits and raisins for extended periods, he still managed to set up scientific collecting procedures for himself and perfect a job that he actually had to create. There was no one to tell him what to do or how to do it or what to interest himself in. The crew members were initially amused by him, later respectful, as he continued to work no matter how awful he felt. At sea he trailed trawls astern and gathered a variety of marine vegetation and small organisms. He did a great deal of sorting, labeling, and bottling. When he was too weak for that he read — mostly the first volume of Lyell's *Principles of Geology,* just published. Lyell, it will be remembered, was the man who expanded on and popularized Hutton's earlier theories about uniformitarianism and the age of the earth. Darwin devoured that first volume. The second and third were published while he was at sea and would be sent out to him during the voyage.

Every chance he got he went ashore to make geological observations and to collect specimens. His energies and spirits soared on land. He hired horses and guides, arranged for camping trips into the interior, climbed mountains, rented bungalows for weeks at a time while Fitzroy was busy backtracking or going up rivers. Always he collected, dazzled by the richness, the strangeness and variety of what he found. Everything was new. At intervals he packed up crates of specimens and shipped them off to Henslow, who not only took good care of all this material but also showed some of it to colleagues. One of Darwin's finds — the fossil skull of a megatherium, an extinct and even larger forerunner of the more recently extinct South American giant ground sloth — created a sensation when Henslow sent it out for exhibit at the British Association for the Advancement of Science. He also read extracts from Darwin's many long letters at meetings of the Philosophical Society, with the result that the young collector, happily observing, happily shipping his treasures home, was, all unknown to himself, emerging in England as a respectable young scientist.

Darwin was uniquely equipped for this experience. He was extremely observant, scientifically oriented, but still innocent enough and unspecialized enough not to be locked into any particular thought channel. Instead, his mind was wide open, and he noticed many things that other more mature scientists might have missed. Lyell's *Principles of Geology* had made a deep impression on him, and as he traveled about South America he was actually more of a geologist than a biologist. Everywhere he looked he found evidence of the slow continuing processes that formed the basis of Lyell's uniformitarian argument: the same ***kinds*** of forces had been at work on the earth's surface for an unimaginably long time. Lyell opened Darwin's eyes to signs all up and down the Andes of geological turmoil: stratified layers of rock laid down by lava from an inferno deep in the earth. Darwin experienced a severe earthquake at Valdiva and again was made vividly aware of the fragility of the earth's crust. He found marine fossils on upland plateaus, indicating that those places had once been ocean bottoms. In the Galapagos Islands, a volcanic archipelago west of Ecuador, he walked over lava landscapes, inspecting huge eroded craters that were thousands of years old and newly erupted pustules that were only decades old. On some of the islands the hardened lava was as fresh and sharp and crackly underfoot as big irregular pieces of cornflakes. It had run, slowed, stopped — like cooling molasses — in waves, crests, tubes, ripples, falls. Treading on it, he could break through into chambers and cells where volcanic gases had been. All this had happened so recently that no plant or insect had yet invaded those lava wastes. Nearby, on another island, there would be no overt trace of volcanic activity at all. The lava would have been ground to coarse black sand along the beaches. Inland there would be a thick mat of odd vegetation. Only by digging through that could one find signs of volcanic origin. And everywhere amidst this evidence of slow geological process were the most peculiar animals: enormous tortoises that weighed up to a quarter of a ton, platoons of blackish iguanas spitting salt from their nostrils as they basked in the sun, bright yellow iguanas farther inland, scarlet crabs hugging the rocks, finches so tame that they lit on his shoulders as he walked about.

Those finches are known today as Darwin's finches because he was the first to collect them and because later they would be important stimulators of the evolutionary theory toward which the Galapagos experience in particular among all Darwin's experiences on the *Beagle* would help propel him. But the commonly held idea that he stood open-mouthed on the shore of one of those islands, like Balboa spellbound by his first view of the Pacific, and suddenly received the lightning bolt of perception is wrong. He was too busy for that, his impressions unformed, his mind too overloaded with what his eyes were pouring into it. He observed and collected constantly, adding his impressions of the islands to the equally potent ones he had gathered on the mainland. He took note of odd relationships and strange behaviors, but it was not until he had returned home and had a chance to reflect on the extraordinary things he had seen that he began speculating about evolution in any purposive way.

The *Beagle* had sailed from England in December 1831 on a

voyage that was originally planned to take two years. It actually took five; the *Beagle* did not dock in England again until October 1836. For the next three years Darwin was busy in London. He rented a flat there, retrieved his enormous haul of specimens from Henslow, and embarked on the long task of sorting them out. He found specialists to work on birds, on reptiles, on fossils, on insects, and other forms. Ultimately he was able to edit and publish a five-volume work: *Zoology of the Voyage of H.M.S. Beagle.* This took organizational ability of a high order and a willingness to make a long commitment to a big task. It established Darwin as one of the coming young naturalists of England. He met other scientists, he joined clubs and scientific societies, published a paper or two. With an income supplied by his father, and with the prospect of considerably more should his father die, the future looked bright indeed to Charles. In 1838, he decided to marry and proposed to his cousin, Emma Wedgwood. She accepted him. *Everything* now pointed to a career of conventional progress in one branch or another of natural science. The young blunderer had somehow blundered into the right field and obviously was marked for success in it.

Except—there was a cloud of impressions gathered on the trip that began to come together in his mind in a disturbing way. He started thinking about them immediately after the voyage had ended. They would not let him alone. They reminded him of the idea of the mutability of species that he had first encountered in his grandfather's writings, encountered later in the theories of Lamarck, and—with his growing ability to examine scientific arguments coldly and analytically—had even noticed in Lyell's monumental work on geology. Darwin had been totally won over by Lyell. Uniformitarianism explained the geological history of the earth with a logic and clarity and a mass of evidence that were devastatingly convincing. He became a fervent uniformitarian. But even as he did so, he could not help noticing something: if conditions of the surface of the earth kept changing—and proof of that could be seen everywhere—would not that mean that animals that had been created to fit certain former earth conditions would fit less and less well as those conditions became more and more different? Would not all of life gradually get out of step with the world in which it lived?

Lyell, it is true, had been aware of this problem, but he had ducked it. Darwin found himself unable to. Mentally he was a very tenacious man. He concluded that the only way animals could continue to exist efficiently in their surroundings was by changing as the surroundings did. How they did this was a mystery to him. He wasn't even sure that his suspicion was right. But he wanted to know. He had already bought a strongly bound notebook; now he secretly began jotting down thoughts and observations about species variability in it.

Following the processes of another mind is never easy, even if that other mind is present to try to explain how it reaches certain conclusions. Indeed, people who have made lifetime studies of thought and the ways in which new ideas are arrived at do not wholly understand it. The brilliant conceptualizers themselves cannot explain where their ideas come from; often they just suddenly appear. In the case of Darwin, dead now for a century, it would seem particularly difficult to trace his steps in working out the theory of evolution. Any attempt is further complicated by his being so badly cast to do it. Why Darwin? He was a retiring man who hated controversy. He was almost pathologically cautious. He lived a conventional life in a sometimes suffocating upper-middle class, in a Victorian world that was itself sublimely complacent and resistant to intellectual change. The sheer willingness to devote his life to working out a theory that would certainly shock the world and strain his relationships with fellow scientists, some of them his close friends whose opinions he valued highly, a theory that would expose him to charges of heresy, materialism, lunacy, and, worst of all, stupidity and bad science— the willingness to risk all that seems wholly out of character. If ever there was a man ill fitted to play the role of Charles Darwin, it was Darwin.

And yet he was the man who did it. The way he did it can be traced more accurately than might be imagined. Darwin was a notetaker. When he observed something he jotted it down. When he thought of something, he captured the thought on paper. He kept a journal. When jarring ideas about the nature of life began to crowd his mind he began filling notebooks with them. He was a great letter writer and many of his letters survive. He also wrote an autobiography. His daughter wrote a detailed account of his *Beagle* experiences. Thus there is a mass of written material available for the reconstruction of Darwin's intellectual progress. But any inquiry must start with what he actually saw.

So, what did Darwin see?

Quite early during the *Beagle* voyage he noticed something interesting: that the species on the Cape Verde Islands, through which the vessel passed on its way to Brazil, were not quite like the species on the African mainland a few hundred miles away. They were similar, but not identical. This made no particular impression on Darwin at the time. Later, when he got to the Galapagos Islands in the Pacific, he noticed the same thing. But there the differences and similarities were with the South American mainland, not the African. For a young man brought up to believe in divine Creation, it must have struck him as odd that God would have gone to the trouble to create two sets of creatures and plants for two sets of islands that might well have done with one set, since the general climate and environment in both places was substantially the same. Darwin's God was the no-nonsense God of a hardheaded no-nonsense society. There was nothing mystical about the Anglicanism of nineteenth-century England. God was a practical Deity, and this was not a practical solution to God's problem.

Furthermore, the resemblance of the flora and fauna of each island group to its next-door continent made it plain that there was a relationship there. Similar, yet different. The same problem in two places. How was that to be explained?

The next thing Darwin became aware of in the Galapagos was that there were differences in species from island to island. He met a man who could take one look at a giant tortoise and tell which island it came from. A different kind of tortoise on each island — why? Particularly why, when some of the islands were only half a dozen miles apart and had virtually identical environments.

Darwin began looking more closely at other species. His attention was caught by the finches there. They had an uncanny resemblance to one another, except that they had different kinds of beaks. These turned out to be specializations for different kinds of eating. One finch had an extremely stout beak designed for cracking tough seeds. To support such a heavy beak the bird's head was comically large, out of all proportion to its body. There was another with a similar, but smaller, beak; it ate smaller, more easily cracked seeds. Another had a long thin beak like a warbler's; it ate insects. Still another had the remarkable habit of using a cactus needle to poke grubs out of cracks in wood. Its beak was different again.

Back in England, cataloging his finches under the guidance of the ornithologist John Gould, Darwin discovered that he had collected thirteen species, all different, all so closely related that it seemed almost that they might have had a common ancestor, and had gradually altered themselves physically to make the most of the varied food resources of the islands. He wrote: "One might really fancy that from an original paucity of birds in this archipelago, one species had been taken and modified for different ends."

The word "fancy" is the one to pay attention to here. The quotation appears in Darwin's published account of the *Beagle* voyage. He was permitting himself a little innocent speculation: "Let's amuse ourselves with this fanciful idea." That the speculation wasn't really all that innocent he kept to himself. It was the kind of unthinkable thought that began filling up the pages of his "transmutation notebooks," as he called them.

To get some idea of the intellectual problem that Darwin was up against at this stage of his career, let us imagine ourselves as archaeologists studying artifacts along the Atlantic coast. We find ourselves working with Coca Cola bottles because they are common in the mud of harbors and also are found on offshore islands. The bottles on the mainland occur in two kinds. There is the old thick, returnable bottle that we remember from our youth, good for a nickel at the soft-drink store (scarce now because as children we returned them whenever we found them). There is also the thinner, nonreturnable kind. It is abundant because it lies wherever dropped; that kind is worthless and nobody bothers to collect it. These two varieties of bottles represent two acts of creation by the Coca Cola Company, a large distant force that, in this analogy, might be compared to God.

What kind of bottle would we find on the offshore islands?

Logically we should find both, in the same proportion as in the mainland sample. But suppose we found something quite different: an enormously bloated bottle that bore only a family resemblance to the shoreside types. How would we account for it — as some kind of an aberration by the main office? Suppose, when we looked at those oversize bottles carefully, we discovered that they differed from island to island. True, they were all huge; they were all Coke bottles; but Martha's Vineyard bottles were not exactly like Nantucket bottles. Both were slightly different from Block Island bottles and different again from Fishers Island bottles. Could we reasonably assume that the home office had manufactured all those different kinds? And to what purpose?

The flaw in this, as an evolutionary analogy to Galapagos tortoises, is that Coke bottles are inanimate. Once they come out of the factory they cannot change. But still they serve here because the prevailing belief in Darwin's time was that living species were as changeless as Coca Cola bottles.

To Darwin, thinking about finches and tortoises—and about a host of other creatures from around the world—it began to seem increasingly reasonable to assume, not that the home office (God) had changed them, but that they had changed themselves. Species, he concluded, were not stable. He *saw* change, or at least powerful hints of it, in the otherwise inexplicable small differences that revealed themselves wherever he looked.

Once the mind accepts the possibility of small change, it opens itself to considering the possibility of larger change: that is, small changes becoming greater over a long period of time. This is really taking Lyell's theory of uniformitarianism in geology and applying it to biology. Darwin's grandfather had taken a similar mental jump. Along with Lamarck he had speculated that all complex things evolve from a few simple things, perhaps only one. If that were true, then it would explain why animals in one place were radically different from those in another. In different environments their efforts to adapt would have led them in different directions.

Gross oddities in geographical distribution inevitably began to gnaw at Darwin. Why were the things that lived on the east side of the Andes so different from those that lived on the west side? For that matter, why were the animals of North America different from those of South America? Why were the animals of the far north similar in all continents? Why were there no mammals at all in the Galapagos Islands except for one species of small rat? Why was there nothing but a skewed population of birds, reptiles, and insects there — to say nothing of an even odder vegetation? Why was there only one kind of hawk, one dove, one owl, one snake? Why one kind of tortoise that had mysteriously become a dozen kinds?

When the *Beagle* reached New Zealand the distribution problem raised its head again. New Zealand turned out to be a bizarre avian ecosystem. The niches that normally would have been occupied by all kinds of mammals from woodchucks to antelopes were instead occupied by a fantastic array of flightless birds — or had been: most had already been exterminated by Maori hunters before Darwin got there. But formerly, when the birds had the place to themselves (as the fossil record showed they had), they dominated it. They ranged from giant moas that were twice the size of ostriches and laid eggs the size of basketballs, down to a host of smaller types like the chicken-sized kiwi, which is one of the few survivors of that extraordinary evolutionary blooming. The kiwi hangs on today only in a few remote outlying islands or in sanctuaries as an inconspicuous nocturnal skulker.

A believer in divine Creation, looking at New Zealand, would have to conclude that God had capriciously decided not to bother with mammals there but had filled it up with birds instead. If a moa could be created to fill the niche occupied elsewhere by a buffalo, a horse, or a camel, why not create it?

But if one were persuaded of the potential for change in living organisms, one might conclude differently: that New Zealand was too far off in the sea for other mammals ever to have gotten there. But birds, able to fly, had managed the trip, and done all their remarkable changing by slow degrees after they arrived. If they later lost their wings — as many had — that could be accounted for by disuse (Lamarck had thought of that); without mammals to prey on them, what use had they for wings? On the tiny Galapagos archipelago Darwin had observed a similar situation. While there was only one swimming, diving bird there that was also found on other continents — a cormorant — it too had lost the power of flight, and presumably for the same reason: no mammalian predators.

Whether Darwin actually had such ideas in New Zealand is debatable. But the facts certainly registered on him. They did again in Australia. Here was another island, a continent-sized one with plenty of mammals — but what peculiar ones, nothing like those of Europe, Asia, or Africa. Australian mammals were nearly all marsupials, pouched animals like kangaroos. Large kangaroos took the place of cows, smaller ones took the place of sheep. Other marsupials filled the niches that on other continents would have gone to rabbits, dogs, wolves, porcupines, monkeys — once again an inexplicable and seemingly capricious explosion of forms like nothing else on earth.

Thinking about peculiarities in distribution, the conscientious zoologist will ultimately wind up asking some rather difficult questions. It cannot escape notice that certain kinds of animals seem to sort themselves out in obvious ways around the globe. Most bears, for example, live in temperate zones, apes and monkeys in tropical zones. How is that to be explained?

It is easy to say that bears have thick coats; they can stand cold winters; they hibernate; they eat the things found in temperate forests and clearings. Where you find such conditions you find bears, just as you find fruit-eating, thin-furred monkeys in places where there is a lot of fruit and the weather is hot. But those responses do not address a more difficult question. There are many kinds of monkeys; where do they come from?

"That's the crux of the matter," said Don as we were discussing how to develop this point. "Where do *different* monkeys come from? I mean, is each kind of monkey a separate act of Creation, all of them capriciously crowded in the tropics? Or are all monkeys related, descended from one or a few tropical ancestors?"

"It's obvious," I said. "They are related by descent."

"So, how do you account for the differences between species?"

"If you believe in Creationism, they were made that way. Bang, all at once. If you don't, then you have to assume that they got that way gradually. You're asking me questions we both think we know the answers to."

"I know," said Don. "I just want to hammer it home for the reader. You're saying that monkeys changed slowly, gradually getting more different from a common ancestor?"

"Yes."

"Why?"

"Mostly because of the environment. As you say, monkeys are tropical animals. They can't stand the cold; there are only a few kinds that live in temperate zones and those have much thicker coats. As for the tropics themselves, there are different environments within them. I know that African colobus monkeys eat nothing but leaves. They spend all their time up in the trees. If they were scavengers that ate insects and frogs and seeds—anything they could find—they would spend more and more time on the ground. Eventually they would be somewhat different from the tree dwellers. They would be more like baboons."

Then Don hit me with something I hadn't known about monkeys. "What about a tree-dwelling, leaf-eating South American monkey, the howler monkey? It lives very much the same life as a colobus monkey. Would you expect the two to be closely related?"

"I would."

"They're not. There's a big group of New World monkeys that are closely related to each other, but they are not closely related to any Old World monkey. The two groups *look* alike, they *behave* alike, they *live* in similar environments. But when you examine them closely you find deep structural differences. Their noses are entirely different. New World monkeys can hang by their tails; Old World monkeys can't. How would you explain that?"

"I'm not sure I could, except that maybe one group of monkeys got split off from another at a time before the continents got separated. Then each could go its own way."

"You're saying that the howler and the colobus were once more or less the same, but have diverged since?"

"Yes."

"They kept their habits but have changed quite radically in the architecture of the nose and the ability to hang by the tail?"

"That's what you seem to be saying."

"Come on, Mait; think. *All* the South American monkeys changed their noses and their tails, but *none* of the African ones did, regardless of their habitats?"

"Well, if you put it that way, it doesn't seem likely."

"I have to put it that way because that's the way it is. Therefore doesn't it seem more likely that there were only two types with somewhat different noses to begin with, and that they gradually split into different kinds *after* the continents separated? That would produce leaf eaters and ground dwellers in both places, each with the proper nose."

"That makes more sense."

"I'll put it another way. Does it make sense to you that some force or influence produced a group of ten different and immutable species of monkeys on one continent, all with the same kind of nose but living in ten different ways; and at the same time produced ten other kinds of monkeys on another continent, with another kind of nose, but living in the same ten ways?"

"I would have a lot of trouble believing that. I'd have to say instead that two different-nosed ancestors started it off in the two different places, and that the opportunities in the environment did the rest—without bothering to alter the noses."

"So," said Don, "monkeys do distribute themselves around the world in curious ways, curious enough to make you speculate that there are relationships between species, and also change."

"Yes, yes—"

"How about bears?"

"The same situation would prevail. Bears are closely related—

you need only look at them to realize that — but there are differences between them."

"Would you say that a large Siberian bear is a slightly different form of a large North American brown bear, or are they separate acts of creation?"

"Don, for God's sake, how many times — ?"

"I want to nail it down. I want to look at this from the point of view of somebody who has never heard of evolution but still has to make sense of a lot of information that is pouring in on him."

"All right. Once again, it depends on how you look at it. But if you think of all the relationships of animals around the world, how they fall into groups, how they arrange themselves geographically, how they vary slightly as you move from place to place, it gets harder and harder to think of them as fixed species."

"Right," said Don. "That was the conclusion that Charles Darwin came to."

Mutability — the dawning likelihood of it — may have been difficult at first for Darwin to accept, but it was not nearly as difficult as the huge question of what made it happen, what *allowed* it to happen. For, opposed to that nebulous, half-acknowledged, timidly sensed idea was the formidable and universally acknowledged truth that species *did not change.*

What was a species, after all, but something that could be expected to reproduce itself reliably generation after generation? Mate two dogs and you always got dogs, never foxes or anteaters. Mate a dog and a cat and you got nothing. Any proposal about the nature of life and how its various forms came into being would have to reconcile two diametrically opposed tendencies: a tendency to vary, and a tendency to stay the same.

During 1837 and 1838 Darwin wrestled with that dilemma. It brought him face to face with the so-called problem of dilution. Suppose (through some strange agency) an animal came along that was quite different from its fellows. If it were mated to a normal one and its offspring mated to another normal one, and so on, would not its original peculiarity gradually get eliminated in the great pool of normal animals? It would be as if an experimenter poured a half-beaker of dark-pink cranberry juice into a half-beaker of clear water. The result would be a paler pink liquid. Pour half of that into another beaker of water — paler yet. After a while an observer would be able to detect no pink whatsoever. That, according to the wisdom of the time, was what enabled species to remain stable. Any aberration that came along would be swamped by normality.

The seemingly intractable problem of variability versus fixity rose again when Darwin turned his attention to selective breeding and hybridization. Again, an apparent and deep contradiction. Darwin knew, as did everybody else, that by careful stock selection within a species such as dogs all sorts of varieties could be created in captivity. This could be done even outside the species so long as the species were closely related, as they are in horses and donkeys. There, a mating produces a mule. The trouble with mules is that they are almost invariably sterile. Stock breeders can produce mules at will, but in nature this does not happen. Left to themselves, horses do not seek out donkeys as sexual partners any more than humans select chimpanzees or gorillas. Even if they did, the inability of mules to reproduce themselves makes it certain that in the wild state there is no way that an ongoing species of mules could come into being.

It was clear to Darwin that the forces that operated to keep species cleanly and consistently within their own shapes and their own habits were overpoweringly strong. This tendency toward consistency in species was made stronger by another factor: weak or deformed animals did not survive. There was a constant culling of the unfit by predators and starvation that ruthlessly confined the species to its normal and proper shape.

And yet, and yet—the strange variations he had observed on his voyage, the capricious distribution of types around the globe, continued to haunt Darwin. Somehow slight differences *did* proclaim themselves in populations. Moreover they settled in and became permanent. He had only to look around him to see the truth of that in humans. He visited a friend's rose garden: hundreds of varieties. Among animals he had seen gradual changes in types as he moved up the South American coast. That was not ruthless adherence to an eternal form. It was flexibility, presumably expressing itself as a response to differences in the environment. Proper shape in one place was improper in another. How did it come about? He did not know.

Two years after he returned home, and while he was still filling notebooks with thoughts about transmutation, he read Malthus's *Essay on the Principles of Population.* In later years he would credit it with a sudden revelation that enabled all his ideas to fall into place: ". . . it at once struck me that . . . favorable variations would tend to be preserved, and unfavorable ones to be destroyed. The result of this would be the formation of new species."

Let us not breeze past that statement too rapidly. It contains a profound and, up to that point, unrecognized truth. Others had noted the Malthusian doctrine of struggle only as a means of keeping a species fit and unchanged; Darwin looked at the situation in reverse and came up with an entirely new idea. If there was variation among individuals, might not that make certain individuals *more fit,* better able to survive? In other words, could not the Malthusian doctrine of struggle lead to change as well as stability? It could, he reasoned, if the fitter individuals could spread their fitness through a population. Those changes would be small to begin with. But, given time, could not the changes build up and become very large, producing whole classes of organisms that had not existed before?

If the fitter individuals could spread their fitness — there was the rub. Could they? What about the problem of dilution? If there was only a handful of the new improved animals — perhaps only one — in a huge population, would not the improvement, like the experimenter's cranberry juice, thin out to the point where it disappeared in a sea of ordinariness?

Here Darwin's observations of islands came to his rescue. The oddities he had been most impressed by were all associated with islands: Cape Verde, the Galapagos, New Zealand. What they had in common was isolation from continental land masses. If an island were sufficiently isolated, particularly if it were small, a new individual appearing there could represent a significant proportion of that island's entire population. One improved animal in a population of only fifty would have a far better chance of expressing itself than one in ten million.

As a laboratory for testing that idea, Darwin could not have hit on a better place than the Galapagos. The creatures there bore telltale resemblances to South American species, suggesting strongly that they had migrated to the islands at some time in the past. They were sufficiently *unlike* their South American relatives to suggest that they had undergone some change after their arrival. Change could have occurred because their populations were extremely small; a few untypical individuals could well have implanted their oddities and made them permanent. That this had happened, not once but several times, was suggested by the fact that the tortoises on each island were different. Why? The geological history of the archipelago itself suggested an answer. The islands were volcanic. They had risen from the sea in pristine, uninhabited form. They were like clean unmarked slates. *Everything* that arrived on them had had to come from somewhere else. Each island, at some point in its history, had been invaded by a couple of tortoises or a mere handful. If each invader was even slightly different from its fellows on the mainland and on the other islands—and no two individuals of any higher-animal species anywhere are identical—then it would have to follow that succeeding generations of tortoises on each island, simply by breeding only with each other, would gradually intensify their differences and assume separate identities. This presumably had been going on for thousands of years. By the time Darwin arrived in the islands, each had its own tortoise.

To repeat, it is impossible to say exactly how and when these ideas appeared in Darwin's head. Exhaustive analysis of his transmutation notebooks by scholars has shown that, like all great intellectual achievements, this one had many seeds from many sources. Even Darwin's own assertion that the light broke when he read Malthus is suspect; he apparently had been thinking productively on the point already and needed only the push of Malthus to bring everything into focus.

Whatever the case, Darwin now had an explanation for transmutation. From there he could take the grand jump and propose that all life had evolved in a similar way from more primitive earlier forms, responding through what he called "natural selection" in the struggle to survive.

One tends to think of survival as depending on the hare's ability to outrun the fox, or, conversely, on the fox's being smart enough and swift enough to catch the hare. Obviously the two live in a precarious balance. Before Darwin, this balance was considered to be fixed, depending on the natural endowments of hare and fox, each species kept within the narrow limits of its "best" shape by unremitting bloody struggle.

What Darwin added to that concept of struggle was that it would probably result in the evolution of swifter prey and smarter predators. Hare and fox need each other to survive. That may be a peculiar way for the hare to look at it, and for an individual hare it is. But for hares in general it means constant pressure by foxes to produce superior hares through the steady elimination of poorer ones, and a similar pressure on foxes by those superior hares. By concentrating on that aspect of the matter, on the ***kinds*** of animals that survived in the struggle, Darwin was able to identify the driving force in evolutionary change. He also understood something else of great importance: that the struggle was not only between hare and fox but between hare and hare — for food, for mates. Both kinds of struggle expressed themselves in differential breeding success. The better animals (bigger, stronger, smarter, more sexually aggressive) lived longer and reproduced more of their kind. Given time, that edge in reproductive success would have its effect on a population. Given enough time, one species could be said to have been transmuted into another. Evolution would have occurred.

Darwin's thinking about evolution was no straight-line affair. It could not proceed easily from one observation to a revealing insight and then on to another. Rather, it was an ingenious and original weaving together of several observations and conclusions, all of which he had to organize in some logical way before they could become an effective theory. No one has explained this more clearly or more concisely than Ernst Mayr in his penetrating analysis of Darwin in *The Growth of Biological Thought,* published in 1982. In it Mayr boils down Darwin's work to five observations that he made and three inferences that he drew from them:

Observation No. 1. Species have great potential fertility. Darwin knew this, of course, having observed the enormous overproduction of pollen and seeds by many plants, the huge outpouring of eggs by molluscs (a full-grown oyster, for example, can produce several hundred million eggs a year), the large hatches of twenty or more wild goslings or ducklings, of which only two or three may reach maturity. Darwin was reminded of this natural fecundity of all life, as an important element in his equation, by Malthus.

Observation No. 2. Populations normally tend to remain the same size. Even those that fluctuate, like lemming populations, do so in regular three- or four-year cycles, never falling below a critical minimum and never rising above a flash-point of extreme overcrowding. Most populations are extremely stable.

Observation No. 3. Food resources are limited and remain fairly constant most of the time.

From these three observations Darwin drew his first inference: in an environment of stable resources and an overproduction of individuals, there will be a struggle for survival among those individuals.

Having established that, and again probably having been reminded of it by Malthus, who made the same point, Darwin proceeded to two further key observations:

Observation No. 4. No two individuals are identical. This is so obvious among humans and higher animals that it scarcely needs mentioning. However, it is true also for the smallest and simplest flowers. Seeing variation wherever he looked, he could assume it to be universal.

Observation No. 5. Much of this variation is heritable. Again, he could verify that simply by looking around him; offspring do resemble their parents.

Darwin's second inference combines insights from all five observations. In a world of stable populations where individuals (all different) must struggle to survive, those with the "best" characteristics will be the most likely to survive, and those characteristics will probably be inherited by their offspring. This unequal survival rate, where the "good" end up outnumbering the "less good," is natural selection.

Darwin's final inference is that the process of natural selection, if carried far enough and long enough, will produce marked changes in a population and will lead eventually to the emergence of new species.

It should be stressed that this summary does not reflect the order in which Darwin arrived at his conclusions. It is merely a short and simple way of showing how he was able to integrate several hitherto unrelated ideas after long thought and a great deal of intellectual floundering. When set out as neatly and clearly as Mayr presents them, they seem so easy and simple as almost to lose their power. This appearance of ease and simplicity is far from reality. Nobody before Darwin had perceived the relationships between those five observations, and as a result, nobody had been able to arrive at the final inference that he reached. (One man, as will be seen, did reach the same conclusion at the same time Darwin did.) Once organized, those observations and inferences would change forever how humans regarded themselves and their place in the universe.

Stated as an argument, and buttressed by a mountain of evidence, the theory of evolution carries with it a sense of majesty and inevitability. The other great naturalists who had partially glimpsed it had sailed like full-rigged ships from idea to statement. Now we find Darwin at a similar point, ready to speak but with a far more wide-reaching and difficult statement to make. There he sits, hunched over the small microscope he would use throughout his life, scratching down notes, proceeding doggedly from observation to insight, from insight to proof by example — back and forth, methodically inching along from step to step until all is worked out in his mind. What next? Publication, obviously.

What actually happened was quite different. By 1839 or 1840 Darwin had filled several notebooks with his thoughts on transmutation. In 1842 he organized these thoughts into a coherent argument and wrote a short essay outlining his theory. In 1844 he wrote another, much longer one. He published neither. Naturally cautious — many have called him pathologically timid — he preferred to sit on his barrel of biological gunpowder in secret while he piled up more evidence.

Meanwhile, important events had occurred in Darwin's personal life. His cousin Emma Wedgwood turned out to be an ideal wife for him, sensible, patient, loving, a good housekeeper and mother. She bore him ten children. Darwin needed those domestic props and more. He had begun to suffer from a strange assortment of ailments as early as 1838 or 1839, and during the next several years they grew worse. He was plagued by bouts of indigestion that were sometimes extremely painful and left him exhausted. He came down with blotchy skin rashes. He suffered from spells of vomiting and from blinding headaches. The onset of these symptoms had occurred in London, a place that he — a country boy — hated. He blamed his bad health on the smoky, fusty air and decided to move. After many months of search, he and Emma found a suitable house in the village of Downe in Kent, only twenty miles from London.

It could have been a thousand miles. Increasingly sickly, always more comfortable with his own thoughts than with the company of others, Darwin had begun to live a rather secluded life even in London. Now, with his withdrawal to Downe, he became a virtual hermit. His attacks of illness grew more severe. At one point he became so enfeebled that he was confined to a wheelchair. He suspected that he was going to die but was afraid to say anything to his wife for fear of upsetting her.

Ultimately he put himself in the hands of a practitioner of hydropathy, a new medical fad. Darwin was given cold footbaths, cold compresses that he had to wear under his clothes and that made his joints ache, and rough rubdowns with coarse cloths that scoured his skin until it itched intolerably. He tried a variety of odd diets and at one point even experimented with electric shocks by wrapping himself with alternate strands of copper and zinc wire. When these were moistened with a weak acid (he used vinegar) they became a battery and gave off a feeble current—but no relief.

In truth the hydropathy treatments did make him feel better, but this improvement was not caused by the cold baths and scourings. It occurred because he was away from home at a spa and away from his work. His ceaseless labors drained him as they drew him down a perilous track that he knew would alienate his friends and distress his wife. The latter circumstance caused him particular anguish. Emma Darwin was a devout woman and had viewed his growing religious skepticism with some pain. Not long after they were married, she wrote him a long letter that did not directly criticize his work but made it clear that she could never be entirely happy with it so long as he insisted on believing only what he could prove. That, she said, would take him farther from God. But, she added, she would never blame him for anything that he did in the full belief that he was seeking the truth. She ended with an eloquent protestation of trust and love that caused him to write in the margin: "When I am dead, know that many times, I have kissed and cryed over this." He kept Emma's letter all his life. It is now among his collected papers.

Much medical and psychological speculation has gone into the subject of Darwin's health. It seems probable that he did suffer from some digestive weakness. But it also seems likely that, whatever it was, it was vastly aggravated by emotional stress. He was working on the Devil's turf, and his work gave him no peace. At one point he spoke of terror.

A person engaged in heretical speculation needs friends: for support; for discharge during the day, through amicable argument and discussion, of the doubts and terrors of the night. Darwin, ill and hag-ridden at Downe, lacked this outlet because of his reluctance to share his increasingly dangerous ideas with men he suspected would disagree with him. Most would have, and it must have been an added source of anxiety that most were scientists. All his life Darwin had been drawn to such men. During his years in London he had met many of them and become friendly with them in his way: amiable and outgoing on the surface, reserved inside. Over the years he managed to retain his ties with large numbers of them through correspondence. This was easier than direct confrontation, safer too. It enabled Darwin to engage in a practice that he turned into a high art: getting people to share data with him, exchange information, *give* information, without revealing what he was doing. In that respect, whatever his isolation was costing him in other ways, it had one plus: he did not have to waste his carefully husbanded strength in the exhausting effort of socialization; he needed every ounce of that for work.

The result was a swaddled life, smoothly and efficiently run by Emma, but a life nearly empty of close professional association. He lost personal touch with the man he most cherished as a friend and teacher, the man who, by recommending him as the *Beagle* naturalist, had brought him to the narrow ledge to which he now clung. John Henslow was as devout a believer in divine Creation as Emma. Darwin could not talk to him frankly, and during the critical fifteen-year period in which Darwin went from speculating about evolution to a determination to publish a large-scale work on the subject, the two men saw little of each other. Henslow was a man whom Darwin truly loved. He later wrote of the "transparency and sincerity of Henslow's character . . . his kindness of heart. . . . When principle came into play, no power on earth could have turned him one hair's breadth." It was that sense of principle that Darwin respected (and feared), and it must have shaken him to realize that by his own beliefs he had become isolated from such a man.

Darwin never "lost" Henslow entirely; they continued to be friends right up to Henslow's death. What he did lose was intimacy and continued contact. He compensated for that by acquiring a substitute, another botanist named Joseph Dalton Hooker whom he had met through Henslow—he was Henslow's son-in-law, a generation younger, less tradition bound, more open-minded. Gradually Darwin confided to him what he was doing. As early as 1844 he wrote to Hooker: "I am almost convinced . . . that species are not (it is like confessing a murder) immutable." Of all the scientists whom Darwin either met or corresponded with during the years before publication of the *Origin,* Hooker was the only one who was aware right along of the direction in which Darwin was headed or of how far he had gone. Hooker read his notes, argued the subject with him, and encouraged him, although skeptical himself. He visited Downe often, the only person who, outside of other Darwins and Wedgwoods, was always welcome there.

Darwin kept one other close friend, Charles Lyell, whom he had met in London. He made no secret of the profound effect Lyell's *Principles of Geology* had had on him during the *Beagle* voyage. Lyell, in turn, grew extremely fond of Darwin. Both he and Hooker wound up with a protective feeling for their retiring, industrious, sickly friend. Lyell also admired Darwin for the tenacity of his logic and the keenness of his observations, both evident in a matter that might have ruffled a smaller man than Lyell: the pupil had dared question the master.

In his *Principles of Geology* Lyell had explained the existence of coral atolls, those low-lying circular islands with central lagoons that were scattered across the Pacific, by stating that the coral had built up on the edges of submerged volcanoes. That was true, Darwin knew; coral did that. But, thinking further about it, he concluded that Lyell's idea needed a better explanation, the reason being that coral can only grow in an environment that goes from the surface of the water downward for about one hundred and eighty feet. Below that there is not enough light or food for it to survive. Since the Pacific is strewn with coral atolls, it seemed statistically most unlikely to Darwin that there could be a host of submerged volcanoes whose tops were all within that narrow one-hundred-eighty-foot range. Furthermore, there were coastal reefs to be accounted for and barrier reefs like the twelve-hundred-fifty-mile one off the Australian coast. They were not atoll shaped and obviously had different histories.

Taking a uniformitarian leaf out of Lyell's own book, Darwin speculated that, as continents rose, there would be a counterbalancing sinking of the ocean bottom in appropriate places. If that were so, certain islands and island chains would gradually begin to go down. As they disappeared beneath the water, the coral found along their shores would grow upward, generation piling on top of generation in an effort to keep close to the surface. Where it had succeeded atolls had been formed, large crescents of reef and sandy beach, most of the sand consisting of surf-ground coral, and all of it resting on a coral base that went down for hundreds or thousands of feet before it hit solid rock, depending on how slowly and how long the original island had been sinking. Presumably, somewhere in the wastes of the Pacific would be submerged volcanic cones topped by cities of long-dead coral that had not managed to grow fast enough and instead had been drawn downward into the darkness and expired.*

His coral researches were important to Darwin in several ways. He published a paper on them; it was widely admired and enhanced his reputation as a serious scientist. He also noted that selection would have to operate on the corals through their need to adapt to changing water depths. When he actually began finding differences in corals at different depths, this observation helped in the development of a vital cog in his evolutionary machine. It strengthened his conviction that the forces that nudged evolution (in this case, slow, steady sinking) would result in very small changes gradually brought to a point of observable difference over many generations. This principle of the heaping up of minute changes through selection, a cornerstone of Darwinian theory, is known today as gradualism. Finally, his work on corals helped cement his friendship with Lyell, who quickly saw the logic of it and amended his own ideas accordingly.

The two became closer. Darwin dedicated a book to him and later wrote, "Amongst the great scientific men, no one has been nearly so friendly and kind as Lyell." But there was a thin veil between them: Lyell was a temporizer about transmutation, and Darwin knew it. He never quite achieved the easy uncontaminated affection for him that he had for Hooker, although eventually he confided in Lyell too. He needed them both. Given his cautious nature and the uneven state of his health, it is conceivable that without the support of those two men, he might never have published the *Origin of Species*. In the words of his biographer Peter Brent, he "might have sunk into the self-indulgent intellectual torpor of a provincial dilettante, collecting his facts forever, their collection his hobby and their publication his myth."

This, of course, did not happen. Darwin continued his labors, however intermittently. When the time came to be counted, both Hooker and Lyell stood up.

---

* This assumption was shown to be correct in the 1950s and 1960s by the discovery of many such sunken peaks by oceanographic vessels studying the sea bottom. Darwin's theory about the buildup of coral had already been proved at Eniwetok Atoll. U.S. naval engineers had bored through more than four thousand feet of coral before hitting bedrock.

The coral reef paper was published in 1842, the same year in which Darwin's first outline on evolution was written. His energies were now wholly committed to transmutation. He had the theory; he needed more evidence. In 1844 he wrote his second outline, larger and more detailed than the first. It is interesting to speculate about what he might have done next, had he not been derailed by an event that he could not have anticipated.

In that same year a book was published anonymously: *Vestiges of the Natural History of Creation.* Darwin read it, of course, and found it to be a shrewd but scientifically inaccurate and rather jumbled effort to explain transmutation. The author of the book was Robert Chambers, a publisher and amateur scientist. He dared not put his name to it because he feared to risk the reputation of his successful publishing business by identifying himself with such a dangerous work.

The book is an interesting one. It recognizes the great age of the earth as explained by Hutton. It notes that in the oldest known rocks there are no fossils, meaning that those rocks were presumably older than life itself. It notes that the earliest fossils (found in somewhat younger rocks) are invariably those of simple marine forms; that increasingly complex ones, fossils of fishes, reptiles, birds, and mammals, appear sequentially in the geological timetable. It concludes from this that there has been progress upward, in short, evolution.

As to how this had taken place, Chambers had a number of ideas. He was acutely and painfully aware of variability in nature; both he and his brother had six fingers and toes. He was impressed by the malleability of species under domestic breeding and insisted that the same thing could take place in the wild state by various means, including the Lamarckian principle of inheritance of acquired characteristics. He even went so far as to suggest that complete new organisms as complex as flowers could be created spontaneously and that those various events of creation were still going on. He believed that there was a purpose in evolution and that the recent emergence of human beings could have been predicted before it happened—had there been a conscious observer on hand capable of making the observation.

Such disparate ideas could not be composed into a single theory of "how" except by invoking a higher agency, and there Chambers reached for God. His book is laced with references to divine agency. What it all boiled down to was a strong pitch for evolution as a fact, with God as the motor.

Vestiges was enormously popular, going through ten editions in a few years. Although it was a much more serious book than modern ones on astrology, or something like Erich Von Daniken's *Chariots of the Gods,* it caught the same kind of undiscriminating audience. But it also got the same kind of professional roasting. Without exception, all the prominent scientists whom Darwin knew savaged the book. His old Cambridge geology teacher, Adam Sedgwick, wrote a sulfurous review of it. T. H. Huxley dissected it unmercifully. Added to this din was a bellow of outrage from the Church. All this was actually far worse than anything Von Daniken got, because the Von Danikens of today are largely immune to social and religious ostracism. To Darwin, who saw *Vestiges* excoriated from pulpit and scientific lectern as a wicked and seditious work (Von Daniken escaped with mild epithets like "silly" and "ignorant"—and large royalties), the reaction was horrifying. The message to him was clear: talk about evolution in public and you get your head taken off. It didn't matter so much that Chambers had failed to find the right answer to how evolution worked; the espousal of the idea alone was deemed outrageous, even though he attempted to involve God in the process. Now here was Darwin, without God, engaged in the same unspeakable act.

In October 1844, Darwin turned away from transmutation to a far less hazardous task, a study of barnacles. He wore himself out on barnacles, learning things about them that no one had known before. He discovered that they were crustacean, not molluscs, as others had thought. He expended his fitful strength on them for eight years that culminated in the publication of a monumental four-volume work on those humble creatures that still stands as a landmark in its field. Recently the biologist and historian of science Stephen Jay Gould, in a charming essay on Darwin's last published work—a study of earthworms—made the point that, whatever Darwin wrote about, he always managed to relate that subject, far from the mark as it might seem, to the consuming interest of his life: evolution. That is certainly true, although it may not have been Darwin's conscious purpose at the time. In the case of barnacles, they probably came as a blessed relief from the study of evolution. They were something he was interested in; they were an immense, largely untouched subject that could be approached in the slow, painstaking way in which Darwin liked to work; they were spread in a great variety of types all over the world; many of their vital processes were poorly understood; it was not even known whether male barnacles existed among certain species (Darwin found minute males). But high on his list must have been the desire to lay down the frightening burden of transmutation for a while and tackle something easier and safer. Barnacles certainly were safe.

3

A Shock from the Spice Islands. The Shocker: Alfred Russel Wallace.

V*estiges* may or may not have blown Darwin off course on an eight-year barnacle odyssey, but another publication, released in 1855, must have hit him like a line squall. It was a paper in a scientific journal, the *Annals and Magazine of Natural History*. Its title: "On the Law that has Regulated the Introduction of New Species"; its author: Alfred Russel Wallace; its thesis: life was not constantly being created anew, but new forms were gradually evolving out of old ones. As Wallace put it: "Every species comes into existence coincident in time and space with a preexisting closely allied species." He was unable to explain the "how" of this, but he was right on target as to the "what."

This paper had come floating out of the blue from the Spice Islands of Southeast Asia. Darwin read it with alarm. He knew of its author only as a far-traveling professional collector of botanical specimens and insects. He spoke to Lyell and Hooker about it; had they read it? Both had. He fretted and dithered. Cautious and private as he was about evolution, it was *his* child. Indeed it was so much his child, his reason for existence, that he had already left instructions that on his death his wife was to spend up to four hundred pounds to publish his 1844 manuscript, the second, longer version of his ideas. If he could not bring himself to speak out on evolution during his life, he would do so from the grave. Now was some stranger going to stumble in, get the credit, grab his prize?

Yes, said Lyell, if Darwin didn't stir himself.

By this time Lyell was fully aware of Darwin's long and secret labors on evolution. Interest in the subject was growing. A book had recently been published by an Oxford geometry professor; it covered much the same ground that Wallace had. Lyell pointed out that the container had begun to spring dangerous leaks and might soon burst. All Darwin had left was his patent on natural selection — and he had not yet applied for that patent.

If he wanted the field to himself, Lyell said, he had better publish. Reluctantly Darwin agreed to put together "a very thin and *little* volume" that might take him a couple of months to write. But when he sat down to the job he realized that a little volume would not do. Ferocious blasts at Chambers's *Vestiges* were still being fired off eight years after publication. Darwin did his best to ignore the ecclesiastical ones, although they frightened him. He was more intimidated by professional charges that Chambers's arguments were not scientifically tight. If Darwin were to venture onto that battlefield, it would have to be with a case that was bulletproof from every angle, backed by an entire arsenal of evidence. In May 1856, he set grimly to work on a mammoth book that would wrap up the subject for good and all.

Meanwhile he found himself involved in a curious, almost coy correspondence with the man who had stirred him into action. It began when he received a letter from Wallace, who wanted to discuss the very things Darwin most wanted to avoid discussing.

Yes, Darwin replied, we think alike. Then he went on to stake out priority in that field: "This summer will make the 20th year(!) since I opened my first note-book, on the question of how and in what ways species & varieties differ from each other. I am now preparing my work for publication . . . but do not suppose I shall go to press for two years." After explaining how long he had been at work, Darwin went on to emphasize how complex the task was, how much more deeply he had gone into the subject than Wallace had, and that his views could not be summarized in a letter. Clearly he wanted Wallace browsing in another pasture.

Having, as he thought, disposed of Wallace, Darwin returned to work, ill, exhausted, and distracted. One of his daughters was ailing. His youngest child, a boy, was wasting away from an undiagnosed physical condition he had had from birth. Even the spirits of the indomitable Emma were low. Darwin worked on. By June of 1858 he had written a quarter of a million words.

On June 18 he heard from Wallace again. This time the letter was accompanied by a manuscript. Darwin read it with horror. It was about evolution and so nearly matched Darwin's own ideas on the subject that he might have written it himself. He finished it in a state of paralyzed despair, rousing himself finally to write to Lyell:

> Your words have come true with a vengeance — that I should be forestalled. You said this, when I explained to you here briefly my views of "Natural Selection" depending on the struggle for existence. I never saw a more striking coincidence; if Wallace had my MS. sketch written out in 1842, he could not have made a better short abstract! Even his terms now stand as heads of my chapters. . . . So all my originality, whatever it may amount to, will be smashed.

Lyell immediately got hold of Hooker, and the two men paid a call on Darwin, prostrate in Downe. After some discussion, to which Darwin was able to contribute almost nothing beyond the opinion that it would be dishonorable for him to suppress Wallace's paper while rushing something of his own into print, he was persuaded to dig up his 1844 manuscript and also an abstract of it that he had sent not long before to the American botanist Asa Gray for comment.* His two friends studied these documents, boiled them down into a summary statement — Darwin was incapable of doing it himself — and made a joint presentation of it *and* the Wallace paper two weeks later at the next meeting of the Linnaean Society, a London scientific organization to which they and Darwin belonged. They gave to both papers the following introduction, which they had written:

> The accompanying papers, which we have the honor of communicating to the Linnaean Society, and which relate to the same subject, viz. the Laws which affect the Production of Varieties, Races and Species, contain the results of the investigations of two indefatigable naturalists, Mr. Charles Darwin and Mr. Alfred Russel Wallace.

---

* Gray later became a champion of Darwinism in the United States.

> These gentlemen having, independently and unknown to one another, conceived the same very ingenious theory to account for the appearance and perpetuation of varieties and of specific forms on our planet, may both fairly claim the merit of being original thinkers in this important line of inquiry. Neither of them having published his views, though Mr. Darwin has for many years past been repeatedly urged by us to do so, and both authors having now unreservedly placed their papers in our hands, we think it would best promote the interests of science that a selection from them should be laid before the Linnaean Society.

Meanwhile the Darwin family was swept by scarlet fever. The damaged little boy died. Darwin continued in a state of collapse, unable to go to London for the Linnaean meeting. Only after it was over did he learn from Hooker that the two papers had been received respectfully but without comment—or even much comprehension—by the members. The reaction, in fact, was surprisingly muted. The president of the society noted in his annual report that the year past had not "been marked by any of those striking discoveries which at once revolutionize, so to speak, the department of science on which they bear."

In due course Hooker informed Wallace of the reading. Wallace's reply indicated complete satisfaction with the arrangement; he asked only that copies of his paper be sent to some of his friends.

Who was Wallace, and what had he said?

Alfred Russel Wallace was born fourteen years after Darwin, in conditions conspicuously different. His father was an idle, deeply religious, apparently gullible and stupid man. He had lived comfortably in London on an inherited income of five hundred pounds a year until, after marriage and six children, he found that this sum would no longer cover expenses. He moved to a remote and little-traveled part of England on the river Usk. There the living was cheap and the society primitive. Coracles, small boats made of wickerwork and skins, were still used on the river, and there was a great deal of local subsistence fishing. In Usk the senior Wallace proceeded to lose half his inheritance on a foolish publishing venture, and the rest in a building swindle that also gobbled up the small inheritances of his children. Those disasters broke up his family. The father pottered about, went to church, read his books, played the role of a bewildered, decaying gentleman, and watched his gently bred children go out to work, one by one, as soon as they were old enough to go.

In Alfred's case the day came when he was fourteen. He had been an excessively shy little boy, but stubborn and determined to go his own way. Without playmates, he made his own toys, invented his own games, came early to a strong belief in simple living (the toys you make yourself are the best), later in simple societies (natives of Borneo are nicer and on the whole more honest than Englishmen), and in everyone's right to self-respect and an independent life.

Wallace's first job was in London, working for an older brother who had made a start as a carpenter. From the age of fourteen he was self-supporting, but for many years nearly penniless. Everything he learned—optics, mathematics, surveying—he learned by himself, from books and from going to night courses for workingmen. It was at one of those that he first came in contact with the ideas of the social reformer Robert Owen and drew from him the belief that there was a mix of heredity and environmental influence in all men—and that the ultimate destiny of each man was determined not by himself but by the working of those two forces. Free will was an illusion.

He speculated about God:

Is God able to prevent evil, but unwilling? Then He is not benevolent.

Is He willing but not able? Then He is not omnipotent.

Is He both willing and able? Then whence all the evil in the world?

These thoughts, plus others about hellfire and punishment, convinced Wallace that all religions were degrading. He became an agnostic.

He was a gangling, bony, almost inarticulate young man, lonely, diffident, highly intelligent, highly observant, a dry sponge ready to soak up anything and everything. As a professional surveyor he encountered his first fossils and realized with a rush of joy that there was such a science as geology; he read Lyell. He stumbled over botany; he had not realized that it was a discipline that could be learned, one that men gave their lives to. He discovered beetles and became an obsessive collector of them—his *first* parallel with Darwin—and as with Darwin they led him eventually into a life of science.

He read Chambers's *Vestiges.* He read Malthus. He read Darwin's *Journal of the Beagle* and hankered to go to South America himself and collect exotic specimens. He spoke to a man at the University of Bristol about it and was told that if he collected enough valuable specimens in South America they could pay for a trip there. He met a beetle expert named H. W. Bates and through him found an agent to handle specimens. In 1849 he sailed for the Amazon.

Like Darwin before him, Wallace was overwhelmed by the beauty and richness of the South American jungle. He was a stoic man, at home with his own company. Collecting assiduously, he traveled thousands of miles by canoe in the Amazon basin, slept in native villages, got dysentery and fevers, recovered. He compared the "tame" Indians from around the towns—those who wore clothes and worked as laborers—with the far nobler ones that lived in the interior. The latter were totally independent (that word keeps cropping up with Wallace), made their own clothes and weapons, got their own food, ignored the white man. Those were the ones he admired; they were in tune with their environment.

He collected more than a hundred new species of fishes and innumerable plant and insect specimens; the totals will never be known because most of them perished in a fire at sea when Wallace was bringing them home. His ship went down and he spent ten days adrift in a small rotting boat that barely made it to Bermuda before it also sank. But he had sent home several shipments previously and had already identified himself to scientists and museums as a meticulous and reliable collector, one who could live on almost nothing in the roughest of places and come out with perfectly preserved, perfectly described specimens.

Back in London, he met professional scientists for the first time, among them T. H. Huxley, whose erudition staggered him. He published a small book, *Palms of the Amazon and Rio Negro,* putting up the money for a tiny edition of 250 copies and barely getting his money back. He learned that the largest blank spot on the collecting map was the Malay archipelago and went there.

He spent eight years in Borneo, Sumatra, the Celebes, and many of the smaller islands. His observations of the distribution of species were keen. He noted that there was one kind of flora and fauna on one side of an imaginary line that threaded its way north through the islands between Bali and Lombock, and an entirely different population of things on the other side. He deduced that it marked an ancient geological separation, and he was right: the division is still recognized as Wallace's Line. Always he collected: 100 reptiles, 300 mammals, 7,500 shells, 8,050 birds, 13,100 butterflies, 83,200 beetles, plus 13,400 other insects. His total haul was 125,660 specimens. He became the greatest collector that Southeast Asia had ever known and a legend back home.

He became a friend of James Brooke, the White Rajah of Sarawak in Borneo. He lived comfortably with Dyak headhunters who complained to him of the loss of the good old days when anyone, feeling restless, could go out and take his chances on collecting a head, a practice that the Rajah had abolished. Wallace himself collected orangutans and shot one out of a tall tree one day. When it came crashing into a swamp he discovered it to be human, a "wild woman" of the woods. She was dead, but the baby that fell with her was not. He took it home, learned to feed it, found a monkey playmate for it. It thrived.

He cruised the Moluccas, a chain of remote islands between Borneo and New Guinea, suffering intermittent bouts of fever as his health gradually gave way. He went by outrigger canoe from Goram to Waigiou and Ternate:

> My first crew ran away in a body; two men were lost on a desert island and only recovered a month later after twice sending in search for them; we were ten times run aground on coral reefs; we lost four anchors; our sails were devoured by rats; our small boat was lost astern; we were 38 days on a voyage that should not have taken twelve; we were many times short of food and water; we had no compass-lamp . . . and to crown it all, during our whole voyage occupying in all 78 days, we had not one single day of fair wind.

Collecting in the Celebes was more than just walking cool jungle paths. Between 1821 and 1851 eight of Europe's most distinguished zoologists either were murdered or died of disease in the East Indies. In Gilolo Wallace came down with such a severe bout of fever that he had to lie over there for several weeks while he recovered his strength. It was from Gilolo that he mailed the outline on evolution that had sent Darwin crashing.

Like Darwin before him, Wallace, in his meticulous descriptions of the multiplicity of specimens he was collecting, could not escape the dazzle of variation in nature. He became acutely sensitive to the geographical distribution of animals and plants, and particularly to the differences between forms in the tropics of the Eastern and Western Hemispheres. He had explored both and realized that those profound differences — in similar environments — had never been properly explained. He concluded that an evolutionary force was at work in the world; it became the core idea of the paper that he sent from Sarawak to London for publication in 1855.

As a result of that publication he received Darwin's first letter to him and wrote his beetle-collecting friend Bates that Darwin agreed "with almost every word" of the paper. He continued by dutifully reporting to Bates the message that Darwin had been at such pains to transmit: "He is now preparing his great work on 'species and varieties' for which he has been collecting material for 20 years."

In further exchanges with Darwin he did not get much out of him, and the correspondence languished. Three years later, recovering from fever in Gilolo, lying in bed with nothing to do but think, he again turned his mind to the subject of species. He later wrote:

> At this time I had not the least idea of the nature of Darwin's proposed work nor of the definite conclusions he had arrived at, nor had I myself any expectation of a complete solution of the great problem to which my paper was merely the prelude. . . . My paper written at Sarawak rendered it certain in my mind that the change had taken place by natural selection and descent, one species becoming changed either slowly or rapidly into another. But the exact process of the change and the causes that led to it were absolutely unknown and appeared inconceivable.

With the clarity of mind that sometimes follows a fever, Wallace addressed himself to the mystifying problem that Darwin had faced: how do species change and yet always retain their identity as species? He wrote:

> The great difficulty was to understand how, if one species was gradually changed into another, there continued to be so many quite distinct species. . . . One would expect that if it was a law of nature that species were constantly changing so as to become in time

new and distinct species, the world would be full of an inextricable mixture of various slightly different forms, so that the well-defined and constant species we see would not exist.

He asked himself three tough questions:

How and why do species change and yet retain their specificity?

How do they become so precisely adapted to distinct modes of life?

Why do all the intermediate grades die out and leave only clearly defined and well-marked species?

As Darwin's had, his mind wandered to Malthus and to what he had read therein some years before. The answer fell into his open consciousness with a clang. Unlikely as it may seem, two men on different sides of the world, each following his own thought processes, each bringing those thoughts into focus through the agency of Malthus, had hit on the same answer to the riddle of evolution: selection through survival of the fittest. One had gnawed at it for twenty years and was still laboring to get a torrent of ideas, arguments, and evidence down on paper. The other captured it as effortlessly as he might have plucked a ripe fruit from a tree. He wrote it all up in a few days. To cap the incongruity, he mailed it off to the other man. This is one of the most extraordinary coincidences in the history of science.

Darwin's version, spelled out in detail in the *Origin of Species* and available today in numerous editions throughout the world, is well known. Wallace's is not. It is worth repeating here what he said later about his theory and the revelation that brought him to it:

> Why do some die and some live? The answer was clearly that on the whole the best fitted lived. From the effects of disease the most healthy escaped; from enemies the strongest, the swiftest or the most cunning; from famine the best hunters. . . . Then it suddenly flashed on me that this self-acting process would *improve the race,* because

(and here is the kernel of it)

> in every generation the inferior would inevitably be killed off and the superior would remain — that is, the fittest would survive. Then at once I seemed to see the whole effect of this, that

> when changes of land or sea, or of climate, or of food supply or of enemies occurred . . . and considering the amount of individual variation that my experience as a collector had shown me to exist, then it followed that all the changes necessary for adaptation of the species to the changing conditions would be brought about. . . . In this way every part of an animal's organization would be modified, exactly as required . . . the unmodified would die out, and thus the *definite* characters and the clear *isolation* of each new species would be explained.

We can only imagine the moment of utter breathlessness that must have seized Wallace as the implications of this colossal insight hit him — his vision streaming backward through the ages like a comet to a world of simpler beginnings. There he lay, his body wasted by fever, his strength almost gone after eight years of unremitting hardship, but cool and refreshed now, recovering in the small shack he had rented, as the unbelievable tendrils uncurled in his head.

> The more I thought it over the more I became convinced that I had at length found the long-sought-for law of nature that solved the problem of the origin of species. For the next hour I thought over the deficiencies in the theories of Lamarck and of the author of the *Vestiges* and I saw that my new theory . . . obviated every important difficulty.

In this recollection he made no mention of Darwin. The reason is obvious: at that time he did not know that Darwin had also found the key to the cipher. Wallace believed that the idea was original with him. He would not learn the contrary until a year later, not until after Darwin had scrapped his enormous work on natural selection and embarked on a streamlined version. Darwin was now desperate to get into print and managed it in 1859, a year after both his and Wallace's papers had been read before the Linnaean Society. This shorter version was the *Origin of Species,* and Wallace was sent a copy. He wrote of it: "It will live as long as the *Principia* of Newton. Mr. Darwin has given the world a *new science,* and his name, in my opinion, stands above that of many philosophers of ancient or modern times. The force of admiration can no further go!!!"

Science is a competitive business, and it is getting more so. Its history is studded with examples of men losing all sense of dignity as they crowd into the limelight, jostling each other for credit they believe is theirs alone, trampling on each other to crush competing ideas. By contrast, the relationship between Darwin and Wallace has a sweetness about it that almost makes one smile. The two met for the first time in 1862. They corresponded regularly thereafter, differed on some matters—but never on the central one—and remained friends until Darwin's death. Wallace always deferred to Darwin. Darwin, having gotten over his original protective feeling about the theory, leaned over backward to make sure that Wallace was not overlooked. When Wallace was having a hard time financially, Darwin applied to the government for a pension for his distinguished colleague. He had the pleasure some time later of being informed by Prime Minister William Gladstone that it had been granted.

Neither man pushed the other—ever. Luckily for them, they saw their joint discovery in somewhat different ways. For Darwin it was his life's work. Although he was slow and timid about it, there was a doggedness about him that transcended illness, public obloquy, and the alienation of friends. He had seen a vision and was determined to claim it. Everything else in his life was small by contrast.

For Wallace it was a sudden glorious inspiration interrupting a life that was devoted to something else: collecting. If he could get credit from his peers for sharing the inspiration, that was enough for him, and he got it. Darwin seemed humble but truly was not. Wallace truly was. He was awed by the years of thought and investigation Darwin had put into the subject, casting up one idea after another, turning the theory around and around, examining every facet of it for flaws.

"I never could have approached the completeness of his book," Wallace wrote. "I really feel thankful that it has *not* been left to me to give the theory to the world."

After returning to England Wallace spent a couple of years recovering his health while he went about the slow job of organizing and cataloging his enormous haul of 125,000 specimens. It was so great that he had to give out sections of it to be worked on by others. He wrote a number of papers on birds and insects, gave exhibits of his unparalleled collection of bird skins, and in 1868 published his most popular book, *The Malay Archipelago*. All these things he did slowly, modestly, and in a small way. He had now turned forty, was still poor, but was becoming more and more celebrated as a scientist. His knowledge of what he had collected was uncanny. Once, looking over the collection of the Reverend O. Pickard-Cambridge (another of that seemingly endless procession of English clerics interested in natural history), he exclaimed: "Why, that's my old Sarawak spider!"

Pickard-Cambridge had been greatly troubled by that spider. He did not know what it was or where it had come from, and he was extremely dubious when Wallace told him he had collected it a decade or two before and had not seen it since.

"If it is my old spider," Wallace continued, "it ought to have my own private ticket on the pin underneath." They looked; it was there. On the spot Pickard-Cambridge named it after him: *Friula wallacii.*

Meanwhile Wallace had met a Miss L____ (her name is not given in his official biography) and fallen in love with her. He called on her formally for a year, too shy to declare himself, and at the end of it proposed to her in a letter. She replied equivocally: no, but not necessarily a crushing no. So he continued his formal calls for another year and finally spoke to her father. An engagement was announced, but at the last minute Miss L____ abruptly broke it off. Wallace had been so silent, so mysteriously uncommunicative about himself that there must be something in his background that he wished to conceal, a shady past, another woman. She refused to see him when he called to explain. He never saw her again.

He became friends with another quiet man, William Mitten, a world authority on mosses, met his daughter, and this time succeeded in marriage. Wallace was forty-two, his wife eighteen. He had been earning enough from his collections and by writing articles to scrape along as a bachelor. But as a married man he needed a steady income. Modestly, as always, he put in for a post with the Royal Geographic Society; it went to his beetle-collecting friend Bates. He tried for the directorship of a new museum; it fell through. He put in for the conservatorship of Epping Forest and failed again. Each time this happened he shrugged it off, saying that it was probably for the best. He described himself as a lazy man who would never have done anything but garden if he had not been forced out into the world to make a living. In financial matters he was incredibly naive, and the sharklike world of English commerce must have made him wish more than once for the society of Dyak headhunters. The proceeds of his collection had been invested by his agent in safe securities that brought him three hundred pounds a year, but, like his father, he was relieved of it all by bad investments. He built a house and was swindled by an unscrupulous contractor.

His worst experience by far came at the hands of a "flat earth" fundamentalist named John Hampden, who had offered a prize of five hundred pounds to anyone who could prove the roundness of the earth by demonstrating curvature on the surface of a canal, lake, river, or railroad track. For Wallace, a surveyor, this was a simple exercise. He located a six-mile-long lake in Wales with a bridge at one end and a dam at the other. He painted a bull's-eye on the side of the bridge exactly six feet above the surface of the water, set up a telescope at the same elevation of the dam six miles away, and halfway between them lined up a stake with another six-foot-high bull's-eye painted on it. As Wallace knew it would, the stake stuck up higher than it should, proving that the surface of the lake was curved, and he claimed the prize.

Hampden, however, proved to be mentally unbalanced. He bombarded Wallace with a fusillade of vilification and court actions. Bewildered, Wallace tried to defend himself. Hampden was put in jail for slander, came out, resumed his attacks on Wallace, went to jail again. In all, Wallace was beleaguered by Hampden for nearly fifteen years and ended up paying court and legal fees at least as large as the prize he had gotten.

His financial affairs a shambles, he was rescued by an elderly female relative who had planned to leave him one thousand pounds. Instead she gave it to him outright, and it kept him afloat. To this was added the civil service pension that Darwin had secured for him, plus fees from a lecture tour of the United States (where he met President Grover Cleveland and was disgusted by the Americans' universal habit of chewing gum). He was able to live out his life in modest but comfortable circumstances. He became an ardent socialist and produced a stream of political pamphlets, along with papers on a great variety of other subjects such as phrenology, hypnotism, and spiritualism.

Despite his preoccupation with head bumps and table rapping—which now seem silly but then did not to great numbers of educated people—Wallace slowly became one of the grand old men of British science. Awards cascaded in on him. He was a pallbearer at Darwin's funeral and later won the Darwin Medal and the Medal of the Royal Society. He became head of the Entomological Society and President of the British Association. Up into his eighties he continued to be very active, taking long walks, always filling his pockets with insect and plant specimens. He horrified his housekeeper one day with a "great disgusting" beetle. Assuring her it was harmless, he put it on his nose where it promptly bit him, to the great amusement of his children. At ninety he was still vigorously writing pamphlets on social issues until, catching a chill one day in 1913, he was put to bed and died almost immediately. Wallace had long outlived the tumult and the shouting attendant on the publication of the *Origin of Species*. By the time of his death the theory that he and Darwin had jointly explicated had been accepted by scientists the world over.

4

The *Origin* Is Published. The Reaction.

Even before the appearance of the *Origin,* Darwin's publisher, John Murray, tried hard to get him to abandon his absurd ideas about species variation and concentrate on a more popular subject: his breeding experiments with pigeons. Darwin ignored him, ploughed ahead, and finished his manuscript, which was published on November 24, 1859, in an edition of 1,250 copies. Its full title was

ON THE ORIGIN OF SPECIES
By Means of Natural Selection
or the Preservation of
Favoured Races in the Struggle for Life

There had been a certain amount of prepublication rumor about it as the word went around that the reclusive genius at Downe was about to release his findings after twenty years of work. On the day of publication the edition sold out. From that moment on Darwin was in the eye of a social, ecclesiastical, political, and scientific hurricane. Some scientists were on his side. Many definitely were not.

Conspicuous among his supporters was T. H. Huxley, who has already been mentioned twice in this book and should now be identified. He was perhaps the most brilliant, widely informed, restlessminded, quick, aggressive person among all the professionals involved in the Darwinian controversy. The others tended to be academic types, steady, scholarly, devoted to research and to their teaching, courtly more often than not in their exchanges with one another, even in disagreement. But Huxley was downright pugnacious. He was the son of an obscure schoolteacher, and had made his way up the scientific ladder by sheer mental power and enormous industry. Always short of money, he was forced throughout his life to expend so much of his energies on teaching to support a large family that he never succeeded in the kind of pure scientific achievements that immortalized men like Darwin and Lyell. It is said that, on reading the *Origin,* he exclaimed, "How obvious. How stupid of me not to have thought of it myself!"

What was obvious to all was that if Darwin's extraordinary idea were true for the corals and barnacles and birds and plants that he used as examples in the *Origin,* it would have to be true for humans also. Magazines were flooded with cartoons of monkey ancestors sitting in trees, long-tailed, scratching themselves, but with the high-domed bald head and beetling eyebrows of Charles Darwin. He was abused in Parliament, mocked in music halls, ridiculed in the press.

Those were the easy things to ignore. Scientific blasts were not. Darwin's old friend and geology teacher, Adam Sedgwick, wrote to him about the book ". . . with more pain than pleasure. Parts of it I admire greatly, parts I laughed at till my sides were almost sore; other parts I read with absolute sorrow, because I think them absolutely false and grievously mischievous." He went on to charge Darwin with ignoring the link between science and God, a point that was taken up by the clergy, who, in one voice, rattled their pulpits with denunciation.

Darwin had anticipated the religious burst of outrage and had deliberately downplayed the role of humankind in evolution. In fact, he mentioned man only once in the *Origin*—in the last chapter—and there in a single sentence: "Much light will be thrown on the origins of man and his history." But no careful reader missed it, and the tempest grew.

Through it all Darwin never emerged from Downe, hiding there like a badger in its burrow. He noted with dismay the opposition of Sedgwick and was saddened that neither Henslow, the botanist with whom Darwin had "walked" during his student days at Cambridge, nor Lyell, though they supported him, would go the last ideological mile. He brightened at an extremely laudable review that appeared in the *Times,* and discovered that it had been written by Huxley.

Huxley, indeed, had decided that Darwin, who could never be brought to defend himself, needed defending, particularly against Richard Owen, an able anatomist but a bitter and unscrupulous adversary. Owen wrote long, nasty, and misleading essays about the *Origin* and, of all its detractors, was the one who most angered the usually deferential, perennially ingratiating Darwin. He and his friends detested Owen for having degraded science by deliberately misunderstanding the *Origin.* The most inflammatory charge being made against it was that the book asserted that man was directly descended from living apes or monkeys, a horrid thought to most, and best expressed by the Bishop of Worcester's wife, who, hearing about it, said:

"My dear, let us hope that it is not true, but if it is let us pray that it will not become generally known."

Darwin, of course, never said that humans were directly descended from living apes, nor did the *Origin* imply it. On the contrary, his theory was that the two had a common ancestor from which both, by gradual differentiation through selection, had evolved along their separate paths. There is a vast difference between the two concepts. The former derides and confuses evolution; the latter explicates it. But this was by no means clear to a public consternated by the appalling idea of ape ancestors, ape brothers, ape cousins—the exact relationship did not really matter. Owen did nothing to clear it up. Instead, he muddied the waters by giving a quick and venomous cram course on the book to a popular clergyman who was scheduled to attack evolution at an upcoming date.

The clergyman was Samuel Wilberforce, Bishop of Oxford. He was a great public favorite, the producer of a stream of such smooth and oily sermons that he was nicknamed "Soapy Sam." He knew nothing of science, but Owen would take care of that. The event at which he was slated to demolish Darwin was the annual meeting of the British Association for the Advancement of Science, set for a Saturday in June 1860. This would be no routine, droning scientific exchange. It would be the first direct confrontation between the forces of the new evolutionary view and those of the entrenched view held by the Church and by catastrophist, fixed-species men like Owen. Both factions had had six months in which to declare themselves, and they had been doing so with increasing passion. Now they would meet.

The Association had a peculiar position in British scientific and intellectual circles. It was the forum in which confrontations of this sort could take place. As a result it often produced fireworks. With this reputation the meeting often drew large public crowds.

Friends of Darwin, aware that Wilberforce planned to speak, tried to make sure beforehand that his side would be represented. John Henslow decided to attend, although he found himself in a tight spot. He did not believe in evolution, but he was devoted to Darwin and was determined to see that he got a fair shake. He recruited Joseph Hooker, his son-in-law and Darwin's closest friend, to appear and, at the last minute, the fire-eater Huxley.

Huxley had planned not to attend the meeting, believing that the circus atmosphere of a mixed scientific and public debate would clarify nothing. His mind was changed when, crossing one of the quadrangles at Oxford, he met Robert Chambers, the anonymous author of the notorious *Vestiges.* By this time most people in the natural history field had learned who the writer of that infamous book was; Huxley knew him well. Although Chambers's ideas had been made obsolete by the superior ones of Darwin, he was still a fervent believer in transmutation. When he heard that Huxley planned to pass up the debate, he begged him with great earnestness not to. Huxley agreed to attend.

Actually, Huxley was already in the scrap. The day before he had faced down Owen in another confrontation, a separate debate over the relationships of humans, gorillas, and monkeys. Owen's chief contention was that it was totally wrong even to think about men as being closely related to any of the other primates. He regarded humans as belonging in a separate genus, order, and subclass from all the others. As it happens, he was right as to genus but dead wrong about the others; modern classification puts humans and apes, together with monkeys, in the order of Primates. Separating humans into a different subclass is preposterous.

Rising to defend his view, Owen made the fatal error of basing his argument on the differences between the human brain and the gorilla brain. He maintained that those differences were far greater than the differences between gorillas and monkeys—and even lower submonkeys like lemurs. Huxley, as it happened, had devoted two years to a detailed comparative study of humans and gorillas. He knew as much as anyone in the world did at that time about their differences and similarities. He had the backing of papers by a renowned German anatomist Friedrich Tiedemann. With great vigor he flatly contradicted Owen: the gorilla's brain was *closer* to that of a human — far closer — than to that of *any* monkey. Lashed so unexpectedly and so devastatingly by this passionate younger man, Owen — although he was generally regarded as one of England's leading comparative anatomists — had nothing to say in reply. It was a hideous embarrassment for him, made worse by Huxley's having used Owen's own data to demolish him.

Having disposed of Owen, and on the next day having met and been persuaded by Chambers to stay over for the main event, here was Huxley, on a beautiful Saturday in June, on the platform again, this time with Hooker, to defend — if need be — the cause of evolution. The need became apparent very early. A crowd of nearly a thousand had tried to jam its way into the lecture room, necessitating its removal to a much larger hall. Now that hall was stuffed; many of those present were noisy partisans of Bishop Wilberforce, impatient for the bloodletting to begin. Owen had been scheduled to chair the meeting but did not show up — presumably still smarting from his savaging by Huxley two days before. His place was taken by the gentle Henslow, who was scarcely equipped to keep order in so emotionally charged an arena. Early speakers drew catcalls from the audience. When Bishop Wilberforce finally rose there was a flutter of handkerchiefs waved from the windowsills where a platoon of young women, staunch supporters all, were perched.

The nickname "Soapy Sam" was well earned. Wilberforce was an able and persuasive speaker, and he loved the sound of his own voice. But his ignorance of science quickly became clear to Huxley, as did the fact that he probably had not read the *Origin,* or if he had, certainly did not understand it. He talked smoothly and condescendingly on a wide range of topics including Egyptian mummified cats. But his main thrust (again an attack on something that Darwin had never said) was that there was absolutely no evidence that one kind of animal could be turned into another; no breeder had ever succeeded in doing so. Then, turning with a courtly gesture he said:

"I beg to ask you, Professor Huxley, is it on your grandfather's or your grandmother's side that you are descended from a monkey?"

There was an explosion of applause, under which Huxley turned to the man next to him and whispered: "The Lord hath delivered him into mine hands."*

Puffed with self-satisfaction, Bishop Wilberforce sat down, and the room quieted for the response of Huxley, a pale, stern-looking man with large features and deep-set eyes. Stating quietly but forcefully that he was there only in the interest of science, and not to distract his listeners with eloquent and emotional appeals to religious prejudice, he then tore Wilberforce apart, point by point, making it clear that he knew nothing about the subject of evolution and had no business talking about it.

As for ancestry, he concluded, turning directly to Wilberforce: "I assert that a man has no reason to be ashamed of having an ape for a grandfather. If there were an ancestor whom I should feel shame in recalling, it would rather be a *man* endowed with great ability and a splendid position who used those gifts to obscure the truth."

Pandemonium! The students, who at the beginning had been noisily on Wilberforce's side, were completely turned around. They shouted and stamped their feet. A high-born woman, Lady Brewster, who had been sitting in the audience, was overcome with the vapors and had to be carried out. Various clerics and would-be rebutters bobbed up here and there—among them, like the Ancient Mariner himself, croaking as it were from beyond the grave and waving a Bible, Darwin's old Captain, Robert Fitzroy, ready to fire one last broadside and sink him forever. But Henslow, with an uncharacteristic show of force, declared that only scientific matters would be allowed for discussion and shut them all off. Hooker then stood up and completed the dismemberment of Wilberforce with a cogent summary of his own.

So ended the first great battle for the new evolutionary theory. In the minds of those who "counted," the scientists and critics who were dispassionate enough and informed enough to make sensible judgments, Darwin had carried the day. Hooker, writing to him afterward, summed it up:

* In an article published in 1986 in *Natural History* magazine, scientific historian Stephen Jay Gould reviewed all the events of that dramatic day and ended up slightly dubious about whether Huxley actually made the famous remark. He also gave more credit to Hooker than to Huxley for demolishing Bishop Wilberforce. Hooker's letter, quoted on page 90, would seem to bear out Gould's conclusion.

> Well, Sam Oxon [as Wilberforce, the Bishop of Oxford, was usually referred to] got up and spouted for half an hour with inimitable spirit, ugliness and emptiness and unfairness. I saw he was coached up by Owen and knew nothing, and said not a syllable but what was in the reviews; he ridiculed you badly and Huxley savagely. Huxley answered admirably and turned the tables. . . . The battle waxed hot. Lady Brewster fainted, the excitement increased as others spoke; my blood boiled. . . . Now I saw my advantage; I swore to myself that I would smite the Amalakite, Sam, hip and thigh if my heart jumped out of my mouth. I handed my name up to Henslow, the presiding president, as ready to throw down the gauntlet. There I was cocked up with Sam at my right elbow, and there and then I smashed him amid rounds of applause. I hit him in the wind at the first shot in ten words taken from his own ugly mouth; and then proceeded to demonstrate in as few more: (1) that he could never have read your book, and (2) that he was absolutely ignorant of the rudiments of Bot. Science. . . . Sam was shut up—had not one word to say in reply, and the meeting was dissolved forthwith, leaving you master of the field after 4 hours' battle.

Would that it were so easy. The Owens and the Wilberforces retired from the stage, but they continued to mutter in the wings. They found many listeners. An idea as shocking as Darwin's, presented in a society that for hundreds of years had believed literally in Adam and Eve, and whose monarch was sworn to defend the Faith, could not be swallowed without a great deal of gagging. Darwin, for the rest of his life, was regarded by many people as the most vicious man to have appeared in the Christian world since Pontius Pilate.

He resigned himself to that and continued to live the life of a semi-invalid and semi-recluse at Downe. In due course most of his friends came over to his side, one of the last to do so being Lyell. By this time both Lyell and Hooker had so distinguished themselves that they were knighted. That honor never came to Darwin; he was far too controversial. He went on working in the way that most appealed to him: long, slow, thoughtful examinations of natural history matters that had a bearing on evolution. He published a two-volume work, *The Variation of Animals and Plants Under Domestication* in

1868, and *The Descent of Man* in 1871. Both are, in a large sense, continuations of the *Origin.* The first makes an effort to elucidate the causes of variation, the second to apply evolution to the human species. Darwin, being the kind of thinker he was, determined to get down to unassailable, rock-bottom, logical first principles, realized only too well that the theory of natural selection did not really satisfy unless an explanation of variation and heredity was added to it. How, in short, had the first primordial microbe spawned a second that was not exactly like it? He was similarly plagued by the rise of intelligence in humans. He had made a long study of instinct in animals and had come to the conclusion that since it varied, it, too, was subject to selection. The same, he reasoned, must hold for intelligence in humans. But how had it first come about?

In asking that question, Darwin was inching toward a problem that would trouble evolutionists a hundred years later: how much of our behavior is inherited—instinctual—and how much of it is learned? Whatever the mix should turn out to be, Darwin was convinced that human intelligence had evolved, just as the brain that did the thinking had. Then, by logic, if evolution had brought us to where we are now, there was no reason to suppose it would stop.

Here Darwin and Wallace parted company. Wallace believed that the process had ended with humans and that no further evolution could take place. He held that the human species had achieved an intellectual and moral plateau that must have originated in "the unseen universe of the spirit." That was his rather fuzzy way of acknowledging some kind of first cause without bending to ordinary religious belief, which he had abandoned in disgust as a young man. The "universe of the spirit" might also explain his fascination with spiritualism; for him it could have been a way of linking conscious, breathing man with that unseen world.

Darwin took a much harder, more rationalist view of human evolution. He had stuck a toe into the swamp of behavior—instinct, intelligence, emotions—and had come away convinced that there was a continuum of sorts between human beings and the lower animals. He found reasoning ability in dogs, and useful behavior of one kind or another in all sorts of lower animals all the way down to worms. Those abilities, according to his theory, must have evolved from even lower, even simpler beginnings. Put more bluntly than he ever put it, behavior (including the ability to think) had to start somewhere. That being so, and accepting the theory of natural selection, human behavior could be traced backward to those simpler beginnings.

There are two different, but related, points here, and it is important to note their difference and their relatedness. The first is that *all* of evolution depends on variation. The second is that behavior is as variable as skin color and thus must be equally amenable to the laws of natural selection. Darwin was astute enough to recognize that the second point was a special application of the first, and that the real key to it all was variation.

Darwin broke his lance on variation. In his book he looked at it from every imaginable angle. He measured it, weighed it, tracked it, but he could not explain it. The reason is that he was looking in the wrong places and in the wrong way. The answer to that enormous question of *how* it happened was locked up inside the cell and was destined to remain locked there until the next century. In Darwin's day cytology—the study of cells—was in its infancy. No one had even known that cells existed until they were discovered by Robert Hooke in 1665. The microscope had been invented not long before and Hooke was one of the first users of it. He was astonished when, slicing an extremely thin sliver of cork and putting it under his primitive instrument, he saw it to be made of an enormous number of minute rectangular boxes—more than a billion of them in a cubic inch of cork.

Later improvements in optics, and the revelation that all living tissue was made of cells—as the tiny boxes came to be called—were followed by the discovery of one-celled creatures, myriads of them, heretofore invisible. The human skin crawled with them. The human mouth was infested with them. The human intestine swarmed with them. There were hordes of them in every drop of brackish water: a teeming universe down there, a subvisible world of fantastically variable, wriggling, jerking, snatching, gobbling life that no one had ever dreamed existed. Darwin's closest approach to that world came on the eve of his departure on the *Beagle* voyage, when he paid a call on the botanist Robert Brown.

Brown had had the inspiration to concentrate, not on the little boxes (that is, the cell walls) but on the contents of the boxes, the jellylike liquid that seemed to contain dots of matter moving about in it. He also discovered a larger dark spot in the cell. He named it the nucleus, assuming that it had some central role in the cell's function. But Brown was a strange, secretive man. He refused to publish papers about his work. He refused to speculate. When Darwin was shown some green particles streaming about in a plant cell under Brown's microscope and asked what they were, Brown refused to tell him. He refused to tell anybody much about anything, and died without ever having revealed what—if any—were his thoughts about how cells worked.

No one else knew much about cells. Darwin, struggling to find a cause for variation and heredity, never tried to poke into cells. In frustration he was drawn closer and closer to the Lamarckian belief that acquired characteristics could be inherited. There, perhaps, in the changes that animals wrought in themselves during life, could the secret of variation be found. He had dismissed that idea as a younger man. Now, in desperation, he grasped at it.

Outwardly his life became increasingly serene. His reputation as the greatest living naturalist flew around the world. He was taken up with enthusiasm in America. One faction at Harvard embraced him. Another, led by the renowned geologist and glaciologist Louis Agassiz, fought him bitterly. At home philosophers and politicians and even some clergymen learned to live with his ideas. Indeed, they exploited him. Confronted by the dilemma that has always existed in Christianity—why does a benevolent God permit all the cruelty and suffering in the world—they solved it by turning to the struggle for survival and finding in this evolutionary mechanism a justification for that suffering. It also justified a stratified society of dukes and earls, landed gentry, ordinary working people, paupers, madmen, and, in some societies, slaves. If the world was indeed evolving in response to the constant struggle that Darwin identified as its driving force, then it was clear that the people at the top were the winners and deserved to be there, just as the paupers deserved to be at the bottom.

That is the essence of social Darwinism, an intellectual force that became widespread toward the end of the nineteenth century. Just as was done by Owen and Wilberforce when they misinterpreted Darwin in their attempts to destroy the very concept of evolution, so does social Darwinism misinterpret him in its efforts to justify the inequities in society. Darwin never meant his theory to be applied that way, and—once again—he never really said the things that social Darwinists were saying to excuse themselves. His "struggle" starts with the appearance of superior traits in an organism, traits that make the organism more apt to survive and reproduce itself during its lifelong attempt to cope with *all* the elements of its environment. He did not see "struggle" as something to be engaged in purposefully by humans, singly or in groups, smashing other people in their efforts to conquer them, enslave them, kill them, or even just to hold them down by laws and customs. Social Darwinism did see it that way. It was a pervasive influence for several decades. But, as sociology, anthropology, and biology have expanded and become better acquainted, social Darwinism has become an anachronism, cropping up only in the fevered claims of demagogues like Hitler for the purity and superiority of certain groups, certain races, certain nations.

This application is a queer turnaround. A few hundred years ago science was regarded with huge suspicion by the Church and by established society. Since then it has been embraced in scores of ways to justify repression. And it goes on. It is sad but true that anachronisms do not disappear. They just become increasingly odd and illogical. But there always seem to be people willing to believe the illogical and willing to perpetuate its damage.

One good thing came out of social Darwinism. It solidified Darwin himself. Once authority found him useful and began quoting him, he ceased to be the great threat to society that he first seemed. Nevertheless, he was far from out of the woods. Science now began to turn its slow but penetrating gaze on natural selection. This is in the best tradition of science, but it gave Darwin and his theory some dreadful moments.

We have already emphasized that the concept of natural selection depended initially on the acceptance of two fundamental ideas: the idea that the world was old and the idea that species were not fixed, that there had been change between living and fossil forms. By Darwin's time, both ideas had been accepted by most as proved fact, and the arguments about them were essentially over. Their acceptance enabled him to postulate that selection based on extremely small increments of change could, over enough time, account for all change.

Fair enough. At first tiny change confers some kind of tiny advantage. Other minute bits of advantageous change accrete to it. In time the webbed foot of an aquatic animal like an otter replaces the unwebbed one of its less aquatic ancestor. The more webbing a foot has, the better a swimmer its owner will be, and the more apt it will be to spend an increasing amount of time in the water. Thus behavior is linked to physical change and evolves along with it.

But what about something like the eye? How can there be any advantage at all in the host of preliminary evolutionary steps that must be taken before the usefulness of the eye becomes apparent? To be useful, an eye must see. But if the steps that precede the ability to see — the gradual construction of an infinitely complex system of lens, iris, photoreceptors, and special nerves — if those steps are so minute and so primitive that they have nothing to do with seeing, cannot even predict it, what good are they? How will they be selected for?

Darwin was sorely taxed by the problem of the eye. It was not so much that he regarded it as invalidating his theory, but that he was afraid that this particular application of it would be so hard to swallow that the entire theory would be unacceptable to others. He was also aware that attempts to think through this problem could produce a chain of logic that would actually lead away from evolution.

For example, a solution to the eye problem would be to postulate, not a host of tiny changes, but a few larger jumps. In that way a dim but detectable future optical advantage could arise all at once and pave the way for others to be laid on top of it. A process like that could more easily be accepted as leading to the development of a complex organ like the eye. Unfortunately Darwin had no evidence of jumps of this sort. Moreover, he had to deal with the widely held dictum: *natura saltum non facit* — nature does not make jumps. It followed that if one were to depend on jumps that could not be accounted for in nature, they would have to be credited to God. And if one had to credit God with jumps, why not credit Him with the whole bundle? Why bother with evolution at all?

What Darwin did not know was that nature sometimes does make jumps, albeit small ones, a phenomenon that would be recognized later when mutations were discovered. Those small jumps, accumulating over time, would account for variation (an insight that Darwin would have been everlastingly grateful for). They would help explicate the eye argument, but they *would not* contradict natural selection; they would accommodate to it. The later knowledge would strengthen, not weaken, his basic argument, one reason being that the jumps, now known as mutations, are usually small — not unlike the minute variations on which he based his theory.

In short, Darwin's hunch was right; he just didn't have the mechanism down pat. Extraordinary change *can* eventually grow out of simple and apparently aimless beginnings. A tiny initial change may be neutral, neither good nor bad. If it is not bad, it will not necessarily disappear in the evolutionary struggle. It may hang on in numberless individuals for many generations until it is nudged by another inexplicable, neither helpful nor unhelpful, change. If, in time, that kind of slow accidental piling up kindles only the faintest glimmer of responsiveness to light in a few cells, then the whole miraculous optical apparatus that produces keen vision can follow without any intellectual gulping at all. Eyes, in fact, have appeared independently several times — and in several forms — in the course of evolution.

In the last analysis, the shaping and development of the eye is different only in degree of complexity from the shaping of the nose. Humans have noses that come in a bewildering variety of shapes, and it is safe to say that no shape is intrinsically superior to another so long as all provide a good flow of air to the lungs. And — looking at the wide sample of shapes that surrounds us today — it is also safe to assume that we have enjoyed a similar variety for many thousands of years. BUT — out of that variety has come a nose that *is* useful in a particular setting. With a choice between high thin noses and low broad ones, Eskimos tend to have low broad ones, presumably because those are less apt to freeze in extremely cold environments. Did they arrive in the Arctic with broad noses? Or did they develop them there? Who can say? It is likely that such noses evolved in the distant past, when human beings began to spread out from warm habitats into increasingly cold ones where they became tundra hunters and, through selection in response to the environment, developed the noses that best suited them. In any event, the development of the eye and the differences in noses both stem from small mutations. As with what we now know about the age of the earth, this is an instance where a later scientific discovery would bolster Darwin's theory in an area where it at first seemed weakest.

Another case that Darwin wrestled with was the habits of social insects. A honeybee, for example, defends its hive by stinging an invader. In the process the stinger is ripped out of the bee's body and it dies. As bee experts can testify, the behavior is instinctive (that is, developed through evolutionary selection and not learned by the bee). Yet, how can the urge to sting continue if the bravest and most loyal bees are the ones to die? Logic would suggest that the cowardly or lazy ones would be the ones most likely to perpetuate themselves, and that the urge to sting would disappear. This is an exceedingly complex question, to one aspect of which Darwin found an answer. The bees that sting are sterile females in whom selection cannot operate because they leave no offspring. But their behavior does benefit the queen, the egg layer, whose genes make it into the next generation.

Another soft spot in the theory had to do with the most inflammatory corner of it: evolution as applied to humans. If humans had evolved, the criticism ran, they must have evolved from more primitive ancestors. And if our closest living relatives were apes, then those ancestors probably looked more like apes than we do. Conceding that, and conceding that other living animals were represented by earlier fossil forms, where were the apelike human fossils? Embarrassingly, there weren't any.

More accurately, there was one, the first Neanderthal skull, discovered in Germany in 1856. But so locked-in were the minds of the scientists who examined it and spoke about it, that it was uniformly accepted as a deformed or degenerate example of a modern human. For Darwin, publishing in 1859, there were no fossil ancestors to fall back on. That is one reason why he limited himself to his one unobtrusive sentence about humans in the *Origin*. He wrote about that to a friend: "With respect to man, I am very far from wishing to obtrude my belief; but I thought it dishonest to quite conceal my opinion." Many thought that, in being so timid, he *was* dishonest. Loren Eiseley, in reviewing the matter, thought not; he believed that Darwin was simply being prudent: "There can be no doubt, considering the temper of the times, that Darwin's caution was well justified and probably had the salutary effect of broaching what was then an unpleasant topic by successive doses that were found assimilable [that is, a first small dose in the *Origin,* a second larger one in *The Descent of Man*], rather than as Lyell was accustomed to say, 'going the whole orang' all at once."

Had Darwin lived another sixty years he would have seen a miraculous enrichment of the supply of human fossil ancestors, the oldest and most primitive of which go back four million years. The knowledge gleaned from those fossils, much of it gained only in the last couple of decades, would have stunned Darwin's contemporaries.

The human fossil criticism, irritating and frustrating though it was, was a negative one, based on a lack: "You have no fossils, therefore you must be wrong." It could always be countered by the weak but reasonable answer: "Wait, we'll find them." Darwin scarcely could have hoped for the profusion of fossils that now rest in museum drawers around the world and constitute the first step in an overwhelmingly powerful confirmation of his original timid suggestion: man has evolved from lower forms.

On the whole Darwin managed to deal fairly well with all the aforementioned questions and criticisms, although he did write to Asa Gray: "To this day the eye makes me shudder." He anticipated those difficulties in the *Origin,* he thought them through, argued them cogently, and gave many examples to support his contentions. More importantly, in the years that followed he remained for the most part undisturbed about his theory. It satisfied, as far as he could tell, any objection that could be leveled at it. Yet there were a couple of pure scientific shots launched at him to which he could find no answer. Then the doubts went very deep.

He received from Scotland two rounds amidships—right on the waterline—and the sea came pouring in. One was in the form of a closely reasoned attack published in the *North British Review* in 1867. Its author was a hardheaded, mathematically sophisticated, logically tenacious engineer, Fleeming Jenkin (misspelled Jenkins in many books). Jenkin went after Darwin in the admittedly soggy area of blending. Darwin had gotten around blending by specifying small populations as a way of getting variety established. Jenkin demolished that argument with a barrage of statistics and mathematical logic. He pointed out that a single variation introduced into a population of any respectable size would be swamped almost immediately. He cited the hypothetical example of a white man cast away on a tropical island populated only by black people. No matter how industriously the castaway impregnated the native women, Jenkin pointed out, the population would never become white. Indeed, his presence there would, in a few generations, be obliterated. It would be as if a trout had swirled its tail in a running stream, tossing up a few grains of sand, which, settling again, would leave no trace of the fish's passing.

The only way total dilution through blending could be overcome, Jenkin went on remorselessly, would be for *a large number of identical variations to be introduced simultaneously.* That, of course, was so unlikely—other than by divine intervention—that it threw a wrench into the whole evolutionary scheme.

Darwin was thoroughly shaken by Jenkin's argument. Had he but known it, an answer had already been provided two years before by an obscure Austrian monk. That monk, Gregor Mendel, had delivered a paper to a local scientific society, demonstrating that traits did not blend, and that swamping did not take place. But his proof, based on numerical ratios, was not understood by his audience, nor was it by others when he subsequently distributed his paper. Neither Darwin nor Jenkin ever heard a whisper from Mendel. His theories would lie forgotten for thirty-five years before surfacing, rediscovered by other botanists. Once again a punishing criticism would be destroyed by a later, more sophisticated scientific insight.

The second Scot to level at Darwin was a true dreadnought, William Thomson, Lord Kelvin, and he nearly blew him out of the water. Fleeming Jenkin had been bad, a trained engineer and a penetrating thinker, but Kelvin was far worse. He spoke from the bastion of physics, that most precise and respected of all the sciences, in which hypotheses could be weighed to the thousandth of a gram or measured to the thousandth of an inch. Kelvin stood in the very parapet of that bastion. He was the greatest physicist of the day, an expert on heat. He had made extensive calculations about the nature of the sun and the rate at which it was cooling. He had also studied the earth, measuring the rate at which it became hotter as one went deeper and deeper into mines, and from that calculating the speed at which its crust presumably had been cooling. Putting sun and earth together, and applying Newton's second law of thermodynamics, Kelvin came to some horrifying conclusions about the age of the earth as a habitable surface: not more than twenty million years. Before that it would have been far too hot to sustain life.

In that statement one hears an echo of a pronouncement made years before by Buffon to explain the demise of the dinosaurs: they became extinct because it had gotten too cold for them as the earth continued to cool. Buffon's figure for a habitable earth—70,000 years—was an inspired hunch, plucked out of the air, so to speak, in an effort to get across the idea of an ancient earth. To eighteenth-century listeners bound to a 6,000-year-old earth, it was a mind bender. But to a more scientifically oriented nineteenth-century England, inoculated against such shocks by the now familiar theories of Hutton and Lyell, for whom time on earth stretched endlessly, it had become increasingly easy to bind hundreds of millions of years into the book of the earth, and thus increasingly easy to accept Darwin's requirement of that huge expanse of time to account for evolution through gradualism.

—Until Kelvin. What a wrecker he turned out to be. If it came to a contest between physics and geology (that nest of fumblers scrabbling in the earth, measuring the depth of deposits by thumb and ruler, and trying to guess how long it had taken them to accumulate), sharp, precise physics would trample all over sloppy geology. And it did.

Kelvin was a strong believer in the biblical account of Creation, and he wasted no time in going for the evolutionary jugular. Darwin, as usual, was silent, although he did admit privately to Lyell: "I take the sun much to heart." Indeed, he took it so much to heart that he was driven to wonder if gradualism was really the answer. Take away time, and how could gradualism survive?

What a dreadful shock to the old man. A lifetime devoted to a theory based on an apparently safe geological assumption, only to see it crisped in the flame of Kelvin's calculations about heat. Failure to fathom variation was bad enough. This was devastating. The geologists were in disarray. Even Huxley, Darwin's public bulldog, who had staked his career on an almost obsessive defense of evolution, was put to ducking and dodging. Other physicists weighed in, elaborating on Kelvin. The outlook for evolution became grimmer. By 1900 there was a strong backwash against it.

Darwin did not live to see the peaking of this counteroffensive, although he must have sensed the possibility of its occurring. Publicly, at least, he was still the wonderful old man, the discoverer of a starry new idea, worthy to stand alongside Newton and Copernicus. He looked the part: deep, hooded eyes under shaggy, heavy brows, a long white beard, a mystery because he was so reclusive. A Merlin in his cave. The sage of the century. That is how he was generally regarded when he died in 1882. His family had wanted him buried quietly at Downe, but in an outpouring of public respect, it was decided to honor him with a place in Westminster Abbey. There he was laid to rest in a solemn and noble ceremony, among his pallbearers three of his oldest, closest scientific friends: Hooker, Huxley, and Wallace. Huxley had suggested that a line from Emerson be put on his stone: "Beware when the great Gods loose a thinker on this planet." But Darwin had left his own instructions: nothing more than his name. And so he lies, in a corner of the Abbey, close to Isaac Newton.

There is no record of Lord Kelvin's coming to Darwin's funeral, although many scientists did attend. He could afford to sit at home in triumph, having macerated the pernicious theory — or so he thought. His calculations had given the earth an age of one hundred million years at the very most, with only the last fifth of that time providing a surface temperature low enough to support life. Darwin had thought the earth to be at least ten or twenty times as old as that; he based his calculations on what he had learned about geological processes, on his observations of fossils, and on his theorizing about evolution. Evolution had happened, he was sure of it. He was also sure it had happened slowly. With the almost unbelievable complexity of life, how could it be otherwise? He could not prove physics wrong, but he was quite convinced that he was not wrong either. When two things are in such blatant opposition, one must go. In the end Darwin had enough faith in himself to say: "I feel a conviction that the world will be found rather older than Thomson [Lord Kelvin] makes it."

How right he was. Nuclear physics would prove it so. It explains how the earth generates its own heat by nuclear decay, and how the sun can put out an unvarying amount of heat for many billions of years, and allows for the earth's having basked in that even glow from the moment of its creation. This would restore to Darwin all the time he could possibly need. Kelvin died in 1907, not knowing that his argument would be blown away forever, and that the newly revealed histories of the sun and earth would fix the theory of evolution more solidly than before.

Two Problems That Darwin Could Not Solve

I will not go so far as to say that to construct a history of thought without profound study of the mathematical ideas of successive epochs is like omitting Hamlet from the play that is named after him. That would be claiming too much. But it is certainly analogous to cutting out the part of Ophelia. This simile is singularly exact. For Ophelia is quite essential to the play, she is charming —and a little mad.

—ALFRED NORTH WHITEHEAD

5

Gregor Mendel. The Problem of Blending Explained: Traits Endure.

Ah, but variation — what of it? The mystery of mysteries, the core enigma, the key to a complete and satisfactory theory of evolution. All unknown to Darwin, someone had already given that key a turn. He did it in 1865, the same year in which Fitzroy killed himself by cutting his own throat, six years after the publication of the *Origin,* three years before Darwin's book on variation in plants came out. His name was Gregor Mendel.

As Schubert was neglected in his life by music and Van Gogh by painting, so was Mendel neglected by science. This seems scarcely credible today, and yet it is just one more example of a rule often observed: look at a problem from a totally new angle and people won't so much disagree with you as completely misunderstand you. They won't grasp what you are talking about and will ignore you. Mendel ran into just that trouble. He attacked the problem of heredity through mathematics. He spoke to botanists about numerical ratios. For all they knew, he could have been counting flies in his garden. They paid no attention to his work, and it was completely forgotten.

"If Mendel was hard for his fellow scientists to understand," said Don, "it's going to be hard for our readers, too."

"What ought we to do about that?" I asked.

"We'll have to think up a good analogy, do some kind of experiment with familiar things. You give cocktail parties, don't you?"

"Occasionally."

"Do you serve ham and chicken paté?"

"Yes. I put it on crackers."

"Do you ever smear the two together on the same cracker?"

"No. Why should I do that?"

"I think you should. We can get a good picture of what Mendel was up to by doing a little smearing of paté on crackers. We'll pretend you're giving a cocktail party."

Here follows the imaginary party that Don directed:

Being bored by the same old ham and chicken patés, I decided to mix them together. I tasted the mixture — not bad, although there wasn't much difference between it and pure ham; the ham was so salty and strong that in flavor and color it tended to overwhelm the chicken. At any rate, I made up a large plateful of canapés and served them at the party.

"What will we do with the leftovers?" I said to my wife, Helen, later.

"We'll have them for lunch tomorrow."

"But the crackers will be stale. They're getting soggy already."

"Then scrape the paté off, cover it with some plastic wrap, and put it in the icebox. You can spread it on some fresh crackers tomorrow."

Which is what I did. The next day I gave the mixture another stir, spread it on crackers, set it on the table, and we ate it.

But suppose something quite different — something totally inexplicable, resembling witchcraft — had happened. (Here we ask the reader to suspend judgment for a moment and just read, as we pick up the story a second time.)

. . . The next day I gave the mixture another stir, spread it on crackers, set it on the table. . . .

"I see you had some chicken left over," said Helen.

"What's that?"

"Some of these canapés are pure chicken."

"They can't be," I said. "I mixed the whole lot thoroughly."

"Well, you didn't," said Helen. "There are three chicken ones and nine ham ones."

I looked. She was right. "That's impossible," I said.

"No, it's not. You know how absentminded and careless you are. You just forgot some of the chicken yesterday and added it now."

"I MIXED IT ALL!"

"No need to get your back up. You *are* forgetful, you know."

I was not. I knew it. I went over my actions of the day before. I remembered emptying the two cans, throwing them out, spreading *all* the mixture on crackers. Something strange was going on. "I'm going to do it again," I said.

"But we're not having another party."

"I mean, mix some more of this stuff. There's something very fishy here."

"Well," said Helen, "if you're talking about repeating an experiment, it must be a controlled experiment. You must do everything as before." She was right; she has had training in laboratory research. "You must mix two cans of paté, put them on crackers, scrape them off the crackers in a couple of hours, put them in the refrigerator overnight, serve them again tomorrow on fresh crackers." Then she said smugly: "You must be very foolish to do all that, because I know what you'll get—a mixture that tastes like ham."

That irritated me.

"*Looks* like ham, too," she added.

That irritated me more, so much more that I repeated the experiment that afternoon. I got her to watch me empty the cans, mix their contents thoroughly, and spread them on crackers. This time, without guests to eat them, there were forty-eight of them.

"Now, wait two hours, scrape the mix off, and put it in the refrigerator."

I did. It stayed there overnight.

The next day she said, "Where did you remake the canapés the first time?"

"In the pantry."

"But isn't it rather dark in there?"

"Not too dark to smear on some paté. Anyway, you told me to do it exactly as before."

"So, go in and start." She watched me as well as she could in that gloomy pantry where it was impossible to tell exactly what was being smeared on. But when we got the canapés out into the sunny dining room and she looked at them, there were thirty-six ham and twelve chicken.

My wife was flabbergasted. So was I. We repeated the experiment. Same result. "This place must be haunted," I said.

"Not necessarily. Have you noticed one thing about the numbers?"

"What numbers?"

"The proportion of ham to chicken." She has a methodical and scientific mind and would notice those things.

"What about it?"

"It's always the same. Three of ham to one of chicken."

That, in a crude way, was the mystery that faced Gregor Mendel. He stumbled over it, not by mixing paté but by mixing plants, by crossbreeding common garden peas. His "ham paté" was a plant that produced yellow peas, his "chicken paté" was a plant that produced green ones. When he mixed them—that is, when he crossed a yellow with a green—he expected, by all logic, to get an equal number of yellow and green peas from the seeds of that cross. But when the seeds grew up and began bearing peas themselves, he was surprised to find that *all the peas were yellow.* The green seemed to have been swallowed up by the yellow, much as the more delicate flavor of chicken had been swallowed up by the ham.

What to do next? Mendel saved all the "yellow" peas that his cross had produced and planted them the following year: the equivalent of putting them in the icebox and serving them again. Logic told him that yellow would produce yellow. But it didn't. When they grew up, among the yellow peas were some green ones. Where had that green been hiding? It should have gone. He tried the experiment again, this time crossing a plant with smooth round peas and one with wrinkled peas. The result: all round. He planted those "round" peas and again was rewarded with an inexplicable crop of mostly round ones but with some wrinkled peas magically reappearing—just like the chicken paté.

Mendel experimented with peas for seven years in a small strip of garden in a monastery at what was then the Austrian town of Brünn (today it is Brno, in Czechoslovakia). During those seven years he worked with plants that had other characteristics that, when the plants were crossbred, came and went in the strange way that "green" and "wrinkled" did. There were tall plants and short ones, for example. Did he get a middle-sized plant by crossing them? No, he did not. He got all tall ones. But those "tall" seeds, once again, produced a few short plants in the next generation.

The recalcitrance of those pea plants was extraordinary. They stubbornly persisted in doing something that was quite illogical. A cross of yellow and green should, by common sense, have one of two results. Either the resulting peas should be yellowish green (a blend of the two colors), or they should be divided (half yellow and half green). For green to hide, for wrinkled to hide, for short to hide, *and then reappear later,* just didn't make sense.

Except — there was one consistent feature in the otherwise apparently irrational behavior of his pea plants. What might be called the "weaker" traits — green, wrinkled, and short — although they invariably *disappeared* when first crossed with strong traits — yellow, round, and tall — always *reappeared* in the next generation of hybrid peas. After a couple of years Mendel found another consistent feature. The ratio between weak and strong in this next generation was always about three to one: three strong to one weak.

That is, the ratio *seemed* to be three to one. Actually it was not, and the best way to understand that is to follow a labeling system that Mendel himself used. He named the strong yellow trait "dominant" and labeled it with a capital letter *A*. The weak green one he called "recessive" and labeled it with a small *a*. There are only four ways that the sex cells of two individuals in a random group of plants containing A or a can be joined in pollination: AA, Aa, aA, and aa. Since the first three have at least one dominant trait (we would now call it a dominant gene), they will all *look* yellow, although in truth only one of them is pure yellow (AA) with a dominant gene from both its male and its female parent. The next two (Aa and aA) are a mix; they, too, look yellow but contain hidden green. The only truly green one (aa) is the one with no dominant trait. Geneticists have learned to make this situation clear by presenting it in the form of a checkerboard:

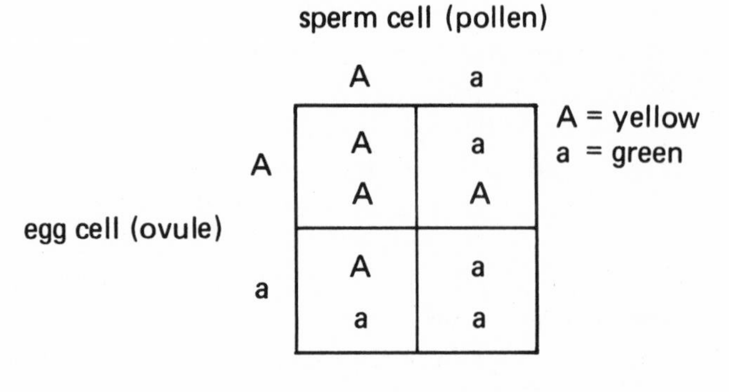

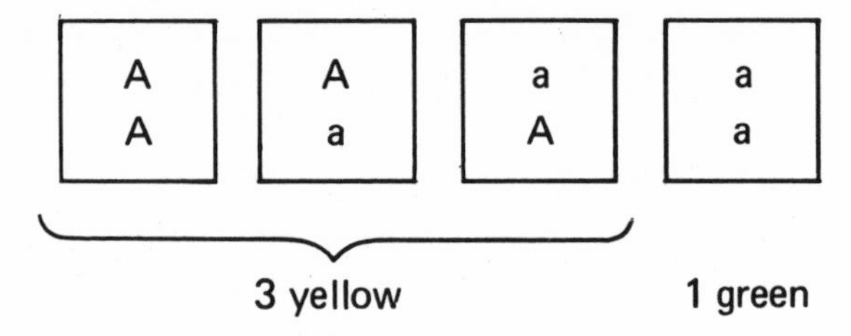

To repeat, three peas *look* yellow, but only the far left one is pure yellow. The middle two are capable, in future generations, of producing a green pea, just as this generation did. Thus the true ratio is 1 : 2 : 1. The *apparent* ratio is 3 : 1.

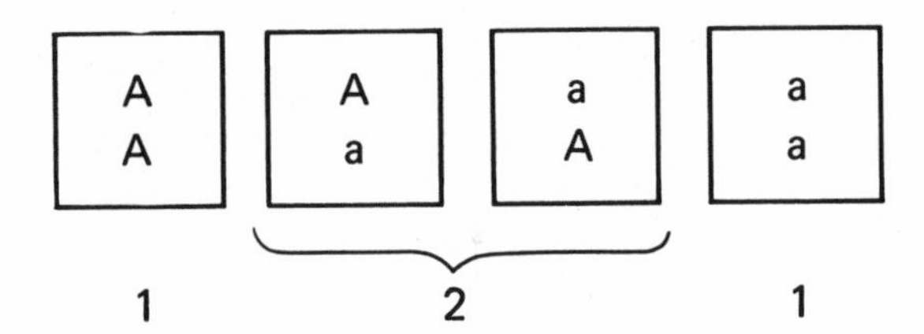

In real life things do not work out that neatly. One cannot take four peas and expect to get three yellow ones and one green one. The

sample will be too small and the probability of error too great. One pod might have two green peas in it; the next one might have none. A sample should have hundreds of peas in it for the statistical probability of a 3 : 1 ratio to begin to reveal itself with any degree of accuracy. Mendel, who had some knowledge of mathematics, understood this and made sure he had plenty of peas. In that hybrid harvest he got 8,023 of them, of which 6,022 were yellow and 2,001 were green —an almost perfect 3 : 1 ratio.

To test his test, Mendel made another try, this time with different traits: smooth peas and wrinkled ones. Following the same procedure, he crossed a plant that he knew produced nothing but smooth peas with one that he knew produced nothing but wrinkled ones. Again his hybrid generation yielded an almost exact 3 : 1 ratio— 5,474 smooth peas and 1,850 wrinkled ones.

Mendel then put his peas to a sterner test. He predicted that if he mixed two traits at once he would get a 9 : 3 : 3 : 1 result: nine that

	AB	Ab	aB	ab
AB	AB AB	Ab AB	aB AB	ab AB
Ab	AB Ab	Ab Ab	aB Ab	ab Ab
aB	AB aB	Ab aB	aB aB	ab aB
ab	AB ab	Ab ab	aB ab	ab ab

A = yellow
a = green
B = smooth
b = wrinkled

This time the peas sort themselves out like this:

	smooth yellow	smooth green	wrinkled yellow	wrinkled green
	9	3	3	1
Of 556 peas in his sample, Mendel's actual figures came out like this:	315	108	101	32

were both yellow and smooth, three that were yellow and wrinkled, three that were smooth and green, and only one that was green and wrinkled. A larger checkerboard with a second letter added, *B* (smooth) and *b* (wrinkled), explains why. Note that there are now four units to each side of the square. This is because there are four ways of combining A or a with B or b, thus: AB, Ab, aB, and ab. Those four possibilities from the male plant are on the top of the square and the same four possibilities from the female plant are on the side of the square; when the two are combined in all possible ways, they provide sixteen possible combinations for the hybrid plants of the new generation.

We can imagine Mendel's excitement when he sorted out those 556 peas and found that they conformed almost exactly to his prediction: 9 : 3 : 3 : 1. This should have convinced him that he had unraveled something dependable about the apparently undependable way in which peas carried on their hereditary traits. But to make absolutely sure, he decided to add a third trait: *C* (gray seed coats) *c* (white seed coats). A test with three traits requires a checkerboard with sixty-four squares to accommodate all the possibilities. Mendel calculated that the ratio should be 27 : 9 : 9 : 3 : 9 : 3 : 3 : 1, with twenty-seven peas showing all three dominant traits and only one all three recessive traits. To his great delight his three-trait harvest closely approached the numbers that his mathematics had predicted. This three-trait checkerboard bears out the ratios predicted by Mendel.

ABC = 27 (The capital letters A, B, and C appear at least once in these squares and signify the traits which will be *visible* in the actual plant)

ABc = 9
AbC = 9
Abc = 3
aBC = 9
abC = 3
aBc = 3
abc = 1

	ABC	ABc	AbC	Abc	aBC	abC	aBc	abc
ABC	ABC ABC	ABc ABC	AbC ABC	Abc ABC	aBC ABC	abC ABC	aBc ABC	abc ABC
ABc	ABC ABc	ABc ABc	AbC ABc	Abc ABc	aBC ABc	abC ABc	aBc ABc	abc ABc
AbC	ABC AbC	ABc AbC	AbC AbC	Abc AbC	aBC AbC	abC AbC	aBc AbC	abc AbC
Abc	ABC Abc	ABc Abc	AbC Abc	Abc Abc	aBC Abc	abC Abc	aBc Abc	abc Abc
aBC	ABC aBC	ABc aBC	AbC aBC	Abc aBC	aBC aBC	abC aBC	aBc aBC	abc aBC
abC	ABC abC	ABc abC	AbC abC	Abc abC	aBC abC	abC abC	aBc abC	abc abC
aBc	ABC aBc	ABc aBc	AbC aBc	Abc aBc	aBC aBc	abC aBc	aBc aBc	abc aBc
abc	ABC abc	ABc abc	AbC abc	Abc abc	aBC abc	abC abc	aBc abc	abc abc

A = yellow
a = green
B = smooth
b = wrinkled
C = gray coat
c = white coat

The twenty-seven wholly dominant squares are heavily outlined. The reader can check the others. What is interesting is the rapidity with which the number of all-dominant squares increases. In a four-trait checkerboard there are eighty-one all-dominants; in a five-square board there are 243. But there is never more than one all-recessive square: the one at the lower-right-hand corner of the board.

"Those things are called Punnett squares," said Don, watching as I laboriously completed the large square on the preceding page. "Boy, are they dull."

"I don't know about that," I said. "This one was fun."

"You *made* this one from scratch?"

"Yes."

"Why?"

"I wanted to see if I could find a formula for predicting ratios for any number of traits. I had to keep making bigger and bigger squares.

I ended up with one that had twelve different characteristics before I could see any rules emerging. It took me an entire day. Here — "

"You must be nuts. A mathematician could have worked out that formula in thirty seconds."

"But I'm not a mathematician," I said. "I wanted to understand what Mendel was up against."

"You are certifiably nuts. And any reader who would look at that big square would have to be as nutty as you are. Leave it out of this book."

Mendel did not have to go through the arduous process I did. Being a mathematician, he could predict ratios and never went beyond a square with three variables. Adding a fourth would have required a checkerboard with 256 squares. All seven traits visible in his peas would have required a checkerboard of 16,384 squares and a test sample of millions of peas before they began to fall into a statistically meaningful pattern. Instead, Mendel shrewdly began backcrossing hybrids with pure dominant and recessive plants, predicting the appropriate ratios. In this way, he was able to use manageably small samples. When they, too, matched his predictions, Mendel was certain that his hypothesis was correct: *hereditary traits of living things come in separate packages; they do not blend; they behave according to simple mathematical laws; some are dominant and "show," while others are recessive and lie "hidden" unless present in the pure state.* This was a momentous insight. It became the keystone for the great edifice of genetic knowledge that would be erected in the following century.

What enabled Mendel to do what he did? Intelligence, certainly, plus a knowledge of mathematics. But many other plant experimenters had both. What they lacked was the ability to make the inexplicable intuitive leap that enabled him to *think* of applying mathematical ratios to the problem of heredity. Mendel also had enormous patience. In that he was not unlike Darwin, who had conducted numerous studies of pigeons and flowers, and who painstakingly and over long periods of time tested his ideas with practical experiments and made careful notes of the results. That was how Darwin liked to work.

Another asset that Mendel had was a lifelong affinity to plants. His parents were peasants in the Austrian hamlet of Heitzendorf, a rural community with a tradition of supplying skilled gardeners to the great landowners. Heitzendorfers, although extremely able plantsmen, were otherwise conspicuously backward. Scarcely anybody could read or write in that somnolent little village, where the buzzing of bees in the gardens and orchards was often the loudest sound on a summer day. It was a stroke of luck for little Johann (he assumed the name Gregor when he became a monk) that an elementary school was opened in the village when he was a child. He was extremely bright and did so well at school that he hoped for a career as a teacher. But grinding poverty limited the decisions in the Mendel household. His father, who had inched his way upward from landless sharecropper to the more elevated status of peasant with a small orchard holding, could not afford a schilling extra for education, so the son became a monk instead, entering an old and prosperous Augustinian monastery at Brünn.

As a novice at the monastery Mendel was somewhat of a failure. He was a friendly, energetic, sympathetic, and concerned man, just what a priest should be. But he proved to be so sensitive to human suffering and to the problems of the destitute in the decaying corners of a decaying empire that he could not perform the regular duties of a parish priest. Unable to figure out what else to do with this so-easily-bruised young man, the abbot of the monastery decided to turn him into a teacher. That is how Mendel ended up doing what he had hoped to do in the secular world.

Mendel was a born teacher. He had a gift for working out simple solutions to problems and explaining them clearly. He did so well at the local high school as a substitute instructor that it was decided to send him to Vienna to take an examination for a proper teacher's certificate. Self-educated in all the sciences, he flunked miserably, returned to Brünn, continued as a markedly popular and successful teacher for five years, and was sent back to Vienna in 1856, this time for a year's cram course in science before trying again.

Austria was then a police state, with a crumbling monarchy doing its repressive best to ward off wars from without and revolutionary plots from within. Indeed, there would be a war with Prussia ten years later. Brünn would be occupied and the surrounding farms devastated by troops trying to live off the country. They would bring cholera with them and two thousand of them would die, along with a thousand citizens of Brünn. Even as early as 1856, every male in Austria had to carry an identity card and show it wherever he went. Mendel had one when he went to Vienna. It identified him as of mean stature, fair hair, gray eyes. Other peculiarities, none. Language, German.

During his year at Vienna, the mean-statured (actually rather chunky) young monk threw himself into a whirlwind of study; physics, chemistry, laboratory equipment and its uses, systematic botany, physiology and paleontology of plants, use of the microscope, mathematics, and morphology. He devoured his courses; here was the lost opportunity for an education. He survived — somehow — by acting as a demonstrator at science lectures and at the end of the year confidently presented himself for another examination. He did well. But then came botany — the one subject he truly felt at home with; he had grown up with it in his bones — and he made the mistake of disputing his examiner on some point or other. He was told he was wrong. He insisted he was right. He was flunked again.

It must have been a dreadful humiliation for him to return to Brünn and report his second failure to the abbot. But humiliation and humility are two different things. Mendel bore the humiliation because he was a truly humble man — but he was also stubborn. He was sure he had been right and his examiner wrong, and he persuaded the abbot to let him have a garden plot in the monastery ground to do some plant hybridization experiments. This small area, 120 by 20 feet (it is still there today, though no longer planted in peas), became the unlikely place where the first and most important step in the riddle of heredity would be worked out.

In his determination to find out for himself whether he was right or wrong, it was natural for Mendel to turn to plants. He had been interested in growing things all his life and had already done some experimenting with hybrid flowers in an effort to learn why they produced the riotous and unpredictable bouquet of colors that they did. He had also bred mice, but they were smelly and kept getting out of their cages; they also seemed to the abbot to concentrate Mendel's mind on sex to a degree that was inappropriate for a monk. So he gave up the mice and turned to peas.

In choosing peas Mendel was unbelievably lucky. Although he did not know it, the seven dominant and recessive traits that he had observed and was determined to trace were each located on a different chromosome in the cell nucleus of the pea. Therefore, each behaved independently of the others. There was no linkage and no perplexing irregularity in the ratios he predicted, which otherwise might have confused him so much that he might never have worked out the principle he did.

When it comes to explaining something like Mendel's law, the writer of a book has a great advantage over the reader in being able to say only as much as he thinks necessary. The reader will swallow it obediently, nod, and go on. However, there is always the analytical and recalcitrant one who will say, "Wait a minute, you haven't really explained anything."

"That reader's right," said Don, as he looked over the draft of this part of the manuscript.

"What do you mean?" I said. "I think I explained it pretty well."

"Then explain this," said Don. "You pointed out a while back that the biggest problem bothering Darwin was variation, and that if he had only known about Mendel's experiments he would have been vastly relieved. From what you're saying here, I suspect the reader will think that Mendel proved that there *wasn't* variation. Traits don't vary. They appear and reappear in predictable ways."

"That's true," I said. "But it's not so much a matter of appearing and reappearing. The important point is that they don't *disappear*."

"Then you should make that point stronger."

"Very well, let's remind the reader of the cranberry juice we talked about earlier, the juice that was increasingly diluted as it was mixed with more and more water."

"What about it?"

"As Darwin and his contemporaries saw it, particularly that pestiferous Fleeming Jenkin, who nearly turned Darwin inside out with anxiety, blending single traits into a large population would quickly cause them to fade out and disappear, just like pink juice. Mendel showed that this does not happen, that variation can be preserved. All sorts of things can be preserved and must be. Consider that a flower has not seven but hundreds or thousands of genetic traits. It needs them all. And you would need such a large checkerboard to take care of all the possibilities of combination that only a huge computer could calculate them."

Don was still not satisfied. "You said there was no blending. Well, I've read a little further into Mendel, and I happen to know that those early flower experiments he made before he got into peas don't indicate that. He crossed a red flower with a white one, and he got a pink one. Mixing red with white to get pink seems like blending to me."

"It may seem so. Actually it is the expression of two traits coming together to produce a fixed result. Both succeed to some extent in expressing themselves. But whether the recessive trait manages to express itself a little or a lot or not at all is not the important thing. What matters is that, seen or not, it is there. In this case two traits express themselves and the result is pink. But, keep breeding those pink flowers, and they will still sort themselves out into the proper ratios, including pure red and white flowers along with the pink. What you don't get is progressively *paler* pink. There is a near infinity of combinations that can arise from a pairing of the genes of two parents. Sometimes the pairing shows. Sometimes it doesn't. But it is never the kind of blending—that would lead to dilution and disappearance—that bedeviled Darwin."*

"Well, in that case I have to agree with you; Mendel was lucky when he chose peas."

He was indeed. Peas, while by no means unique in having a number of obvious traits on different chromosomes, are unusual. The odds are one in 163 that a blind guess among a great number of plant species will come up with one that behaves as simply and as satisfactorily as peas do. It was as if Mendel had gone into a locksmith's shop with a padlock in his hand, turned to a board on which 163 keys were hanging, reached out once, and picked the one that fit his lock.

After that, Mendel's luck as a scientist was all bad.

His first setback came when he released his findings. He had helped found a local organization, the Brünn Society for the Study of Natural Sciences, in 1862. On a cold evening in 1865 he appeared before its members to deliver a paper. Everyone in the Society knew that Mendel had been working hard and long on hybridization—a phenomenon whose thorny intractability had caught all the botanists of Europe by that time—and they were most anxious to hear what he had discovered. Nearly all the Society's forty members showed up. They listened to him in complete bafflement as he

* It is now known that there are varying degrees of dominance and recessiveness in some genes, also that more than one gene may be involved in determining something like skin color. That, not blending, accounts for the variety of tints in human skin.

described the procedure he had gone through and the numerical ratios he had arrived at. This sorting-out of traits "by the numbers" meant absolutely nothing to them. If he had hoped for questions or comments, he was disappointed. Nobody said a word.

A month later Mendel made a second appearance before the Brünn Society, this time explaining the mathematical means by which he arrived at his ratios. Again total incomprehension; he was met by nothing but glazed eyes. Algebra and botany just didn't mix. The members went home, shaking their heads about their nice fat friend, who had apparently wasted six or seven years in the monastery garden counting seeds. Mendel's grand concept, that this counting led to an explanation of the heredity tangle that nobody on earth at that time had the faintest glimmer of an understanding of, drifted out the door along with the homeward-bound members and vanished like a puff of smoke.

The Society did give Mendel its customary courtesy of publishing his paper. Copies of it were duly sent to universities and scientific societies in a number of European cities. Where there was any reaction at all, it was the same: this monk was talking about something that made no sense. It was not quite as bizarre as trying to study astronomy by making abstruse calculations to predict the length of grass stems, but the stretch was almost as great and just as meaningless.

Mendel's paper had been read in February 1865 and was published in 1866. On the last day of that year he made one more try. He sent a printed copy to Karl von Nägeli, one of the three or four most distinguished botanists in Europe. He hoped that this sophisticated and brilliant man might understand him and offered to send him some seeds so that he could test Mendel's conclusions, had he a mind to. Mendel also very modestly nudged himself forward as one willing to do some experiments on hawkweed *(Hieracium),* a plant he knew Nägeli to be interested in.

The great botanist took a long time to respond to the obscure monk. When he did, it was to point out that Mendel's work could by no means be considered conclusive; indeed, it was just beginning. How Mendel felt about that, after having raised more than ten thousand plants and having kept exact records on every one, can only be imagined. However, he was as aware of his own modest pretensions as he was of the splendid achievements and world renown of Nägeli. He apparently swallowed hard and sent the seeds that Nägeli agreed to accept (there is no record that he ever did anything with them). Nägeli, for his part, must have been pleased to find a volunteer researcher who could be counted on for the most meticulous work and the most scrupulous record keeping. He urged Mendel to get to work on hawkweed — one of the worst recommendations one scientist has ever made to another.

Hawkweed is a small wild plant that grows in profusion on the open, dry hillsides of Europe. For Mendel, it meant long walks into the higher pastures and field edges in the hills above Brünn. He had now been at the monastery for more than twenty years, and the sedentary life there had allowed him to grow very fat. But far worse than the arduousness of collecting was the plant itself. Hawkweed does not have separate and easily trackable hereditary units. They come in perplexing clusters and the variation among individual plants is immense. Moreover, the plant sometimes grows as a result of sexual fertilization and sometimes without any fertilization at all. Nägeli did not know this, neither did Mendel — nobody did. Furthermore, the plant's sexual apparatus is extremely small and delicate; any attempt to fertilize it artificially is apt to damage it irreparably. And in that tiny blossom everything is so close together that attempts to remove pollen from a plant will inevitably sprinkle a few of the plant's own grains on its own stigma.

Hawkweed stopped Mendel in his tracks. None of his predicted ratios appeared. During several years' work he succeeded in getting severe eyestrain and in producing only half a dozen hybrids. These departed so puzzlingly from the makeup of their parents that he was utterly bewildered by them.

Those years he could have spent on peas, or on the colors of the blossoms of some bean plants he had experimented with that had wild explosions of unexpected hue. If he had spent his entire time on them, he might well have been able to interpret the seemingly random results. As it was, he did glimpse the principle. It was what he had deduced from peas. The beans, he reasoned, were already hybrids of a complex nature; the apparently irrational eruptions of color that occurred in the various generations were simply expressions of those hybrid complexities. He was right. Had he had the time and the energy to keep separating out a few bean plants from succeeding generations until he got pure strains, he might well have proved it. And that second demonstration, illustrated by an orderly and *predictable* production of blossoms from what had been a fireworks of *unpredictability,* might have been just enough to catch the attention of the scientific world. It never happened. He died utterly unknown to science.

It was not all Nägeli's fault; there were many other botanists just as unperceptive as he was. Mendel, after all, had offered his services on hawkweed. And as long as his eyesight held out, he was apparently willing to keep up the work, sterile as it seemed, in order to maintain a correspondence with the one man in the mainstream of botany with whom he had any contact.

Nägeli may have slowed and frustrated Mendel's plant experiments, but he did not stop them. That was accomplished by the sudden elevation of Mendel to the office of abbot in 1866. His administrative duties expanded. He handled them well. He was given more. He became an important man in Brünn, curator of a local institute for deaf-mute people, chairman of a bank. In 1872 the Emperor made him a Commander of the Order of Franz Joseph "in recognition of your meritorious and patriotic activities." Mendel was now living in sumptuous quarters and had a sizable discretionary income that he donated to good causes in Brünn—saving for himself only enough to set a fine table at the monastery. Its food was locally celebrated and made him fatter yet and nearly immobile. He never collected another flower.

Mendel was only forty-six when he became head of the monastery, a young age for this important post. There was a reason for this. Whenever an abbot died, the monastery was compelled by the state to have its accounts audited and to pay a large tax. For that reason, the monastery had a long history of choosing young men for the post. In Mendel the brothers chose well.

His last act as a teacher had been to divide his salary among his three neediest students, knowing that he would immediately fall heir to the emoluments of one of the richest monasteries in the empire. He could scarcely believe his sudden rise. "When he looked back on his own past [Mendel wrote of himself in the third person], as a peasant lad in Heitzendorf, who had had so hard a struggle to achieve a high school education, often ailing and always poor; and when he saw himself twice rejected as incompetent by the distinguished examiners at Vienna University—he cannot but have been amazed to find himself at forty-six a mitred abbot."

All went smoothly until 1874, when the government passed a stiff new tax bill. Mendel regarded it as unjust. He refused to pay, taking the same position that Winston Churchill, as prime minister of England, would later take: "I was not elected to this post to preside over the dismemberment of the British Empire." Mendel fought the tax and mired the monastery in a long and expensive court battle. All his peasant stubbornness came out. He regarded as bribes (which they were) some hints that if he relented he might be elevated to even higher office in the Church. Other monasteries, which had likewise balked at payment, caved in. Mendel fought on, embittered, without friends, and was finally trapped when the monastery's funds were sequestered by court order. By 1882 a faction within the monastery had decided that his behavior was becoming dangerously bizarre, and a private watch was put on him to see if signs of mental derangement could be detected. None was, other than a mad determination to see a just and godly fight carried to its end, no matter what the cost.

At this impasse Mendel collapsed. He began having serious heart symptoms and also suffered from dropsy. Administration of the monastery was taken out of his hands and a compromise with the state was quickly reached. With heart and kidneys failing, and vastly overweight, Mendel went downhill rapidly and died in 1884. He had a terror of being buried alive and insisted that a postmortem examination be performed on him. It was.

In the course of writing a biography of Mendel, Hugo Iltis visited Brünn many years later. A subscription drive was then being conducted to erect the statue of Mendel that now stands in the city. But few citizens there knew of the world fame of the most distinguished man whom Brünn had ever produced. Those who did were not sure what he was famous for. To help the campaign along, pictures of Mendel were put in store windows.

"Who is this fellow Mendel that everybody is talking about?" Iltis heard one man say to another.

"Don't you know? He's the one who left an inheritance to the town."

That much, and only that, was what the people of Brünn understood of the "inheritance" work that Mendel had done.

Hugo de Vries. The Source of Variation Found: Mutations Occur. Mendel Is Vindicated.

Although Mendel was totally forgotten during the latter half of the nineteenth century, he lived on invisibly in the paper he had written. The Brünn Society had sent more than one hundred copies to scientific libraries and universities throughout Europe. Mendel had sent forty more to botanists he had hoped to interest. No one paid the slightest attention to it. It gathered dust on obscure shelves, its quietly ticking message unheard. Nearly twenty years went by. Then Karl von Nägeli, the man to whom Mendel had sent seeds, to whom he had written so many hopeful letters, by whom he had been entangled in the thicket of hawkweed, published a book. By that time Nägeli had become one of the most famous botanists in the world. His book, a big one, summarized everything he had ever learned about heredity in plants. It said not a word about Mendel.

Sixteen more years went by. Then, in 1900, in a coincidence even wilder than the joint announcement of natural selection by Darwin and Wallace, three men, within a period of months, rediscovered Mendel.

The first to do so was a Dutch plant physiologist, Hugo de Vries. Like many others in the last decades of the nineteenth century, he was bothered by the uncomfortable legacy left by Darwin: a persuasive new theory (evolution) that most experienced students of natural history had been forced by the evidence of fossils and geology to accept, but whose method (natural selection) was not swallowed nearly as easily. There was no known biological mechanism to show how natural selection worked. Worse, variation, on which natural selection depended, was inexplicable.

It was in this dilemma — a good theory without a good final proof — that de Vries and a number of other able young botanists found themselves caught. Galvanized by the *Origin,* compelled by its logic to accept evolution, they still looked at it in an old-fashioned way. Most of them believed in blending, having no knowledge of how cells worked and no explanation of how traits were carried from one generation to the next unless by some vague process of blood mixing. That traits *were* carried on was a concept that had been known since antiquity. Sex was recognized very early as the vehicle for making babies, as was the role of semen in impregnating the female. But as to just what semen was, or what it did when it entered the female, no one had much of a clue.

Botanists had a somewhat better idea than most. They had recognized for at least three thousand years that pollen was needed to fertilize a seed. After the invention of the microscope they had even learned how this happened: by the pollen grain's putting down a long tube to unite with the seed and — in some miraculous way — stimulating it to grow.

For people who had been bringing bulls to cows and stallions to mares from time immemorial, it was clear that something comparable took place among animals. It was equally clear that fertilization produced offspring that were usually a pretty fair mix of the characteristics of the parents. That knowledge — that one could carry forward to a new generation of children the traits found in the father and mother — underlay the crude but successful principle by which all stock breeders worked. They did not know *how* it worked, but it did work; they had different kinds of cows and sheep than their ancestors, and better ones.

All sorts of wonderful ideas have been promulgated over the centuries to explain how traits are carried down. The father, with his strength and his powerful sexual fluid, was thought by many to be the influencer, and the mother merely the vehicle. Others thought that the mother supplied the traits and that the father merely triggered them off. Still others believed that a mother, by entertaining several lovers during the time she could conceive, would manage to pass on to her baby some of the traits of all those men. And it had long been believed that the blood of a woman impregnated by a man of another race would be forever tainted by that alien influence. Some dog breeders still believe that if a purebred bitch has a litter of puppies by a mongrel she is useless for future breeding purposes. Somehow, some of that mongrel's "bad" blood will stay with her.

Hugo de Vries grew up in that sea of ignorance, encouraged to a belief in blending through lack of a better alternative. He was also exposed as a student to the concept of "soft," or Lamarckian inheritance. It will be remembered that Darwin himself, unable to deal with some of his critics, had moved reluctantly in the direction of Lamarck from a position of "hard" inheritance—that is, one in which the environment or striving on the part of the individual played no part. For two decades after Darwin's death a wave of Lamarckian thought continued strong in Europe.

De Vries swam in that wave—and not very comfortably. The rigorous logic of the *Origin* (he had read it in 1866) told him that Lamarck was not the answer. So, what was? Variation had to be explained in some other way.

Like Mendel, de Vries sought the answer in flowers. But he went about it quite differently, and actually with a different aim in view. Mendel was looking for the rules that explained the passing on of variable traits. De Vries was looking for the source of variability itself. He could not accept Lamarck, but on the other hand he could not quite accept Darwin either. How could Darwin's multitude of small changes, themselves caused by an unknown mechanism, piling up generation after generation, produce the miracle of a new species? The experience of animal breeders was against it.

True, different and better sheep could be created, but there was a limit to which such improvement could go. No sheep, for example, would ever stand six feet tall, although what a boon that would be: much more wool on the hoof, fewer animals on the hillside, less wear and tear on the shepherd and his dog, and a quicker and more efficient shearing. But sheep—and all other animals—are stubborn. Bred for size, for thickness of coat, for anything, there comes a time when they no longer cooperate. Improvement ceases. You simply can't get a bigger sheep. If you do, it has other undesirable qualities. It may be sterile, it may be feeble, it may be clumsy and snap its legs. Its coat may be thinner. Get too far away from the classic, essential sheep, and things break down. In the end "sheepishness" asserts itself.

Knowing this, de Vries concluded that Darwin's gradualism, though it might explain the differences between sheep and sheep, was not sufficient to explain the greater differences between sheep and goat. But—and a very big but—those differences had to be explained if natural selection was to be believed, for logic demanded that there be a common ancestor of sheep and goat—indeed of sheep and cow, of sheep and wolf, ultimately of sheep and fish.

Larger evolutionary jumps, de Vries concluded, had to be called into play. Somehow, he speculated, nature was making those jumps. He decided to go looking for them, and chose to look in a field of primroses* near his home in Amsterdam. He chose that field because it was crowded with a great burst of plants showing a wide range of variability, all of them having escaped from a cultivated garden in a nearby park. In that new and unfettered environment he hoped to find some plants that were not merely different in scale from the others but truly different: new species.

He found them. In a mammoth experiment lasting twenty years, de Vries raised more than fifty thousand primrose plants. Among them were a few that he concluded were new species. Plants appeared with traits their parents did not have. Those traits appeared suddenly, apparently out of nowhere. What was more, the plants bred true. Their differences appeared consistently generation after generation. There, before his eyes, was the proof he had been looking for. A jump, a sport, a saltation had occurred, not gradually but all at once. He decided to call the phenomenon a mutation.

De Vries' choice of primroses, like Mendel's choice of peas, was one of those extraordinary strokes of luck that occasionally crop up in science. His luck was to discover mutations in plants that were not mutating. What looked to him like mutants were actually hybrids with "new" features cropping up and then persisting in succeeding generations. Those features, however, were not really new. They already existed in the genetic systems of the plants but did not

* These were specimens of the evening primrose, *Oenothera lamarckiana,* named, by a lovely coincidence, after Lamarck.

express themselves until the right combinations of traits happened to come together during crossing experiments. What enabled the primroses to behave in this way—seeming to produce new forms that endured—is that they flourish only in the heterozygous (hybrid) state. They will not grow in the homozygous state (that is, the "pure" state which arises when both parents are identical with respect to a given trait). Thus, in every new strain of hybrids that de Vries found, what he got by breeding them was more of the same: apparent mutant forms. His theoretical conclusion was right, but he had used the wrong flowers to arrive at it.

De Vries was an odd and difficult man. He came of a distinguished Dutch family of merchants, lawyers, professors, and statesmen. His father, prominent in Dutch politics, once was honored by the king, who asked him to form a government. He did so, becoming Minister of Justice in that government. Much was expected of young Hugo, who was a brilliant boy but somewhat of a loner; he spent his time wandering all over Holland collecting flowers. By the time he was old enough to enter a university he was already a competent botanist and had accumulated a complete collection of the flowering plants of the Netherlands. He was a ferocious worker but an opinionated one. An early convert to evolution, he had so many acrimonious disputes with his anti-Darwinian professor that he left the University of Leyden to study in Germany. He became an expert on the interior structure and behavior of plant cells and was appointed a lecturer at the University of Halle. But his determination to lecture only about what interested him produced almost no listeners. Again at odds with his superiors, he returned to Amsterdam and, already emerging as a world authority in his field, began an illustrious career as a teacher and researcher there. In 1877 he went to England, visited Darwin at Downe, and had a long talk with the old man. That intensified the interest he already had in evolution. By 1890 he had abandoned his work on plant cells and was deep into studies on variation. When he found a clump of strange-looking primroses in a field of ordinary primroses, the course for the rest of his life was set.

He now wanted to devote all his time to research, and although his decision again provoked storms he more or less got his way. He paid so little attention to his doctoral students that, during a period of nearly ten years in the 1890s, not one of them earned a degree. Despite this, his reputation continued to attract brilliant students, who put up with his contentious character and his obsessiveness with his own work, and struggled to their degrees willy-nilly through sheer exposure to their neglectful but inspiring teacher. He was a prodigious producer of scientific papers, writing hundreds of them as well as numerous books. Typical of everything he did was a series of articles he wrote for a Dutch agricultural journal on the improvement of domestic crops. Its approach was evolutionary, highly scientific. While it helped de Vries sort out in his own mind some of the tremendous problems he was working on, it was utterly useless to the practically minded farmers at which it was aimed. His scientific reputation, however, continued ever upward. He collected medals and honorary degrees from all over the world. It was a poor scientific society that did not elect him to membership. As early as 1889 de Vries had begun to suspect that organisms had hereditary traits that acted independently of each other. Attempting to identify these, he suggested the existence of minute units that he named pangens. From his earlier study of cells, de Vries was better suited than other botanists to think small. He was familiar with some of the biochemical processes that went on inside the cell and had worked out some extremely sophisticated principles of osmosis, the ability of cell walls selectively to admit or expel molecules. He also had discovered that small bodies inside the cell had linings of their own. Therefore it was not hard for him to imagine something exceedingly tiny, working at the molecular level to control development.

Pangens were de Vries's theoretical way of disposing of the problem of blending, of arriving where Mendel had arrived with his pea experiments. Unaware of Mendel's work at that time (he subsequently claimed), he had to find another mechanism that explained what Mendel had discovered: you get either yellow or green peas, never yellowish green ones. In short, you get discontinuous variation, either one thing or another, and not things in between. Unfortunately the world seemed full of in-between things. Among humans there is not one crowd of short people and one crowd of tall ones. Within limits we come in every size, with differences measured in fractions of a millimeter. It is now known that those mixed sizes are the result of a great many genes operating in combinations to produce the results they do, as well as the result of influences like diet.

Although surrounded by what looked like blending, de Vries was still clever enough to see through it to the purer situation that lay back of it. Discontinuity—his botanical insights were telling him

—was the true state of affairs, and blending was a special application of it that could still be explained in pure terms. To do that, he invoked pangens.

He visualized pangens as minute clusters of atoms or molecules. If there were a lot of pangens for brown hair in a nucleus and only a couple for yellow hair, the resulting individual would be brown-haired. Take away some of those dark pangens and add a few light ones, and the individual would be lighter-haired. Thus, by judicious flavoring, as one might put pepper in a stew, one could achieve what looked like blending, although the ingredients themselves did not mix.

The principle is well explained by a colored magazine illustration, which is made of blue, red, and yellow dots. All shades of the spectrum are achieved by how the dots are arranged and how numerous they are. What seem to be green areas in the picture are actually great numbers of blue and yellow dots next to one another, as a magnifying glass will reveal. There are no green dots, no brown dots, no purple ones—just the original three: blue, red, and yellow.*

For de Vries, pangens acted somewhat like those dots. It was a brilliant concept that turned out to be wrong. (It is not the *number* of pangens—or genes—that determines the quality of a trait.) But it did lead him to a conviction about the separation of traits, and to the long series of experiments by which he hoped to demonstrate the truth of that idea. He had hit on the same principle that Mendel had, and by 1900 he was ready to announce it to the world. It would be the fruit of a decade of intensive work, the crowning ornament to a life of outstanding research. Realizing that this paper was probably the most important that he would ever write, and sure to cause argument, he decided to submit it simultaneously to both the French Academy of Science and the German Botanical Society. Both societies would print and distribute it in their respective languages.

What happened as a result of this plan was hideously embarrassing to de Vries. He had no way of knowing it, but two other plant hybridizers had also been working on inheritance ratios: a German named Carl Correns and an Austrian named Erich Tschermak. What is more, *both had read Mendel.* Both had had to face the fact that they were merely doing research that confirmed what Mendel had

* In most color printing black is also added to strengthen the illustration. It does not, however, create the illusion of new colors, as the other three do.

demonstrated more than thirty years before. Each, independent of the other, was ready to publish.

What each might have said, in the absence of the others, will never be known. What is known is that de Vries's plan came completely unstuck. His French paper was published a few days before the German one. Correns got hold of a copy and read it. He immediately pointed out to de Vries that it contained no reference to the true discoverer of the laws of inheritance, Gregor Mendel, whose original paper de Vries certainly must have read. De Vries, as it turned out, had read it, at least a year earlier, perhaps two or three years earlier. He hastily inserted a reference to Mendel in his German paper just as it was about to go to press. Even then he did his best to claim a share of the glory by saying in a footnote about the Mendel paper: "I first learned of its existence after I had completed the majority of my experiments and had deduced from them the statements communicated in the text." In other words, "I made my discovery independent of Mendel."

How true this was is debatable. Certainly de Vries had come a long way on his own. However, in his teachings and writings prior to the publication of his French-German paper, he had not closed in on the majestically clear 3 : 1 ratio that lay at the heart of Mendel's insight. De Vries's plants, with complex gene arrangements, gave him other ratios. Only in 1900, presumably *after* he read Mendel, did he grasp the core point. And it appears that he was not planning to cite Mendel at all in his paper. The suspicion remains that he was prepared to claim full credit for being the discoverer of Mendelian law until Correns caught him with his hand in the cookie jar.

What about Correns? He may have had the same idea, and he may have been stopped only by learning that de Vries was ahead of him. In that case, the best he could do was spike de Vries's wheel (which he did) and then go on to a statement about how he, too, had made the discovery independently; it came to him "like a flash" as he was lying in bed (shades of Wallace bedridden by fever in Gilolo) only a few weeks before he got around to reading Mendel's paper.

Correns, more than de Vries, had reason to have heard early on of that paper. He was not only a student of Nägeli (to whom Mendel had sent a copy years before) but also a friend—he had married Nägeli's niece. Whether Nägeli ever mentioned Mendel to him or not, Correns never said, but the possibility exists that he did.

That leaves Tschermak, the Austrian. He came in on the freight a few weeks later. By that time both de Vries and Correns had publicly admitted their debt to Mendel. There was nothing for Tschermak to do but rush his own paper into print, also giving full credit to Mendel, and at the same time pointing out that he, too, had arrived at the principle independently; he had *then,* he claimed, checked it by repeating some of Mendel's own experiments. In Tschermak's case there is some reason to suspect that he did not truly understand the import of Mendel's law. However, he did validate it and help in the promoting of it. All three men were distinguished scientists, and it may be invidious here to cast doubt on the purity of their motives. But for more than eighty years other scientists have spéculated about that, among them the American Nobelist George Beadle, who hinted at it in his book, *The Languages of Life,* written in 1966.

Pure or not, de Vries, Correns, and Tschermak together constituted a formidable array of botanical influence. With a suddenness that was absolutely electrifying, they drew Mendel out of the shadows and put him on the pedestal he had been denied so long. He has not budged from it since.

The year 1900 was a milestone in the evolution of evolution. De Vries had discovered mutation, Darwin's missing source of variation. Mendel, as it was only then learned, had demonstrated that inheritance was hard (as Darwin first claimed), and not soft (as the Lamarckians claimed). Both insights were tremendous affirmations of Darwin's theory. Presumably that theory could now sail triumphantly into the twentieth century.

Eventually it did, but it would run into some severe squalls first. The entire thrust of evolutionary studies would change. Whereas Darwin and Wallace had looked at whole organisms, how they were distributed over the world, how they were related to other fossil organisms, how they differed from one another, individual by individual, population by population, the new men would go into the laboratory. There they would divide. Some of them would study the inheritance patterns of certain very small creatures with an intensity never before attempted. Others would dive into the heart of the cell in an effort to understand the nature of newly discovered extremely small structures that lay within the cell nucleus. Those entities were chromosomes. And the probing did not stop with them. Perhaps the chromosomes consisted of still smaller entities. If so, what were they and how did they work? Finally, might those entities—still only hypothetical—consist of entities smaller yet?

There is a principle in science called reductionism. It holds that a good way to a better understanding of things is to take them apart and examine their components and then take the components apart and examine them. A scientific rhyme illustrates this:

Big fleas have small fleas
Upon their back to bite 'em.
Small fleas have smaller fleas—
And so, ad infinitum.

The first half of the twentieth century would be devoted to the examination of those smaller fleas—not quite ad infinitum, but down to the molecules that de Vries had so prophetically mentioned when he was speculating about pangens. Molecules—a mere handful of them—strung together in simple patterns, would turn out to be the directors of the shape and substance of every living thing and also the moderators of how those things behaved. Evolution, for all its intricacy, for all its strange twists and turns, for all the apparent exceptions to the rules, would ultimately be found to turn on the workings of a few simple chemical compounds. That discovery, finally puzzled out in the 1950s, is surely the most astonishing, the most profound, and the most humbling insight about humans that has ever been stumbled across.

IV

Inside the Cell: Chromosomes and Genes

There is nothing alive that is simpler than a cell, and nothing can start to get more complex without first being a cell.

—MAHLON HOAGLAND

Put off your imagination, as you put off your overcoat, when you enter the laboratory. But put it on again, as you put on your overcoat, when you leave.

—CLAUDE BERNARD

He who takes no interest in what is small will take false interest in what is great.

—JOHN RUSKIN

7

The Role of the Chromosome

There is a story dating from the Tin Lizzie days, when a Ford Model T cost only a few hundred dollars, that a Kentucky mountaineer, making a rare visit to town and seeing automobiles chugging up and down the street, decided to buy one. The idea of being able to ride up the long dusty mountain road to the hollow where he lived was irresistible. He had saved some money from moonshining. He bought a car on the spot, was shown how to operate it, and drove home. He made several trips to town, taking his friends there on visits, but one day the car stopped. He complained to the man who had sold it to him. The man examined it and said that there was nothing wrong; it had simply run out of gas.

"Run out of what?" said the mountaineer.

"Gas. This car goes only a hundred and fifty miles before you need to buy more."

"You never said nothing about that."

"Expected you to know."

"Not me. Mister, I think you crawled a little up the back side of my trust in you."

"No, sir."

"Well—you going to *give* me some more gas?"

"Can't do that. It costs a dollar a tankful."

"You mean to stand there and tell me that after I shell out

hundreds of dollars for this dang thing I have to pay another dollar every time I want to ride it?"

"About the size of it."

"Well, to hell with that." He gave the tire a mighty kick, shot the dealer dead, left the car in the street, and walked home.

This book seems to be full of shaky analogies. But we can't resist comparing what the mountaineer knew of the innards of his car with what people, not so very long ago, knew about their own innards.

Everyone today is more or less aware of things about the body that would have made our ancestors gasp with astonishment. We know that oxygen is taken into the lungs and then transferred to the blood, where hemoglobin carries it to all parts of the body to be used by the cells in all sorts of ways: to help repair themselves; to make new cells; to produce enzymes, proteins, and hormones, some for breaking down the food that is eaten, some for turning that food into useful materials for body function, some to act as triggers of bodily activity, some as inhibitors of it. In their first biology classes youngsters learn that the cell is a wondrously complicated chemical engine, furiously going about its business, endlessly performing extraordinary chores of production, change, division, waste elimination, growth. (The number of individual chemical reactions that have occurred in the reader's body during the short time it has taken to finish this paragraph is in the millions, probably far higher.)

It is hard to remember that not many generations ago no one knew what oxygen was, or understood something as fundamental as the circulation of the blood. No Pilgrim who landed on these shores had ever heard of those things, or that the heart acted as a pump. What a preposterous idea! Did the heart suddenly send blood squirting out in all directions as far as the fingers and toes and then mysteriously suck it back again? And if it did that, how did it do it? And why? The answers are clear enough today, but they weren't to the Pilgrims. It took William Harvey, surgeon to two English kings in the 1600s, James I and Charles I, to find that out. He proposed for the first time that blood went round and round, that the heart was a powerful pump, that it shoved pulses of blood out into the arteries every time it contracted, and that the same blood managed somehow to get through small capillaries and the walls of individual cells into the veins, where it trickled back to the heart again, being prevented from reversing itself by a series of one-way valves.

The details of this and other bodily functions were gradually worked out. As each became clear, it also slowly became clear that all depended on the orderly working of cells, on intricate chemical reactions taking place inside those tiny boxes. Science, during the nineteenth century, had to face up to the proposition that if it was going to understand life, it would have to first understand the cell. Much of that understanding would have to be arrived at by inference, for there were limits to what the microscopes of the day could see. It was becoming increasingly apparent that cells contained minute unseen objects whose functions were as unidentifiable as the objects themselves.

Cytology, the study of cells, became a science in the nineteenth century. As noted, the cell nucleus was discovered in the 1820s by Robert Brown. Soon came the discovery that the cell wall could selectively allow some liquids to seep inside and at the same time prevent others from seeping out: a process known as osmosis. That ability had been as mystifying as witchcraft. Finally understood, it revealed a great deal about how cells functioned: they sucked in through their skins the things they needed to do their work, without at the same time losing other things already there that they needed to keep. The fields of metabolism and nutrition began to open up. Meanwhile speculation had been mounting about how organisms grew, how they assumed their final shape of ladybug or walrus from bits of apparently uniform protoplasm. That process was utterly baffling. Attempts to solve its mysteries would lead eventually to the science of embryology.

Earlier ideas about growth were not satisfying. One, a couple of centuries old, was that the fertilized egg contained a model of the complete individual down to the finest detail. This was called a homunculus, a miniature man, miniature bird, miniature crab, somehow executed in microscopic size so that it would fit inside a cell and then slowly grow to its proper dimensions.

Later studies of cells failed to reveal anything like a homunculus existing at any point along the line of development. In fact, as knowledge of cells increased, it was learned that growth occurred in an entirely different way. The fertilized egg split in two, then split again into four, into eight, into sixteen apparently identical cells. What would ultimately become a ladybug or a walrus always started in the same way: as one cell, then two, then four, and so on. It was only after a good many such divisions that individual cells began to behave differently, some splitting more often than others, assuming different shapes, moving about in the initially amorphous blob of splitting cells that constituted the earliest life stage of every living thing.

In the 1870s, with better microscopes and with newly developed dyes that enabled them to stain tissues, scientists were able for the first time to see what was going on inside a cell. Dyes concentrated themselves in different intensities in different parts of the cell, revealing its architecture. Before that, cells had appeared to be nearly transparent, and the most devoted examiner could scarcely distinguish one part from another.

It was through staining that the cell nucleus gave up a surprising secret. It was full of tiny threads all jumbled together, lying every which way. Those threads were particularly good at picking up dye, hence the name given them: chromosomes, "colored bodies," from *chromo* (color) and *soma* (body). Chromosomes apparently had something to do with cell division because, just before the cell divided, they went through a complex dance of their own. First they moved to the center of the cell. Then they became plumper, divided into pairs, and made copies of those pairs.

Whatever the chromosomes might turn out to be made of, and whatever the purpose of their remarkable activity, there were now two sets of them, each a duplicate of the other, in the cell. Their next step was to separate, one set going to one end of the cell, and the other set going to the other end. At that point a wall would begin to be built across the middle of the cell, ultimately dividing it in two. What, an hour before, had been one cell was now two, and each contained an identical set of chromosomes in its nucleus.

Another important discovery was soon made about chromosomes. Each animal or plant had its own number, and always the same number. Fruit flies have eight, Mendel's garden peas have fourteen, humans have forty-six, some kinds of butterflies have more than three hundred. And they always come in pairs — twenty-three pairs for humans.

AUGUST WEISMANN: "YOUR CHROMOSOMES DETERMINE WHAT YOU ARE."

THE extraordinary behavior of chromosomes, whose very existence had been unknown before the staining technique was developed, led to some potent speculation about its meaning and its relationship to the development of organisms. If a ladybug always develops into a ladybug and a walrus into a walrus — and if each starts in the same way, as a tiny blob of splitting cells — what makes them different? Equally mystifying, what makes some cells grow into a ladybug's eye and some cells into its wings and legs? An answer to those profoundly puzzling questions was proposed by a German biologist, August Weismann, in the 1880s. After years of deep thought, he came to the conclusion that the shaper of growth in organisms was chromatin (the stuff that chromosomes seemed to be made of) and that there had to be small units of chromatin, each of which was responsible for a certain kind of growth. If, he reasoned, one bit of chromatin was designed to control the development of the eye and another bit designed to control leg development, then the shapes and the sizes and the complexities of all living things could be accounted for.

The question that Weismann had to answer was: did the cells that would grow into eyes get a special bit of chromatin and leg cells get another special bit? Or did *all* cells get *all* the chromatin? From what was said previously about how chromosomes duplicate themselves during cell division, it would seem that all cells got all kinds. At the time Weismann asked himself the question, no one had undertaken experiments to see which idea was right. He had to make a guess. He thought it highly unlikely that each cell in the body contained the instructions to do everything it was required to do, *as well as instructions for the growth and behavior of every other cell.* This seemed an unnecessary complication. Also, it would further complicate things by requiring the existence of regulator mechanisms in the chromatin that would say "make eyes" in one cell and "make legs" in another. Thus, Weismann decided that it was more plausible to assume that the chromatin existed in smaller units that were *not* exactly alike, that different units got into different cells, and that later cell growth into different organs and different kinds of tissue was explained that way.

Weismann's guess was sensible, but it was wrong. It is now known that the precise pairing up and duplication of chromosomes during cell division is a way of ensuring that each cell gets a full set. Inconceivable as it may seem, a blueprint for the growth and functioning of his entire body existed in every one of the several trillion cells that made up August Weismann, as it has for every organism, large or small, that has ever existed.

Although he guessed wrong as to how units were parceled out among the chromosomes, Weismann was triumphantly right about what they did; they controlled heredity. Furthermore, they did consist of specific particles that were responsible for specific traits. He could not see those particles. In fact he could scarcely see the chromosomes themselves under the strongest magnification of his microscope. They were visible enough during their plump gather-and-duplicate stage, but after cell division they reverted to being threads so thin as to be virtually invisible. Some investigators even insisted that they broke up and disappeared, only to be reconstituted before the next cell division. It is to Weismann's everlasting glory that he recognized the permanence of the chromosomal material, that it was responsible for inheritance taking place at the subvisible molecular level inside the cell, and that it was not subject to the environmental influences of the later life-activities of the animal or plant. At a time when neo-Lamarckianism was widely popular, Weismann came out strongly on the other side. He ended up where Mendel had: traits were distinct; they were passed on whole. What he added was the mind-bending proposal that those traits were programmed by invisible units somehow attached to nearly invisible threads located inside the cell nucleus.

Weismann did not stop there. Noting that chromosomes were paired, he reasoned that one of the pair represented the male contribution during sex—during the fusion of a sperm cell with an egg—and that the other strand represented the female contribution. This put him face to face with another problem: if an egg is united with a sperm there will be twice as many chromosomes. And if that is true, the next mating will produce four times as many—and so on up into the thousands in no time at all. Weismann got around that difficulty with another profound speculation. He predicted that sex cells—sperm and egg—would be found to contain only half the requisite number of chromosomes. This prediction was proved right.

THEODOR BOVERI: "YES, AND YOU NEED THEM ALL."

WEISMANN'S belief that chromosomes were the carriers of inheritance was quickly confirmed by another German, Theodor Boveri, who began picking apart the eggs of sea urchins (he used sea urchins because they have unusually large eggs, making his job easier) and inserting bits of similar sea urchins and even bits of other kinds of sea urchins. In this way he succeeded in scrambling up the normal number and nature of chromosomes that the growing fertilized egg had. Sometimes it got more chromosomes, sometimes fewer, sometimes the wrong ones. The result was always deformed urchins of some kind: half-urchins, double ones, dwarfed ones, sometimes no urchins at all — too much playing around with their chromosomes would kill them. In this experiment, Boveri not only confirmed Weismann's belief that chromosomes played a central role in heredity; he went him one better. He demonstrated that an organism needs all of its chromosomes. Only when a sea urchin had its full and proper set could it grow into a proper sea urchin. There is a profound implication in this. If you need all of your chromosomes, that means that they each must do different things. And if each chromosome must be complete, that means that it must be made of smaller units, each one important to your development. Later biologists would seize on Boveri's insight. They would learn that chromosomes did consist of smaller things, genes.

Although the evolutionary implication of Weismann's and Boveri's work would not be firmly anchored for several decades, that work did, nevertheless, aim directly at the great question left unanswered by Darwin: HOW was the inheritance of variation carried out? Darwin would have been mesmerized by what Boveri had learned.

The last decades of the nineteenth century were heady ones for biologists. The work of Weismann and Boveri came like a thunderclap, followed almost immediately by the equally resonating rediscovery of Mendel's law and de Vries's announcements about mutations. People interested in evolution watched those developments with amazement and delight. Might not the new knowledge help them decide — at last — whether Darwin's theory of natural selection was right or wrong? They now had, for the first time, a clue to where variation came from. They also, thanks to Mendel, could bury the old bugaboo of blending. *Traits did endure.* True, they endured in strange ways, not always in the neat ratios that Mendel postulated. Sometimes they lay dormant for generations before reappearing, sometimes they combined to produce totally unexpected results. But, the reasoning went: track the traits, purify them, understand how they were linked together; ultimately it should be possible to predict what they would do. Equally important, they could be counted on—sometime, somewhere—to reappear. When they did, they would have good, bad, or neutral effects on the creatures that possessed them. If all that were true—and for the first time there seemed to be some scientific basis for its being true—then natural selection could work.

Those were hopeful days for the Darwinians, who had been living reluctantly, peevishly, and restlessly under a persistent cloud of Lamarckianism. Despite having been exploded by one scientific experiment after another, Lamarck would not go away, even though he had been challenged by a better idea proposed by Darwin. The major flaw in Darwin's idea was that it could not explain the origin of variation. Lamarck had an explanation. It was not a very good one, but it was good enough to keep him breathing because of the many "sort of" Lamarckian events that were occurring all the time. Creatures did come in different shapes and sizes; surely the environment and their own efforts had something to do with that. Good food made some people larger than others. Exercise made them stronger. Too much shade and crowding made plants weaker and spindlier. And—whether spindly or strong—traits were handed down. That was obvious.

The problem, of course, lay in distinguishing between inherited traits and acquired ones. We make a clear distinction between the genotype—the genetically ordered constitution of an individual—and the phenotype—the body of that individual and what happens to it during the course of its life: the scars, stains, sunburns, skills, languages, immunities, and so forth, it acquires. Stated crudely, Lamarck thought that stains and scars could be inherited. Science today knows better. It regards the genotype and the phenotype as related but essentially separate systems. They can be seen as two containers, one inside the other, each dependent on the other, each doing its own thing. The cell is such a double system. It is a body, a working instrument, being fed, nourished, growing, subject to the buffeting of life, and being strong or weak according to that buffeting. But its basic shape is ordained by the nucleus at its heart. The nucleus and its chromosomes design the cell. It must live with that design.

THE BICYCLE FACTORY: DESIGNERS AND WORKERS.

PICTURE a small totally enclosed office inside a large factory building that makes bicycles. Inside that central office is a group of designers working on a complex set of blueprints. Out in the factory are a crowd of workers and piles of supplies. The designers' job is to issue to the workers instructions for keeping the building in repair and for making bicycles. The cell and the cell nucleus bear that relationship.

The workers in the bicycle factory never speak to the designers. Every once in a while a small grille in the office is opened and some instructions are handed out. But before those outside can say anything, the grille is slammed shut again. The workers can smoke, they can pollute the air. Flies can buzz around. Sand and dust can accumulate. The food that the workers eat can be delivered on time or it can be late. It can be good or bad. Even disease germs can get into the factory and kill some of the workers. All those things can have an effect on how well they carry out the blueprint instructions given them. ***But they can have no say about the blueprints.*** They might like screens on the windows to keep out flies, but if the designers don't order screens, the workers have to endure flies.

Darwin sensed a separation between the designers in the central office and the rest of the workers. Lamarck did not. He thought that they communicated with each other. He envisioned a factory in which the workers might say to themselves: "We need a bigger factory door so that the supply trucks can get in more easily." They would build one and then instruct the designers to change their blueprints accordingly. If the designers did so, that would be an example of the inheritance of acquired characteristics.

For strict followers of Darwin there could be no such interchange. Central office (genotype) never got any feedback from factory building (phenotype). The only way a larger factory door could be built was for a designer, ***on his own,*** to decide to turn out a different kind of blueprint for a door.

In one factory (call it a human body) a designer might put out a blueprint for an excellent door, the workers would get the supplies they needed, and construction of products could continue on schedule. But a similar factory down the road might have a poorly designed door. Trucks would not be able to get in; production schedules would falter. Ultimately the factory would fail. That, according to a strict Darwinian interpretation, would be an example of survival of the fit. Variation (different-sized doors) would have occurred—he didn't know how—and selection would have resulted.

After 1900 scientists began looking at the factory door problem much more intently than they had before. Mendel had a great responsibility for that. He had shown that a factory could have high doors or wide ones (yellow or green peas). He didn't speculate about which was best, or why there were two kinds. He just satisfied himself that there were two kinds, and, as far as he could see, always would be.

It was de Vries who came up with the observation that a designer, suddenly and for no explainable reason, might capriciously produce a design for a totally different kind of door. That new design (a mutation) would then have to be faithfully followed by the factory workers whether they liked it or not.

It was this idea of *instructions coming out of a central office* that led investigators toward the cell nucleus and to the chromosomal material inside. A whole new set of questions immediately sprang up: who are those designers? Are they really making blueprints in there? If so, how are they doing it, and how are their instructions being carried out? Why do they make the same blueprint over and over again and then suddenly switch to a different one? What kinds of experiments can we think of to answer those questions?

Boveri had taken a big step in that direction already. Through his fiddling with sea urchin eggs he had demonstrated that the development of the sea urchin depended on its chromosomes, and that it needs *all* of them to make itself into a complete sea urchin. To repeat, there is a very important idea buried in this discovery. It is that chromosomes are not all alike! If they were, the loss of one or two would not matter to the sea urchin. Also, if you take away part of a chromosome, and if this has an effect on the sea urchin, it means that *not all parts* of the chromosomes are alike! They must consist of smaller elements. Those invisible smaller elements, speculated about by both Weismann and de Vries, and called pangens by the latter, were in 1909 given the name genes.

That word, a contraction of de Vries's old word, was coined by a

Dane, Wilhelm L. Johannsen,* who, oddly, did not regard the gene as a real thing. It was just a word, like "inch," used to help describe something else. There is no such thing as an inch. An inch is merely a concept, invented as a handy way of measuring other things. So with the gene. Johannsen insisted that it, too, was a mere concept and had no reality, no physical body or structure of its own.

Others were not so happy with that idea. If Johannsen wanted to use the word gene to represent some sort of measurement of hereditary difference, that was all right, but it did not dispose of the fact that there had to be something lurking in the cell—not just a word, but an actual force—that did affect inheritance. The word gene began to attach itself more and more securely to something, albeit an unseen something, that was incorporated in the chromosome.

Increased knowledge of chromosomes, and ongoing speculation about them gradually coalesced into what became known as the chromosome theory: chromosomes are responsible for heredity; small units on them do specific jobs. The theory also contained another extremely important concept originally suggested by Weismann: if an individual is the equal product of two parents and if chromosomes always come in pairs (except in sex cells, which combine and end up as pairs), then should it not follow that one half of a chromosome pair represents the father and the other half the mother?

In the great froth of experiment and theorizing that surged up in the early 1900s, biology itself began to become increasingly complicated and specialized. With the coining of the word "gene" there now could be geneticists whose job it would be to try to prove that those elusive things actually existed, and, if so, what they really did. Geneticists were not quite the same as cytologists, who were still trying to work out everything that went on inside the cell. Nor were they quite the same as embryologists, whose job was to figure out how organisms developed. But all three disciplines were linked in one way or another to the chromosome theory, which gathered strength throughout the decade and would become the arena for evolutionary inquiry for the next eighty years. It still is.

---

* Johannsen also introduced the terms genotype and phenotype.

The Fly Room

T. H. MORGAN (EARLY): "FORGET GENES; THEY DON'T EXIST."

DESPITE the growing acceptance of the chromosome theory, there were pockets of dissent. One who had little use for it was a rangy Kentuckian, Thomas Hunt Morgan. Morgan was a dyed-in-the-wool skeptic. If you couldn't prove something by experiment, he said, forget it. Ideas were a dime a dozen, and worthless until hammered out as true or false in the laboratory. He was scornful of the chromosome theory because it depended on genes whose existence could not be substantiated.

Morgan doubted everybody. He even doubted Mendel. He had run some early breeding experiments with mice; their varicolored and spotted coats defied Mendelian prediction. He then remembered Mendel's own puzzling pink flowers, neither red nor white. He shrugged off the idea that two or more genes might be linked to explain pink. "Once crossed," he stated flatly, "always contaminated."

Morgan elaborated on that at a meeting of the American Breeders Association, gathered to discuss some of the finer points of Mendelism. He stopped the breeders in their tracks with a famous statement: "In the modern interpretation of Mendelism, facts are being transformed into factors (genes) at a rapid rate. If one factor will not explain the facts, then two are invoked; if two prove insufficient, three will sometimes work out. . . . We work backwards from the facts to the factors, and then, presto! explain the facts by the very factors we have invented to account for them."

Having never seen a gene, nor having met anyone who had, he was forced by the rules of his own logic to deny that genes existed. It is ironic that, after this ringing denunciation of genes, Morgan would go on to become the father of modern genetics.

Although he did not realize it at first, Morgan was already caught up in a current that would draw him irresistibly to genes. As a young man he had experimented with frog embryos and found out things about their development that he could not explain. He had studied in Europe and been impressed by the work on cells being done there. He became obsessed with the problem of how and why organisms grew into the consistent and specific shapes they always did: the old enigma of ladybug into ladybug, walrus into walrus. He would eventually go looking for an explanation of the shapes of factory doors himself.

In 1886, four years before de Vries started his work on primroses, fourteen years before he would announce the rediscovery of Mendel's law and lay before the world his own discovery of mutations, Morgan was a gangling twenty-year-old, just graduated from the State College of Kentucky. He would be the first—and to date the only—Kentuckian to be given a Nobel Prize and the first from any state or any nation to get one in the field of genetics.

Kentucky was a bloody border state during the Civil War, its residents divided between passionate defenders of the Confederate cause and equally passionate defenders of the Union. The Morgans, a prominent family from Lexington, were Confederates. Thomas was the great-grandson of Francis Scott Key, author of the "Star Spangled Banner." A distant cousin who elected to stay in the north and become a banker was J. P. Morgan. An uncle was the notorious rebel raider John Hunt Morgan, known to supporters as the Thunderbolt of the Confederacy, and to enemies as the King of the Horse Thieves. John Hunt Morgan commanded a fast-moving cavalry outfit that led daring forays north into Ohio, causing panic and confusion behind enemy lines. He robbed banks, drove off horses wholesale as remounts for his men, was captured, put in jail, escaped, resumed his raids, and was shot and died in 1864. He lives on as a glamorous hero of the Confederate cause, part guerrilla genius, part foolhardy adventurer, whose exploits succeeded only in getting himself and a good many other men killed.

Young Tom had to grow up in the glare of this flaming figure, sometimes a heritage to be proud of, sometimes one to be endured. At college he had to take a French course under a potentially vengeful ex-Union soldier who had been captured by his uncle near Cincinnati and made to ride backward on a mule for more than two hundred miles to Lexington. Never particularly good at languages, Morgan owed it to the passage of time and the forgiving nature of his teacher that he passed the course.

The Morgans had lost everything but a fine house in the war. Conditions at the struggling new state college were even worse. It had been established only six years before and was operating in a few crowded rental buildings, the chemistry department nearly a mile away from the rest, and the administration torn by a bitter dispute over the competence of the science faculty. Morgan, one of three graduates in his year, received the first science degree granted by the tiny college (which would metamorphose gradually into the University of Kentucky). His scientific preparation was sketchy in the extreme, suffering as it did from the poverty and turmoil of the time.

Morgan's initial exposure to disciplined, first-rate science was at Johns Hopkins, where he went as a graduate student. Later he became a biology professor at Bryn Mawr. Like de Vries, he resented the time he had to spend as a teacher. "Neglect your teaching," he would mutter to his assistants. One day he said to them: "Excuse my big yawn, I just came from one of my own lectures." He was a superb teacher, nevertheless. An easy-going, informal, rumpled man, he attracted brilliant students, who were enormously stimulated by him. He not only helped them, but they helped him. After transferring to Columbia, he assembled a group of young men in his laboratory, the so-called Fly Room, who were essential to the long and arduous series of experiments in heredity that would make him world-famous. It would make some of them world-famous too.

Early in his career Morgan, having given the back of his hand to Mendel, decided that Darwin, too, was wrong. To him, natural selection through struggle made no sense. Looking at a plateful of bacteria, he observed that as long as there was food, all would survive, and that as soon as it was gone, all would die. Where was the struggle there? He became fascinated instead by de Vries's mutations, which had just been announced to the world of science. In

***Charles Darwin** at forty, from a portrait by J. F. Maguire. The artist made him look more robust than he was at the time. Plagued by worry and ill health, he had managed to work out his theory of evolution, but would not publish for another ten years.*

Carolus Linnaeus. *His pioneering work in naming specimens was a powerful propellant in the orderly development of science in the late eighteenth century. The method he used— giving two Latin names to every species— is still followed and is known as the Linnaean binomial system.*

Comte de Buffon. *He was the eighteenth century's great scientific popularizer. An ambitious courtier, he undertook to explain everything in the natural world in a series of enormously influential books that ultimately ran to forty-four volumes and stimulated scientific thought throughout Europe.*

***James Hutton.** His curiosity about the origin of minerals on the earth's surface led him to the conclusion that present-day rocks had been churning and remaking themselves again and again over an uncounted period of time, indicating that the earth was far older than commonly thought.*

***Jean Baptiste Lamarck** proposed the first coherent evolutionary idea by suggesting that acquired traits could be passed on to descendants.*

***Thomas Malthus.** His observation that only the fittest survive helped Darwin to the insight that through such a selection process evolutionary change would occur.*

***Robert Fitzroy**, commander of the* Beagle, *was an unlikely five-year shipmate for Darwin. A molten-tempered aristocrat, he was proslavery, pro-Church, and later fulminated against Darwin's heretical ideas. He afterward became governor of New Zealand, and ended a suicide.*

T. H. Huxley. *Known as "Darwin's Bulldog" because of his tenacious support of the* Origin, *he demolished Bishop Wilberforce in a famous public debate in 1860.*

Cartoons of Darwin *proliferated during his life. Most of them, like this one, showed a bearded old man's head on the body of a monkey.*

***Alfred R. Wallace** became a lifelong friend of Darwin's after both had independently hit on the theory of natural selection. In later years they drew apart ideologically. Darwin stuck to hard science; Wallace drifted into spiritualism and table rapping.*

***Charles Darwin.** In a photograph taken shortly before his death, his face shows the effects of years of drudgery and illness. With its beetling brows and flowing beard, it had become a familiar icon, the archetypical sage brimming with secret wisdom.*

***Gregor Mendel** at forty was a stocky monk, deep into his experiments with pea plants. His epochal paper proving the heritability of specific traits would be published three years later.*

***Hugo de Vries** became famous for his ultimately triumphant twenty-year search for mutations in evening primroses. Ironically, the new forms he found were later proved to be not mutants but recessive hybrids.*

them might be the true source of evolution. Furthermore de Vries's suggestion that mutations were caused by tiny molecular units in the cell was something that could be examined in the laboratory.

The Fly Room at Columbia was a strange and messy place. A small, disorderly laboratory with an overpowering reek of rotting bananas, it was lined from floor to ceiling with small glass milk bottles containing hundreds of thousands of tiny fruit flies. Swarms of escaped flies buzzed around hanging bunches of bananas. The place was teeming with cockroaches and mice.

Morgan, coming to Columbia, had made his decision to investigate the source and shape of factory doors. As Mendel had done, he abandoned laboratory mice as experimental animals; they were too large, took up too much space, and bred too slowly. On the advice of an entomologist friend, Frank Lutz, he switched to the fruit fly *Drosophila melanogaster*. This creature is only a quarter of an inch long, lives happily on mashed-up bananas, is a lusty and almost nonstop copulator, and raises another generation of itself in under two weeks.

Having convinced himself that natural selection, as a theory, was full of holes, and that de Vries's mutations held the answer to evolution, Morgan was determined to find mutations in his flies. If he could, perhaps there would then be a way to link the mutation to the chromosome and thus possibly demonstrate whether genes did or did not exist. He was still extremely suspicious of the idea that Mendel's individual factors could be inherited as regularly and precisely as Mendel said. If that were so, how was sex explained? Was there a gene for men and another for women? If so, where was it? How was it expressed? If dominance and recessiveness were to be believed, was it the male or the female who was dominant? It would appear that neither one was, since men and women did not occur in a 3 : 1 ratio but in equal numbers. The idea that there was a factor (gene) for sex seemed ridiculous to Morgan. His fly experiments, in addition to finding mutant forms, might throw some light on the sex problem.

Morgan's first experiments aimed at unraveling the mystery of sex were made, not with *Drosophila,* but with another kind of fly. They were a disaster on the order of Mendel's disaster when he was engulfed by hawkweed and for the same reason: Morgan had selected a fly that had the unusual ability to produce males or females asexually. How it did this was a mystery to Morgan, but it strengthened his belief that sex and chromosomes were unrelated. When he learned that earthworms are bisexual and that oysters can change sex when the water gets warmer, his disbelief became stronger.

Others were not so sure. With better and better microscopes, and with the keener vision that a hopeful idea can often produce, they had noticed something peculiar about chromosomes. Though they had always seemed to come in identical pairs, an exception to that rule was gradually revealed. In many organisms there was one pair whose two parts were not always identical. In about half the population of fruit flies that pair *was* identical, but in the other half it was *not*. In the latter pairs, one part had a distinct hook at one end—it looked as if it had been bent.

If Morgan noticed this he was unimpressed by it. But when a colleague, E. B. Wilson, with a more flexible mind than Morgan's noticed it, he went on to notice something else: only males had the bent chromosomes. Females always had a pair of straight ones. He had no clue as to how "bent" could have a say in determining sex, but he reasoned that if it was the only visible difference between the chromosomes of males and females, and was consistent for thousands and thousands of flies, it could at least be plausibly assumed (though unproved) that the lottery of sex was somehow connected with those two unobtrusive and atypical little chromosomes.

Morgan paid no attention. He was busy with half a dozen other experiments. He had decided, among other things, to prove that Lamarck—as well as Darwin and Mendel—was wrong. Typical of Morgan; others had done it many times, but he wanted to see for himself. He got one of his students to start breeding fruit flies in total darkness to see if it affected their eyesight. Sixty-nine generations later the experiment was called off—the flies of that generation could see as well as their remote ancestors. Lamarck had been buried once again.

But not Mendel! Morgan was now breeding fruit flies on a large scale in an effort to find mutations. At first he found none. It is not hard to see why. To begin with, he and his assistants were not exactly sure what they were looking for. Differences, surely. But how different does a difference have to be in order to qualify as a difference? Nearly every fly that was put under the microscope had some differences. The little bristles that covered parts of its body were slightly longer or shorter than the bristles of other specimens. The flies themselves varied ever so slightly in size. Furthermore, none of them was more than a quarter of an inch long. Finding anything distinctive in a population of laboratory animals that small was hard. Morgan became increasingly impatient and increasingly discouraged. He tried to induce mutations. He subjected his flies to heat, to cold, to numerous chemicals. He whirled them around in machines. He exposed them to radiation. Nothing. According to Ian Shine and Sylvia Wrobel, who in 1976 published a biography of Morgan,* he said in despair to a colleague who visited the Fly Room in 1910: "There's two years' work wasted. I've been breeding these flies all that time and have got nothing out of it."

It should be remembered that each bottle of flies represented a large investment of work. In it was a pedigreed strain of insects whose exact ancestry had been tracked for generations. Each time a new mating occurred and a bunch of new flies was hatched out, they had to be anesthetized and examined one at a time under a microscope. The young men and women recruited by Morgan for this mind-numbing task were a dedicated lot, and why they were engaged in such a ridiculous undertaking was not understood by others at all. One child, when asked at school what his father did, replied: "He counts flies at Columbia."

T. H. MORGAN (LATER): "SORRY; THEY DO."

MORGAN'S attitude toward genes changed with dramatic suddenness one day in 1910 when a fly with white eyes was discovered. A mutation at last. That peculiarity stood out like a searchlight because normal fruit flies have red eyes. It was hustled off to mate with a normal female, and there followed some moments of anxiety because the white-eyed specimen seemed very listless. It exhibited none of the irrepressible sexual ardor of a normal fly. Nevertheless it roused itself sufficiently to mate once and then expired. The female was sequestered and laid eggs that hatched into 1,237 offspring. All had red eyes.

To a student of Mendel this could be no surprise. The white eye

* The previous quotes attributed to Morgan and the anecdotes about his early life are from their book: *Thomas Hunt Morgan.*

was obviously linked to a recessive gene that would not be expressed again until the 1,237 hybrids were allowed to mate and presumably produce, in the familiar 3 : 1 ratio, a proper number of white-eyed individuals. Morgan watched this first half of the Mendelian experiment with his usual skepticism, but when the next generation appeared his attention was riveted. The white-eyed gene had showed up again.

The new generation produced 3,470 red-eyed flies and 782 white-eyed ones, not quite a perfect 3 : 1 ratio but close enough.* The important thing was that "white-eyed" had not disappeared. Morgan's scathing remark of a few years earlier, "once crossed, always contaminated," had been proven untrue by one of his own experiments. He was forced to the conclusion that Mendel might have something after all.

What happened next turned out to be one of the great moments in the history of genetics. The flies were examined more carefully. *Every one of the white-eyed flies was a male.*

We can say glibly now: "No mystery there. The white-eyed trait was linked to the sex chromosome of the fly." Easy today. Not easy for Morgan. It must be remembered that he was suspicious of the entire chromosome theory. Nothing had really been proven except that chromosomes were essential for development. But did they determine sex? Well—maybe. How? Nobody was entirely sure. As for genes, they were still simply ideas.

Sex linkage and genes did not remain ideas long. In a series of inspired experiments, Morgan literally put genes on the map. To understand how he and a couple of assistants did this, a review of what they knew about chromosomes is necessary.

Every cell in the *Drosophila* body has eight chromosomes, four pairs of two each: that much they knew. Most of the fly's cells were so small that their chromosomes were nearly impossible to study, but a young woman named Nettie Stevens, who had studied under Morgan at Bryn Mawr, discovered that the salivary gland of the fly had unusually large cells and much clearer chromosomes. The difficulty of her work and the work of all those that followed can be

* Statistically speaking, it was impressively close. The nearly perfect approximations to a 3 : 1 ratio that Mendel got in his experiments with what actually were rather small pea samples, have caused a number of statistically sophisticated scientists to question his figures. He is suspected of having doctored them—once he caught onto the basic principle—to make them more convincing.

imagined, when one remembers that she had to first realize that it was this small gland in this midget fly that she would have to occupy herself with, not some of its other cells. Next she had to isolate a minute bit of that tissue, stain it so that the chromosomes would show up, then tease the cells apart so that the contents of each nucleus could be observed independently. Most of the time, of course, the cell under observation would not be in the process of splitting, and therefore there would be no sets of visible plump pairs of chromosomes—only a jumble of nearly invisible threads. So she would have to examine one nucleus after another until she found one whose chromosomes she could count. Rather quickly she identified three "large" chromosome pairs. Later, and with great difficulty, she succeeded in spotting a very small fourth pair. Most important, of the three large pairs, one pair was not a true matching pair. Among males, one of the pairs had the hook shape already mentioned.

By the time this peculiarity was discovered in the fruit fly, it had been fairly well established as a working theory that the odd pair were the ones that determined the sex of the individual. The straight one was named the *X* chromosome and the crooked one the *Y* chromosome. A female always had two Xs and a male always had XY. The diagram on page 154 shows how this works. It is well worth a moment's study, for it shows a number of other important things also:

1) There are four *pairs* of chromosomes in the ordinary fertilized fly cell.
2) Two of the pairs are very large. One pair (the sex chromosome) is fairly large. One is extremely small. The importance of this size difference will become clear in a moment.
3) One pair of chromosomes is XY in the male and XX in the female.
4) After dividing to form sex cells, there are four *single* chromosomes in the sperm or egg.
5) When the male cell divides to form sperm, one sperm gets its X chromosome and the other one gets the Y. In the female, each egg gets an X.
6) When sperm and egg unite in the next generation, the only possible combinations are XX and XY. Therefore, as the diagram shows, half are males and half females.

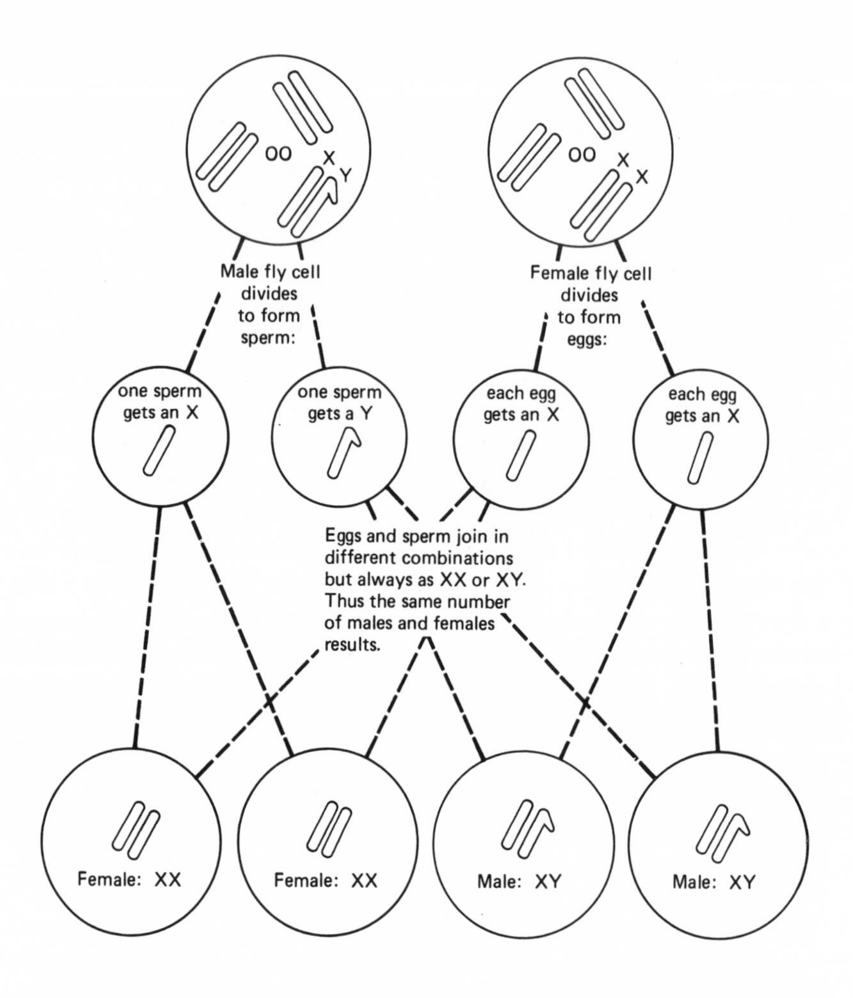

This simple genetic rule for determining sex, universally understood today, was not at all clear when Morgan first tackled it, because there are many exceptions to it. Among birds and butterflies, for example, the situation is exactly backwards: males have the XX and females the XY. The principle, of course, is the same, and again results in an equal population of males and females, but for a while it was confusing, not helped at all by the oddities of a great number of lower organisms that are asexual, bisexual, change sex, or alternate sex from generation to generation.

White-eyed flies, by providing a clear revelation about how sex was determined, riveted Morgan's attention on chromosomes. An obvious next step, of course, was to try to discover where those white eyes had come from. Morgan was utterly in the dark on that one, but he would get help from his team of dedicated and able laboratory workers, notably from two exceptional young men: Calvin Bridges and Alfred Sturtevant. Bridges was an underprivileged orphan from upstate New York, who managed to become a biology student at Columbia and then worked his way into the Fly Room by offering to clean bottles and count flies. Morgan rather grudgingly took him on as a volunteer but quickly changed his status to regular lab worker when Bridges, whose eyes were unusually keen, found another mutant fly: one with a bright vermilion eye instead of the usual darker-red one.

Sturtevant grew up in Alabama and as a boy got interested in the bloodlines of the horses raised by his father, a breeder of thoroughbred trotters. He wrote a paper on the subject that so impressed Morgan that he not only helped Sturtevant get it published but also invited him into the Fly Room as an undergraduate. Sturtevant responded by making an extraordinarily perceptive deduction about the way genes were positioned on the chromosome. He did this when he was only twenty-one. Both he and Bridges worked in the Fly Room for seventeen years. No one has ever attempted to calculate how many flies they examined during that time.

After Bridges's discovery of a vermilion-eyed fly, others in the laboratory began finding mutations, too. Even Morgan found a few. But Bridges was the star. He found dozens. As the flies became more and more familiar to their patient observers, peering at hundreds of little bodies day after day, tiny differences that hitherto had gone unnoticed began to appear. Some of the bristles on one fly's body would be forked: a mutation. The veins in the wings of another would be slightly different from the norm: another mutation. Occasionally a weirdly shaped wing, or no wing at all, would appear: an easy one to spot. Within two or three years more than forty mutant strains were cataloged, about one for every fifty thousand flies examined. The array of bottles mushroomed, for a huge and cloudy concept was beginning to take shape in Morgan's mind: traits *were* located on chromosomes; perhaps they could be located on specific chromosomes. To test this idea, all the mutant strains would have to be crossed with others to see what happened.

The idea was this: convinced now that Mendel was right, Morgan could use that first white-eyed fly as a tracer because he knew he would not lose it. He could follow it, almost like a brightly colored marble in a jar of plain marbles, through many generations of other flies that also bore telltale marks. Those marks, of course, would be the new mutations that his Fly Squad was discovering. The white-eyed fly had already showed up in the second hybrid generation as a male. Did that mean that the mutation was on the sex chromosome? With a vermilion-eyed mutant that idea could be tested again. It was; vermilion, too, seemed to be sex linked. How this works can be illustrated with another diagram of fly chromosomes, this time, for simplicity's sake, showing only the sex chromosomes. In the diagram the X chromosome of the male is marked with a W to indicate that it carries a trait for white eyes. Similarly, R indicates red for a normal red-eyed female. The hooked male Y chromosome has no mark because it is "empty" of color — it has no effect one way or the other on the eye color of its descendants.

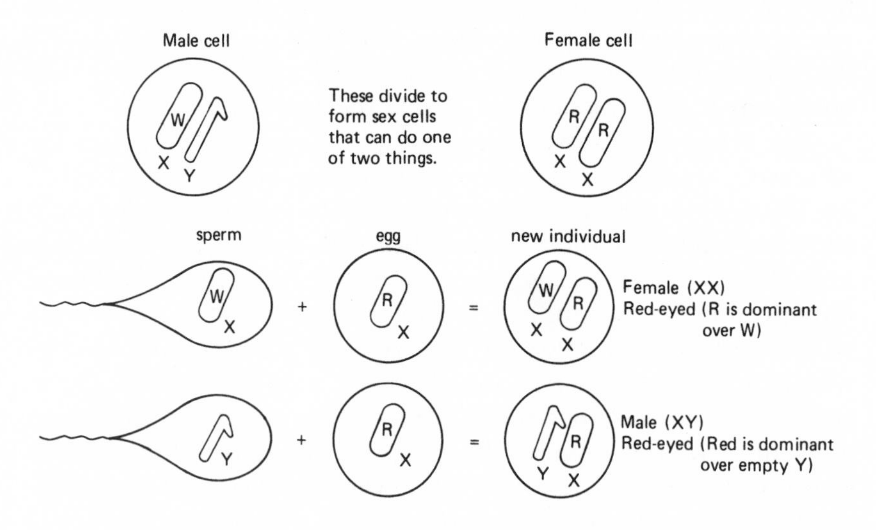

So far, so good. This diagram, like the previous one, explains why half the first hybrid generation flies are male and half female. But it does more. It also shows why all those flies are red-eyed. White has not disappeared, however. As Mendel could have predicted, it now lies hidden in the female, who—though she may look exactly like her mother—is genetically quite different.

"Don't forget to point out," said Don, "that the male is different, too." And it is: it now has a red eye; the diagram explains why. Those differences permit the following results in the next generation, the so-called second hybrid generation.

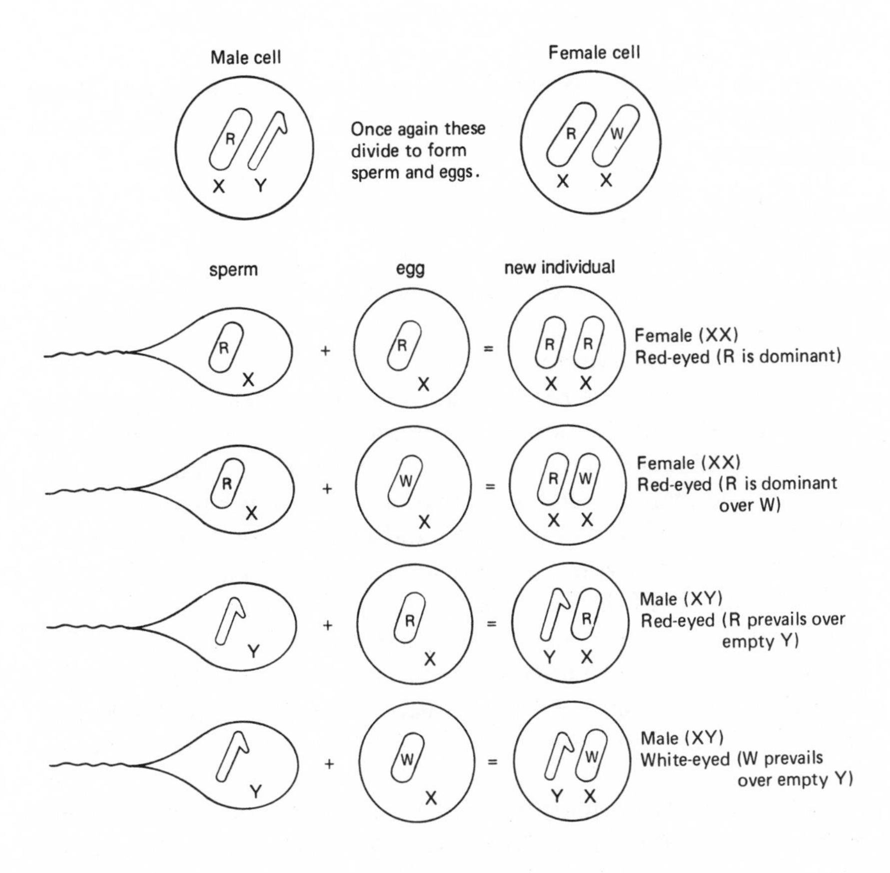

It was this result that finally turned Morgan into a true believer in Mendel. It had been a hard battle for the stubborn scientist, but he could not ignore the results of his own experiments: the 3 : 1 ratio expressing itself so neatly with three dominant red-eyed flies and one recessive white-eyed one. The fact that all the white-eyed flies are males is also explained: the mutation for eye color is on the sex chromosome. If it had been on one of the other chromosomes, things would not have turned out so neatly. There would have been some white-eyed females as well as males, in other ratios.

"Hold it," said Don. "Hold it right there. These explanations are tricky. Are you sure our readers are going to understand the principles underlying this stuff?"

"I think so," I said. "I'm no scientist, but I think I understand it."

"Well, then, how do you explain a white-eyed female?"

"What?"

"We've been talking about eye color being sex-linked. About those big, hot, red-eyed females and those puny little white-eyed males, right?"

"Yes."

"Females are red, males are white."

"That's what we've been saying."

"Morgan got females with white eyes."

"He did?"

"He sure did. His literature is full of them."

That is true, and any reader who is as sharp as Don can probably work out why. For the benefit of others, here is the explanation. It starts with the fact that the first cross between a red-eyed female and a white-eyed male produces a female that no longer has two "red" chromosomes, but rather a red one and a white one. Mate her with a white-eyed fly, and you get the results shown on page 159.

The members of the Fly Squad now had flies with three kinds of eyes: red, white, and vermilion. They went on to discover pink, garnet, cinnabar, and a dozen others. They even found flies with no eyes at all. Some colors were dominant, others recessive — they all followed Mendelian law. And all were linked to the sex chromosomes.

Working out that principle, simple as it may now seem, was a giant step by Morgan and the Fly Squad. It led over the years to something that Morgan himself, not much earlier, would have thought inconceivable: the mapping of genes. Morgan now not only believed in genes, he was rapidly learning what individual ones did, and he was even beginning to figure out where they were. Considering that no one in the world had yet seen a gene — that they were still entirely

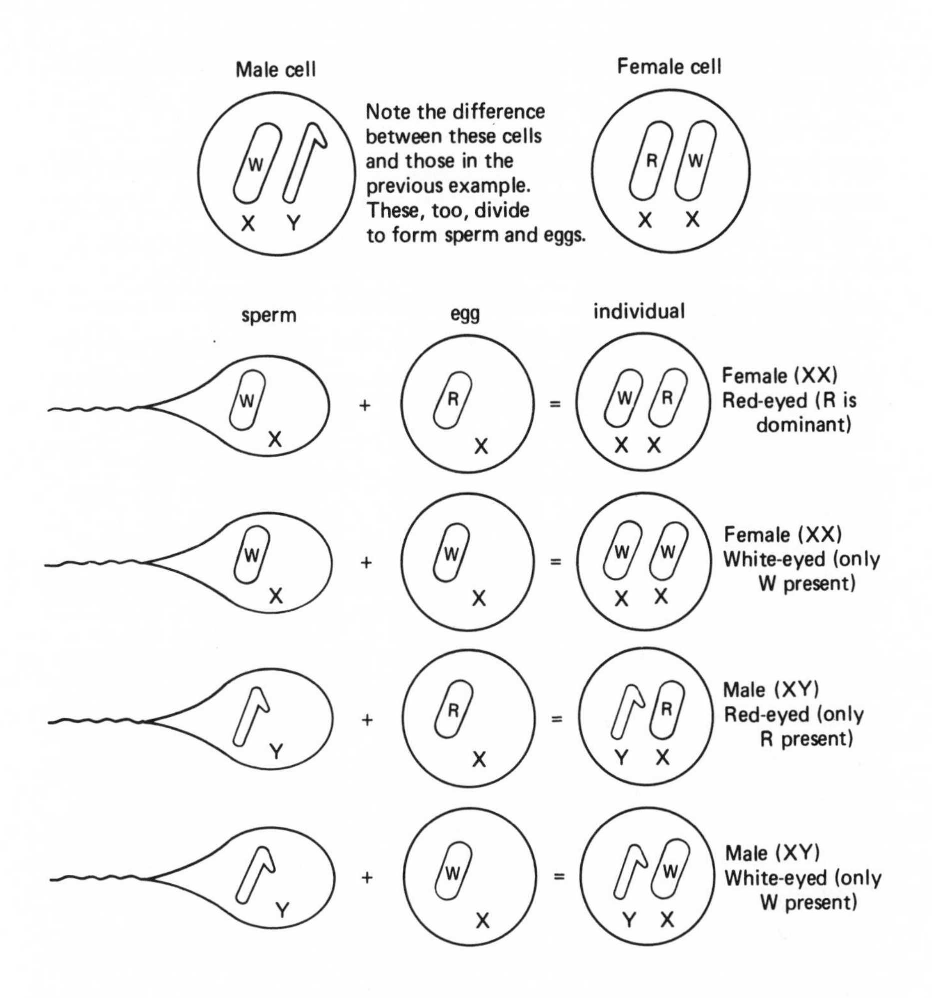

theoretical—this was an extraordinary achievement, made possible, again, in large part by his two enormously talented young helpers; Bridges and Sturtevant.

Going back to the analogy of the bicycle factory with the central office, we should visualize a member of the Fly Squad as being in the position of a man standing in the street, trying to guess which designer was making what blueprint inside. Not only was he unable to get into the central office to observe that, he could not even get into the factory to sneak a look at the blueprints being handed out to the workers. He had to wait until bicycles began coming out of the factory. When he saw blue bikes with white seats coming out and red bikes with black seats (always blue and white, always red and black) he could assume that the designer of blue bikes had stipulated white seats on his blueprint (that is, two genes were linked on one chromosome). Furthermore, if, after a long run of blue bikes with white seats, blue bikes began appearing with brown seats, he could assume that one of those unseen designers had changed a blueprint—another way of saying that a chromosome had undergone a mutation.

Over the years the Fly Squad watched strange things come out of the factory. They saw three-wheeled models, they saw ones with headlights, ones with bells.

"I'm going to test this bicycle factory analogy right here," said Don. "Did those new models always have bells?"

"No. Just some of them."

"Okay. Another question: Do bells always go with blue bikes?"

"No. Sometimes with red. Bells aren't linked to any color. Does that suggest anything to you?"

"Of course it does. I'm not your dumb average student. It suggests that the workers have more than one blueprint to work from. I mean—a worker picks up a blueprint that says 'blue bike.' Then he picks up another blueprint that says 'bicycle bell.' Okay, he puts on a bell. But he might pick up a different blueprint that said 'horn' instead of 'bell.' If 'blue bike' and 'bell' were on the same blueprint, they would always go together; you would never get a horn. But suppose you do get a horn once in a while. Then you have to begin thinking: two blueprints. Maybe a whole lot of blueprints."

"Since we're testing," I said, "let me test you. How many blueprints would you say?"

"For flies?"

"Yes."

"I'd say eight."

"Why eight?"

"Actually four—two sets of four which go to two different workers. The workers are standing on opposite sides of an assembly line. Call one worker a sperm and the other an egg. When the assembly line begins to move, each worker does his thing."

"Fair enough. But what makes you think each has four blueprints?"

"I have a helicopter," said Don. "What you didn't mention is that there is a skylight in the factory roof. When I fly over it I can look right down into that office and see those blueprints lying on the table. Two piles of four each."

"What evidence do you really have—outside of this analogy?"

"A lot. My helicopter is actually a microscope. That office is the nucleus of a fly's cell. And with my microscope I can see four pairs of chromosomes in the cell."

Don was right, of course. Morgan did know that his flies had four sets of paired chromosomes. As the number of mutant forms grew, and as the experiments that crossed them in every possible way grew, there also began to grow separate groups of linked traits. The sex-linked group, the one that controlled eye color and some other things, was called "linkage group one." Soon a second linkage group had to be created; it was made up of traits that behaved independently of linkage group one. It was followed by a third group. These groups gradually took shape on the big charts in the Fly Room as the linkages became clear. Mistakes would be made, erasures made, and traits relocated to fit the evidence of new crossings. But before very long the groups assumed clearer and clearer dimensions. It was possible to visualize them as three imaginary piles of instructions for assembling all the ingredients needed to make a fly (or a bicycle). Two of the piles were somewhat larger than the third. Since their sizes corresponded almost exactly with the sizes of the fly's chromosomes, it became more evident that those piles of traits did indeed represent the chromosomes themselves. The trouble was that there should have been a fourth very small fly chromosome.

Another brilliant young worker in the Fly Room, Hermann J. Muller, went at that problem. Eventually he succeeded in finding a few traits that had no links with those in the other three linkage groups. A fourth very small linkage group had to be created for them. Morgan was now convinced. Chromosomes were the source of heredity. Each had its own set of genes. And since chromosomes were long, skinny, threadlike things, it was probable that the genes were strung together like beads on a necklace, or like buttons pasted on a long strip of paper.

Morgan had, in effect, discovered genes. He still had never seen one. He didn't know what it was made of, or how it did its work, and he had no more than a hunch that the arrangement of genes was linear. Why not in clumps or in circles? Hunches were never enough for him; he wanted evidence. And the only evidence that would satisfy him was that derived from experiments that he had made himself or seen in his own laboratory. It was his assistant Sturtevant who came up with that evidence.

ALFRED STURTEVANT: CROSSING OVER DISCOVERED.

STURTEVANT was a more reflective man than Bridges. He found far fewer mutant flies, but he thought more about them. He would sit, puffing on his pipe, staring at the wall charts, apparently doing nothing for hours at a time. It is to Morgan's credit that he let his assistants go their own way, conduct their own experiments, do their own thinking. Sturtevant was delighted — in a kind of intellectual-esthetic sense — with the way the groups of traits on the wall charts had assembled themselves to match so neatly the chromosomes inside the cell nucleus of the fly. It was one of those wonderful congruences whose sheer logical beauty enchants the mind of the scientist. What troubled Sturtevant was that the congruences were not always so neat. There were some ragged edges to the piles of traits. Things did not always link quite as they should. In effect, once in a while, a black seat would appear on a blue bicycle. And that was wrong because the chart said that black linked with red bikes, not with blue. Endless crossings had proved it so.

Sturtevant chewed over that. What had gone wrong? Had a worker in the factory somehow gotten on the wrong assembly line and picked up the wrong blueprint? More likely, had he somehow torn a couple of blueprints in two, and in pasting them together gotten the two halves mixed? If so, the top half would still say "blue bike" — as it should — but it now might be mistakenly attached to a bottom half that said "black seat."

Sturtevant was not the first to puzzle over this. De Vries, a decade earlier, had noticed that the characteristics of his primroses were usually linked — *but not always*. This did not disturb him. Rather, it explained something that otherwise he could not have explained. Mutations aside (and de Vries knew them to be rare), if all traits were irrevocably locked onto their respective chromosomes, there could never be any more different kinds of flowers than there were chromosomes. Since primroses had only a few chromosomes but hundreds of varieties, that could not be. Obviously there had to be some switching back and forth between chromosomes to provide the great number of combinations he could see with his own eyes.

How did that switching take place? Sturtevant soon found out. Chromosomes were lying across each other in a tangle of threads in the resting stage. That very tangle might cause some of the threads to get mixed up during division and then get reattached in the wrong way like a couple of torn blueprints hastily pasted together by a careless worker. If that were so, something like this would happen:

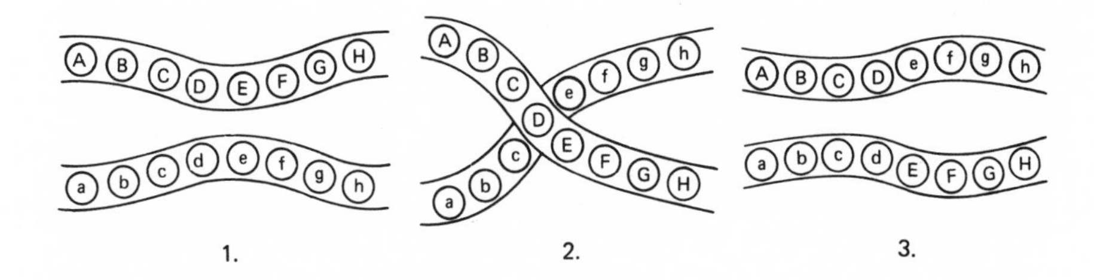

As a result of this crossing over, as it would come to be called, some of the genes would now be on the wrong chromosome and the relationship between certain traits would have changed. If the trait for a blue bicycle was somewhere on the ABCD part of the chromosome, and the trait for a white seat somewhere on the EFGH part, it was clear that as long as the chromosome was intact those traits would travel together—they were linked, as Sturtevant learned in many experiments. However, if crossing over had occurred and resulted in a chromosome like that in figure No. 3, then blue bike and white seat would no longer be linked.

That crossing over might occur had been Morgan's idea originally. Sturtevant turned it into a probability when he began discovering more and more examples of broken linkages. But what really riveted his attention was that certain new arrangements were more likely to happen than others. The more examples of crossing over that he obtained, the clearer it became that there was a statistical frequency in linkage breaks. Some always were more common than others, measurably so.

Such things have to be explained. No scientist, observing a phenomenon that seems to be following some sort of regular principle, can rest until he understands the principle. Mulling over this one night, Sturtevant came to the conclusion that the governing principle was the position of the trait on the chromosome.

If, in the diagram below, "blue bike" is controlled by A and "white seat" by H, then, there are a good many places between them where the chromosome can break and be incorrectly joined again. In other words, the probability of a break somewhere between the two is fairly large. But if "blue bike" is D and "white seat" is E, there is only one way that they can be separated: at the precise point between them—and the probability of its happening is very small.

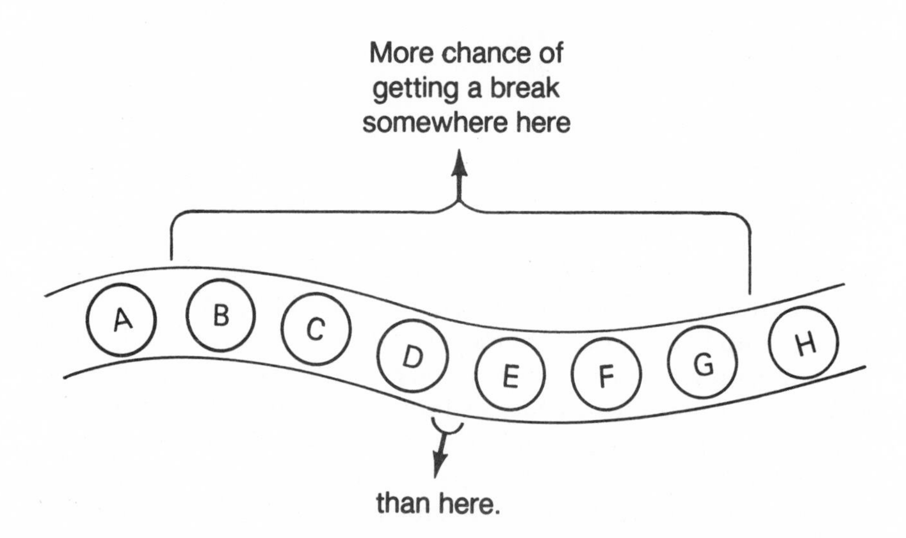

The idea of a linear explanation for the frequency of a genetic event was a profound intellectual leap for the young man. It convinced him—and Morgan—that genes were indeed lined up like beads on the chromosome and could not perform as they seemed to do if they were in clusters or rings or any other configuration.

Since he was now beginning to get regular percentages of crossing over for certain traits, Sturtevant took the next logical step: he used that information to find precisely where on the chromosome a particular gene was located. He learned, for example, that a crossover resulting in flies with a yellow body and a white eye took place not much more than 1 percent of the time but that a cinnabar-colored eye and a vestigial wing crossover occurred about 10 percent of the time; a miniature wing and a white eye occurred 30 percent of the time.

Having decided that distances apart on the chromosome were responsible for those differences in percentages, Sturtevant began drawing long lines on large sheets of paper. The lines represented chromosomes and on them he began placing genes where he thought they belonged—close together for ones that seldom crossed over, far apart for ones that often did. He divided his chromosome lines into units, as a ruler is divided into sixteenths of an inch. Each of his units corresponded to one crossover per hundred fertilized eggs. Thus, the yellow body and white eye, which crossed over scarcely more than 1 percent of the time, were scarcely more than one linear unit apart on the chromosome. Others that crossed over 8 percent of the time were eight units apart, still others ten units apart, and so on.

Morgan became absorbed in what Sturtevant was doing. And when Sturtevant began predicting correctly—from what he knew about chromosomes he had already located—how far apart other combinations would be, it appeared that Morgan's grand concept of gene mapping might be realized after all.

The job was not nearly as easy as it may sound in this simplified description of it. For one thing, there were four separate pairs of chromosomes to be mapped. With traits crossing back and forth, it took many weary hours of fly breeding in increasingly complex combinations to nail down the positions of many genes. Furthermore, it was discovered that crossing over was not always limited to one break. Sometimes there were two, when one chromosome looped over another in two places. Sometimes a broken piece got attached wrong-end-to. Sometimes a small piece floated away and got lost. Each of these anomalies produced flies that didn't adhere to the rules. Each time there was a great deal of head scratching and endless experiments to try to explain the anomalies, reach an understanding of what was going on, and make the necessary adjustments in the charts. Ultimately all exceptions were found to work out according to Sturtevant's basic proposal: genes were lined up on chromosomes; they often rearranged themselves by crossing over; the frequency of crossing over held the key to where the gene was located on the chromosome.

The work of the Fly Room ultimately found its way into scientific publications. It electrified the world of biology. Morgan wrote a book in collaboration with Sturtevant, Bridges, and Muller. He also wrote more than three hundred scientific papers. On his own, Bridges wrote an important paper about chromosomal abnormalities. *Drosophila* became the most famous laboratory animal in the world. Geneticists from everywhere began clamoring for flies to experiment with. Morgan, having accumulated scores of pure strains, began shipping them out, and the work of gene mapping went on. Today, knowledge of the genetic structure of the fly is detailed. The maps that Sturtevant first began making, whose blank spots have been filled in subsequently by many others, have proved to be remarkably accurate. Photographs of the famous fly's genes taken by an electron microscope—with resolving power great enough to make visible what had been totally invisible before—reveal that the chromosome does consist of chains of units, just as Sturtevant said.

Although much of the actual work in the Fly Room was done by his assistants and some of its most original ideas came from them, Morgan was the driving force behind it all. He created the laboratory and had the vision and energy to propel its mammoth experiments, along with the flexibility of mind to change his views when the laboratory evidence forced him to—something that not all scientists are capable of. He came to Columbia with some very stubborn opinions: Lamarck was wrong, Darwin was wrong, Mendel was wrong. He emerged with some profoundly different ideas, some of them hammered into him by his assistants. He was an eloquent and forceful proponent of his views. When he changed them—particularly when he published works that showed why he had to change them—the world sat up and noticed.

What Morgan had produced was not only a magnificently detailed and convincing endorsement of Mendel but also an explanation of how Mendelism worked. All the criticisms of the old abbot, based on experiments that came out "wrong" time after time, were swept away by the work of the Fly Squad. Difficulty after difficulty disappeared when the anomaly that caused each one was fully understood and found to conform to basic Mendelian law. The trouble always turned out to be a peculiar wrinkle in the law, not the law itself.

Paired chromosomes emerged as the bearers of Mendel's traits. The traits themselves were identified as unseen tiny bundles of influence lined up on the chromosomes. The X and Y chromosomes explained how sex was determined. Linkage of traits was established. The location of specific genes on specific chromosomes was worked out, first by discovering the phenomenon of crossing over and then by taking advantage of its varying frequency. Mutations — the things that started Morgan off on his great enterprise — turned out to be rare. True, they were the markers that unbarred the door and made all the subsequent ingenious experiments possible. But they by no means explained the enormous variability of the flies themselves. *That* was mainly the result of crossing over.

What an achievement! It put a firm base under the theory of the chromosome. Morgan's work also was the basis for something more precise: the theory of the gene. Oddly, the insights that launched the theory of the gene were gained entirely by inference — since they dealt with invisible things — in the laboratory of a man who mistrusted inferences. But the flies themselves, hundreds of thousands of them, whose physical characteristics were unpredictable in 1910, had become entirely predictable a decade and a half later. The logic that connected eye color and wing shape to specific locations less than a millionth of an inch apart on the chromosome was overwhelming. A jigsaw puzzle, whose picture was revealed down to the last detail in the anatomy of the fly, had somehow been constructed out of those invisible pieces. Morgan became a passionate advocate of a process whose working and whose ingredients he had never seen. He is generally regarded as the father of modern genetics. All sorts of honors poured in on him. He received the Darwin Medal in 1924, and in 1933 the big one, the Nobel Prize.

There were those who felt that Morgan did not deserve all that acclaim, that he did not share it out generously enough with his assistants. One who felt that was one of the assistants, Hermann J. Muller, the man who had discovered linkages on the fourth chromosome. Muller also regarded Morgan as sloppy (he was), careless (he was), wrong-headed (he often was), and he left the Fly Room to do important work of his own — and win a Nobel Prize himself. But the men who were closest to Morgan, Bridges and Sturtevant, had no criticism of him and were exceptionally loyal. Even Muller, in later years, more than made up for his early criticism. World-famous himself by that time, and less hungry for credit, he was able to regard Morgan more dispassionately:

> That the early findings of Morgan were so quickly followed up . . . was due in no small measure to his having opened the doors of this laboratory and, indeed, his mind to a group of co-workers . . . who chose entirely their leads and who would not have had the opportunity to carry on freely in most European or even American laboratories. Had Morgan been more of an authoritarian and less willing to be an equal member of the group in discussions, the young workers would not have had the opportunity they needed for the further development of the subject, and Morgan's own mind would not have been so opened . . . to a theory of natural selection, now on a more rational basis, and provided with an elaborate mechanism for its operation. Morgan was won to this point of view only against his own very active opposition, yet it is to his enduring credit that he was finally willing to alter his own viewpoint in accordance with the empirical facts.

Thus did Muller pay a fine tribute to Morgan and to the style in which he conducted the Fly Room, while also making clear that Morgan was exceedingly stubborn (he was). On another occasion Muller wrote even more glowingly:

> Morgan's evidence for crossing over and his suggestion that genes further apart cross over more frequently was a thunderclap: hardly second to the discovery of Mendelism, which ushered in that storm that has given nourishment to all our modern genetics.

Having looked Lamarck in the eye and stared him down with an experiment on flies' eyes, and having looked Mendel in the eye and come away converted, Morgan had only Darwin left to deal with. Here he was unconvinced. He had seen the dramatic results that could come from submicroscopic rearrangements of genetic material and was more than ever persuaded that de Vries's mutations held the key to evolution, rather than Darwin's gradualism through selection. Furthermore, Morgan's work had such a powerful effect on the entire field of biology that his views on evolution could not help affecting biologists generally. That is to say, they tended to look at his findings from the same perspective that he did.

Darwin had left a large legacy of skepticism through his inability to explain variation, and after his death that skepticism grew. Now, suddenly, there was a detailed explanation of the mechanics of variation, demonstrated with dazzling clarity, with predictable results, repeatable in tests in other laboratories—and it was all based on mutations.

They were not the whole game, of course; crossing over was a more potent mixer of characteristics. But there would be nothing to mix if mutations did not introduce differences from time to time in individuals. That was the message that went trumpeting out from the Fly Room. Darwin, already somewhat in eclipse, faded further in the early decades of this century. The geneticist G. Ledyard Stebbins recalls that when he was at Harvard in the 1920s two distinguished members of that faculty under whom he studied biology, Oakes Ames and George H. Parker, had little use for natural selection as an evolutionary force. Mutations, they insisted, were the thing. The good ones led directly to change. Natural selection was simply the mechanism for eliminating the bad ones. Stebbins also remembers that the most highly regarded biology text of the time was by the Swede Erik Nordenskiöld, who wrote:

> To raise the theory of natural selection, as had often been done, to the rank of a natural law, comparable to the law of gravity established by Newton, is, of course, completely irrational as time has already shown; Darwin's theory of the origin of species was long ago abandoned. Other facts established by Darwin are all of second-rate value.

Curtains for the sage from Downe? Not quite. Morgan himself, visiting England in 1923, got quite a shock when he was taken by Julian Huxley, grandson of Darwin's "bulldog," to Oxford and shown an exhibit of mimicry in certain tropical butterflies. Half of the specimens in the exhibit were butterflies that tasted so awful that birds, once they ate them, learned never to do so again. The other half consisted of edible butterflies that looked almost exactly like the bad-tasting ones. In fact, mimicry is common among insects, with certain flies resembling wasps, others resembling bees, and so on. But the South American butterfly collection that Morgan examined at Oxford was exceptional. Over time, the various races of the bad-tasting kind had, each in its own region, undergone changes in the shape and patterns of their wings. They no longer looked alike. But each had its mimic, and the mimics had changed along with their protectors with a precision that was uncanny. Whatever the subtle selection pressure or random genetic change that had resulted in alteration of the protectors, it had been reflected in direct selection pressure on the mimics. They became almost identical with their models. According to Huxley, Morgan left that exhibit quite shaken. He said: "This is extraordinary; I just didn't know that things like this existed." By the end of his life Morgan had become reconciled to the possibility that Darwin might have something after all.

But Lamarck — never. In the same year that he saw the butterflies, Morgan attended a symposium in Woods Hole, Massachusetts, and heard an astonishing claim by the great Russian behaviorist, I. P. Pavlov. Pavlov had made himself world-famous by demonstrating that he could condition the behavior of a dog by ringing a bell every time he gave it food. Before very long the dog would drool in anticipation whenever it heard the bell.

This time Pavlov described some mice conditioned to run from one room to another for food whenever they heard a bell. It had taken the first generation of mice three hundred rings before they learned to associate the bell with food. But by breeding those mice among each other, Pavlov had produced a second generation that had learned in only one hundred rings, a next generation that learned in thirty, the next in ten, and the next in five. It was the five-ring stage that Pavlov had reached when he spoke at Woods Hole, claiming a fantastic demonstration of the Lamarckian principle of inheritance of acquired characteristics.

Morgan had a question for Pavlov: were not the mice being selected on a different basis — on their alertness? Might not Pavlov inadvertently have been retaining smart mice and eliminating the stupid or inattentive ones? Absolutely not, replied Pavlov. He declared that when he got back to Russia he would execute the final step of his experiment and arrive at a mouse that had been genetically programmed to respond to the first ring of the bell.

Returning to his own laboratory, Morgan growled that this extraordinary demonstration could only be attributed to the senility of its designer, a judgment that proved right — nothing more was ever heard from the magical mice that had had their genes altered by the ringing of bells.

The mouse anecdote conveys an important message. Having spent so many years poking into genes, and having had it demonstrated to him so many times and in so many different ways that *they* were the programmers of life, Morgan — like many others of his generation — was beginning to make a clear distinction between genotype and phenotype, something his predecessors had not been able to make. Pavlov's dog experiment was a phenotypical one, involving a learned response. The response could not be handed down to the dog's descendants because nothing in its genes had been, or could be, touched by the ringing of a bell. Similarly with his mice. Pavlov discreetly buried them.

HERMANN J. MULLER: THE SOURCE OF MUTATIONS REVEALED. DARWIN IN ECLIPSE.

IF bell ringing or neck straining or tail snipping — or any of the hundreds of Lamarckian tricks that were being tried — could not cause mutations, what could? That was the question that Hermann Muller kept asking himself after he left the Fly Room, where Morgan had tried — and failed — to induce mutations.

Muller was a more careful and precise man than Morgan. Whereas Morgan had several unrelated experiments going on all the time and didn't always remember the details of each, Muller put all his attention on one thing at a time and prepared himself with exquisite care for any program he was about to engage in. He decided, as a starting experiment, to test whether the gene, when it mutated, was being affected by some external force. If it was, he would find out what that force was. Some time later he admitted:

> Animals have been drugged, poisoned, intoxicated, illuminated, kept in darkness, half-smothered, painted inside and out, whirled round and round, shaken violently, vaccinated, mutilated, educated, and treated with everything except affection from generation to generation.

The result: nothing. His test subjects were scorched, chilled, choked, discolored, and often died. But among those that survived, none showed the slightest sign that their genes had been affected. He had either failed to find the correct force, or else there was no force—somehow the gene mutated all on its own.

Muller could not believe that. His scientific training told him there had to be something, some tiny chemical or other reaction that produced a change in the structure or the behavior of the gene. He began thinking about what other kinds of influences might exist, and came up with the idea of using X rays. True, Morgan had tried them, but he had done so in the early days of the Fly Room, before he or his associates had become skilled at recognizing mutations. Perhaps X rays had created a few which, in the slapdash atmosphere of the Fly Room, had gone unnoticed.

X rays made sense to Muller for an entirely different reason as well. He had been trying to bludgeon his laboratory test subjects into mutation by what amounted to brute force, by shaking and squeezing and heating them. It occurred to him that that entire category of change inducers was wrong; it was far too coarse. The reason: if change could be induced in a gene by shaking, would that not change all the genes, since all were being shaken equally? How could there be an alteration in a single gene and not in the one next to it, only a millionth of an inch away? And yet, on the evidence of the Fly Room, that was what happened—only one trait, presumably controlled by one gene, was affected when a mutation took place. Clearly the mutating force was an extraordinarily small, extraordinarily precise, and extraordinarily penetrating one. It would have to be if it was going to pierce the body of the test animal and then the wall of the individual cell, poke its way through the cytoplasm in the cell and then through the wall of the nucleus, and finally strike a single minute spot on the chromosome.

Muller knew that an X ray can do that. It is an extremely shortwave bit of energy. The shorter the wavelength, the greater the power of penetration. X rays will go through practically anything but lead, and it is that property that makes them useful in photographing the interior of the body. They also damage it. Prolonged exposure to X rays causes blood diseases, cancer, and a variety of other disabling ailments that sometimes end in death. X rays possess that potentially lethal effect on living tissue in common with other shortwave, high-energy radiation: fallout from nuclear explosions, radium, cosmic rays spewing from the sun—from all suns—whizzing through the universe in uncountable numbers. Why not use X rays in the hunt for mutations? If an X ray did hit a gene, it was bound to do something to it. Furthermore, X rays had another powerful attraction: they were controllable. Muller could vary the dose, and—if he did get mutations—see if the number varied according to the dose. If it did, he would have excellent evidence to support his idea. For these reasons he decided to go ahead. As his test animals he would use his old friends: fruit flies.

Before actually exposing any flies to X rays, Muller had to have the right kinds of flies. What he needed was pure strains; he didn't want any unexpected types popping up as the result of the matching of two hidden recessive traits whose existence he had not been aware of. So, his first step was to crossbreed flies very carefully in numerous combinations over many generations until he was sure that the strains were pure. Next, he had to make sure that one of the female strains carried a deadly recessive trait which, when matched to a similar bad mutation in a male, would express itself in the males of the second hybrid generation—and kill them all; that generation would contain only females.

As a starter, he was careful to breed his "marked" females only with a strain of males he knew lacked that defective gene. That is, he knew the males lacked the lethal gene *before* he exposed them to radiation. He had tested them and gotten both females and males in the second hybrid generation. Therefore, he reasoned, if he exposed those "good" males to X rays and *then* mated them with "bad" females, and if nothing untoward happened in the second hybrid generation, he would know that there had been no bad mutations in the males as a result of the X rays. But if, by any chance, that second generation produced only females, he would know that a mutation had occurred, and he could assume it had been caused by X rays.

To set this experiment in motion, Muller put a number of flies in small capsules, gave them a hefty jolt of X rays, and let them mate. The X rays did them no apparent damage; they mated lustily and promptly. But damage was done to their genes, just as Muller had hoped. In a significant number of matings there were no males among the second generation of descendants. All had been slain by the introduction of that fatal new mutation. X rays had changed the genetic structure of a particular tiny part of one of the male fly's

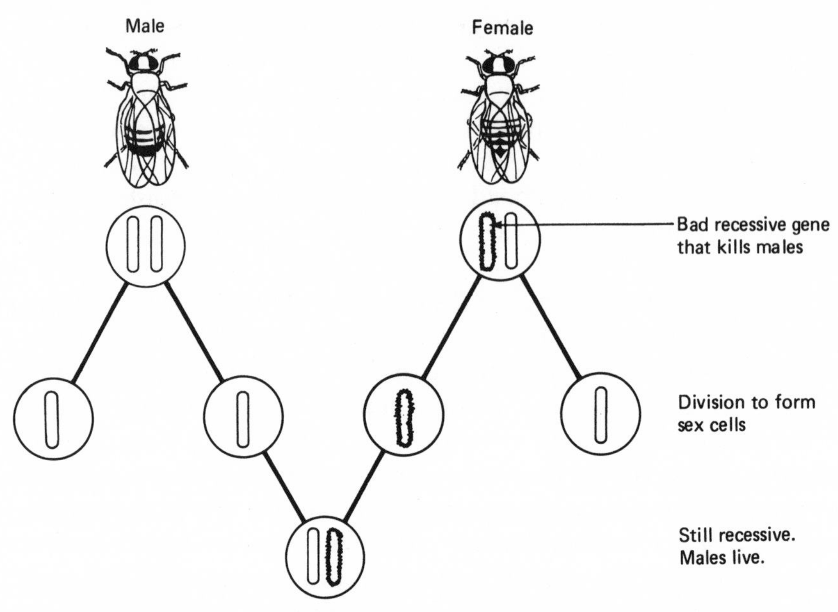
Muller's experiment. Step one, without radiation:
Male
Female
Bad recessive gene that kills males
Division to form sex cells
Still recessive. Males live.

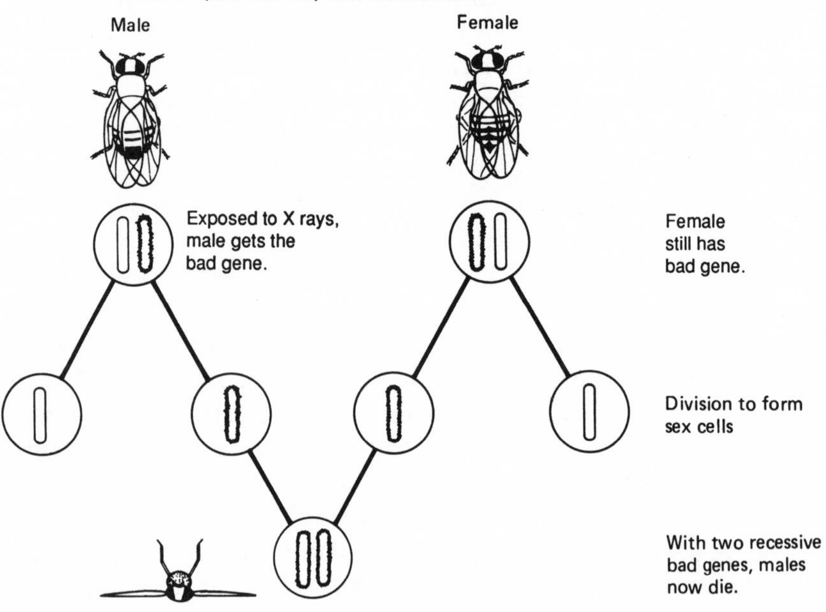
Muller's experiment. Step two, with radiation:
Male
Female
Exposed to X rays, male gets the bad gene.
Female still has bad gene.
Division to form sex cells
With two recessive bad genes, males now die.

chromosomes. The new trait matched the same defect in the female, and this killed the fly.

"That was an extremely subtle experiment," said Don. "I think we should make it a little clearer. Let's go back into the bicycle factory and join the workers on the assembly line."

The assembly line, it will be remembered, is a double one; it has workers standing on opposite sides of a long counter, those on one side representing the female sex cell and those on the other side representing the male sperm. At a particular point on the line a worker is reading from his blueprint: "Put on wheels." He has a very loud voice (he is a dominant gene), and his opposite number across the table, even though his blueprint may read: "Don't put on wheels," does what his partner says. That is because he has a weak voice (recessive). The bicycle gets wheels and passes down the line in good shape.

But suppose the designer of the blueprint, sitting in the central office, is blinded by a flash of sunlight through the skylight and makes a mistake in the blueprint. He might hand it out to the first worker with the instruction: "Don't put on wheels." Now both sides of the assembly line have the same crippling message, and the bicycle frame goes scraping down the line without wheels, to be broken up by an inspector at the end of the line because it is useless.

In this analogy the designer in the office is a chromosome. The sunlight blinding him is an X ray. The changed instruction on the blueprint is the result of a mutation caused by this X ray. The bicycle without wheels is a dead male fly.

Muller is known today for the beauty and precision of his experiments, of which he made many, using "marker" flies of various types to tell him exactly what was happening and where on the chromosome. He need not have been so precise. His original hunch was so right that any observant biologist, sorting over flies descended from those subjected to X-ray bombardment, would have noticed its effects immediately. Flies of every possible description showed up. In the words of the science writer Ruth Moore, there were "flies with bulging eyes, flies with flat eyes and dented ones, flies with purple, yellow, or brown eyes, and even flies with no eyes at all. There were flies with curly hair, with ruffled and parted hair, with fine and coarse hair, and some were bald. There were flies without antennae, flies with broad wings and down-turned wings, with outstretched wings, with truncated wings, and with almost no wings. There were big flies, little flies, active flies and sluggish flies. Some were long-lived, some were short-lived. Some preferred to stay on the ground, others avoided the light, and some had a mixture of sexual characteristics."

This astounding freak show set in motion a great surge of X-ray activity in laboratories all over the world. It was discovered that similar effects could be produced in virtually any laboratory animal or plant. Muller also demonstrated that the rate of mutation varied with the dosage of X rays, thereby providing a precise measuring stick for mutations, for their frequency, and their effects. For his work, a Nobel Prize went to Muller in 1946.

One other insight of gigantic significance surfaced in that menagerie of deformed flies: some of the mutations were identical with those that Morgan had observed to occur naturally. It took Muller and others no time to recognize the implication of this. High-energy radiation is not unique to the laboratory. It also occurs in nature. All living organisms are exposed to it. Therefore, if it produces mutations in the laboratory, it will do the same in the real world. In short, Muller had discovered the missing ingredient in Darwin's theory. In radiation he had identified the force behind variation, the motor that drove evolution.

Natural radiation exists at very low levels. It occurs in radioactive elements in the earth and in the rays from space that are constantly striking the earth's atmosphere. Luckily for all organisms living here, much of that lethal dose is deflected by the atmosphere, and only a small but steady trickle of it hits us. What does passes right through us. Radiation is not a dull knife or blunt needle, ripping apart cells; it is far too sharp for that. Since we are made of atoms arranged in various clusters, and since there is a great deal more empty space in and among atoms than there is matter, a cosmic ray can zip through a good many cells in a body without touching anything. If it does hit something in an ordinary cell, it will rearrange the structure of a few of its atoms but may do the cell no overall harm. However, if it hits one of the chromosomes in the nucleus of a sex cell, the atomic rearrangement there changes the blueprint and causes a mutation. Such hits are rare — which explains the rarity of natural mutations — but they do happen.

Given the great penetrating power of shortwave, high-energy rays, it is easy to see why experiments dealing with such phenomena must be conducted in lead-sheathed laboratories or deep underground to make sure that no unwanted wandering rays interfere. Otherwise, the experiment would be about as useful as measuring the amount of water a man puts out in perspiration during heavy exercise — but doing the experiment in a rainstorm.

It would be interesting to know if the creatures living five or six miles down on the floor of the ocean evolve as rapidly as those on the surface do. The logical answer would be that they do not — because some of the energy of a cosmic ray is lost each time it hits something. As a ray passes through five miles of water, the probability of its bouncing off at least one atom in a water molecule is very high. Therefore it can be supposed that the mutation rate of deep-sea life forms is extremely low and that they evolve correspondingly slowly. Another reason for this, of course, is that the environment is very stable. There is little selection pressure, which helps explain the persistence of certain primitive fishes like the coelocanth. Its surface relatives have been extinct for at least fifty million years, but, to the astonishment of marine biologists, a coelocanth was found in the deep waters off Madagascar in 1938. Others have been found since, proving that the species is alive and well and moving about actively down there — somewhere.

Coelocanths aside, the question remains: how did some of the less active and weirdly different creatures that creep about on the very bottom of the ocean in utter darkness evolve at all? There can be only two answers. First, there may be a very small amount of radiation even there, perhaps a flicker or two filtering down from above, perhaps some more filtering up from radioactive materials underlying the detritus on the sea floor. But certainly far more important must have been the recombination of an elaborate array of genes already in those creatures. Where organisms get their genes in the first place is a mystifying topic that will be discussed later. Here, we need only note that deep-sea creatures have a great many, that they probably got them eons ago, probably from ancestors that were not yet adapted to living in the depths, and that crossing over and the mixing of genes through sex provided more than enough variability for selection in that strange and remote environment.

Getting back to the surface, where there is a significant amount of radiation, it should be noted that the action of high-energy rays with respect to organisms is entirely random. Mutations, therefore, are unpredictable (except in a statistical sense). Since their effect is to make changes in organisms that are already functioning efficiently, those changes are apt to be damaging. If one of the wheels of a bicycle were to be smaller than it should be, or if the handlebars stuck out from the side instead of the front, the bike would not work very well. In that same way mutations that express themselves in dominant form in organisms are usually harmful, so harmful, in fact, that they are more likely than not to kill the individual—and eliminate themselves in the process. Therefore, most mutations that persist in gene pools are either neutral in their effect or recessive. That is: they express themselves in harmless ways, or they are recessive and unable to express themselves. Actually a good many recessive traits are harmful also, but they do not surface unless they happen to match up with others like them. Thus they hang on, and the carrier has no knowledge that he owns a damaging trait until—mated with a similarly marked partner—it shows up in their child. A very few mutations are beneficial.

Muller once compared the effect of mutations to the actions of a man trying to fix a balky watch. If he bangs it on the table to try to get it started again, there is a high probability that he will break a delicate vital part like the hairspring, and the watch will never go again. There is also some chance that he will do neither good nor harm. There is only a remote chance that he will jiggle loose the particle of dust that caused the watch to run badly in the first place.

Why, then, if most mutations are bad, can organisms survive them? To repeat, they are rare. The bad ones often kill off their owners and thus tend to flush themselves out of the gene pool. Those that work their way into the gene pool in the recessive state provide a tremendous variety in the genetic makeup of a population and in that way give it the potential for selection to work later on if there should be a change in the environment.

Flies, those ever-reliable little experimental animals, provide a good illustration of how useful a potentially damaging trait can sometimes be. In a large population of flies it is statistically probable that one fly in every few million will suffer a particular mutation at a particular spot on a particular chromosome that makes it somewhat feebler and less sexually successful than ordinary ones. In a normal environment, flies so handicapped do not survive in competition with their hardier fellows. But that same mutation happens to make the fly resistant to the insecticide DDT. Sprinkle a lot of DDT around, and the huskier, more fertile flies die—well, like flies—leaving the erstwhile feebler ones as the favored form. In a few generations they take over, and the other type dies out as fast as it appears because it cannot stand DDT.

That, incidentally, explains why DDT is useful as an insecticide for only a short time. It doesn't alter flies, as so many people think, somehow making them gradually stronger and more resistant to the poison, as one might gradually acquire a suntan. Rather, it alters the existing ratios in a population of flies. It forces an apparent evolutionary step by eliminating one type and favoring a different one. Put another way, selection pressure in an environment of DDT is overwhelmingly strong. Remove the pressure by removing the DDT, and the fly population will quickly swing back to its former character, with an abundance of the husky, more sexually active type, and only a rare DDT-resistant mutant.

DARWIN REHABILITATED: THE NEW EVOLUTIONARY SYNTHESIS.

BY hindsight, it is not hard to see that the findings of Morgan and Muller supply an ironclad prop to evolution theory. Morgan's experiments in crossing over and Muller's X-ray experiments together provided the explanation for variation long missing from the equation written by Charles Darwin. But as news of the new work circulated through the world of biology, that equation began to seem more and more badly written. If Darwin's idea of gradual evolution through selection had seemed shaky to many before, it was now, with Muller's demonstration of how mutations took place, in danger of collapse nearly everywhere. With lethal mutations cropping up in scores of laboratory populations of flies as a result of radiation, the whole question of selection seemed settled. The argument ran somewhat as follows.

Species well adapted to their environments do not evolve on their own. However, they are prey to a constant but low rate of mutation. Most mutations are harmful and are eliminated as soon as they express themselves. Those that are harmless or recessive may persist in the genes of a population, and are potentially useful because they provide the variability that a species must have if it is to adapt to future changes in the environment. Lacking such variability, it could (and innumerable species have) become extinct. In that scenario, where was the role for gradual selection? Mutations did it all. They were responsible for the bad, and the bad were eliminated; they provided the material for the good.

That was the laboratory scenario, and it was a powerful one for geneticists. However, it did not satisfy some other biological disciplines. The taxonomists, the comparative anatomists, the biometricians, the field naturalists all saw things differently. In their studies of large numbers of individuals of a species, they saw gradual changes in shape, size, color, and habits, as that species was distributed geographically. They refused to be stampeded into giving up Darwin and his theory of gradual change through selection. Mutations caused jumps. Darwinian selection was not jumpy. It was a smooth, slow affair, proceeding by very small degrees, as seen in nature—and in fossils, too. A Mendelian saw a population of flies divided into two types, those that could endure DDT and those that couldn't; there were no in-between ones. A Darwinian gradualist, looking at other populations, saw in-betweenness everywhere and very seldom an either-or situation.

Both sides were right, of course. Each was looking at his own side of the same coin. The two camps became polarized, and during the 1920s and 1930s there was a great deal of acrimonious debate between them. It was only when it occurred to geneticists to begin looking at entire real populations in an effort to discover what the actual frequencies of certain genes were, and what were the factors that might change them, that there could be a possibility of the camps coming together.

Two geneticists, G. H. Hardy and W. Weinberg, had pioneered this work. They demonstrated that a given trait would endure indefinitely in a certain percentage of a population—say one-third of its members—in a sort of equilibrium unless it were changed by any of several influences. Those included mutation *and* natural selection. But if neither happened to jostle the trait, it would just go on and on, appearing and disappearing for generation after generation with statistical probability in a benign and static environment. It would, in effect, find its place among other competitors in the gene pool, and all would be serene.

From the findings of Hardy and Weinberg and others a whole new biological discipline—population genetics—grew up, attracting a number of thoughtful evolution-minded biologists from various specialized fields: Ernst Mayr from ornithology, George Gaylord Simpson from fossils, G. Ledyard Stebbins from botany, to name only three. All sorts of sophisticated mathematics were applied to answering new questions about how far the overall gene mix in a population might drift by pure accident in one direction or another. This was not natural selection, but it did happen, and it raised some interesting issues: Did that drift really matter (did slightly large spots on a thrush's breast have any significance)? Did it have direction (did those spots tend to get still larger or did they drift back again)? Were populations truly stable genetically (despite mutations) unless nudged by outside influences? How important was the size of the population in answering some of these questions?

The dynamics of populations quickly were discovered to be both delicate and complex. They depended not only on size (very small populations *were* strongly affected by random drift, as Darwin's tortoises on the Galapagos Islands had shown), but on many other influences as well. Populations were found to be enormously sensitive to *any* environmental change, and would flow with it, their gene frequencies changing to match that external change. This explained the geographical variation among certain species found in different parts of South America by Darwin. Conditions were not the same in all places; therefore the frequency of genes that regulated certain adaptive characteristics of the animals was not the same either.

Observations like this could not be repeated very often without its occurring to somebody that for populations to exhibit such flexibility there would have to be a much greater variability in their gene pools than anybody had previously thought. In other words, mutations were probably introducing themselves into the chromosomes of individuals at a much higher rate than anyone suspected and were hiding there.

The first to investigate that matter was a Russian, Sergei Chetverikov. In 1926 he caught 239 wild fruit flies. Then, in order to show matchups of recessive genes that he suspected were hiding in their chromosomes, he mated brother with sister for several generations. He found more than thirty such matchups. What he and others had suspected was true: all of us, fly and human alike, lug around in our genes a vast amount of extra baggage that we are totally unaware of unless our spouses can match it—and it reveals itself in our children.

With a better understanding of some of the intricacies of population genetics, it began to be increasingly clear that both mutation and selection were involved in the evolutionary process, and the gap between strict Mendelians and strict Darwinians began to shrink. Then came another revelation. One of the strongest arguments of the Darwinian gradualists had been gradualism itself: the way shape, size, color, or habit slid imperceptibly through innumerable intermediate stages from one extreme to the other in every population. If there were no such thing as natural selection, how was that gradualness to be explained? There is the world's longest alligator and the world's shortest one, and an alligator of every size between, with the optimum-sized alligator the most numerous. Could that distribution be attributed to mere chance? Put another way: could mutations explain it?

Yes and no. It was soon demonstrated by the geneticist R. A. Fisher that such tiny differences among individuals were the result of many genes working in different combinations to produce their microscopically variable effects, and a potential for innumerable different shapes and sizes. *That* provided the raw material for selection; then, in a given environment, the best-sized alligator for that environment could be selected for. Thus, gradualism was explained by Mendel, not contradicted by him, and the field naturalists began to listen more attentively to the geneticists. When the latter began to understand, through their own observations of populations, that gene frequencies were determined by natural selection, the arguments ended.

The final result was a triumph of intellectual organization, a coming together of evidence from many disciplines to produce what came to be known as the New Evolutionary Synthesis. Leaders in the forging of the synthesis were the aforementioned Mayr, Simpson, and Stebbins, along with Theodosius Dobzhansky, Bernhard Rensch, and Julian Huxley.

What the new synthesis says is this:

Mendelian genetics, with its concept of discrete heritable traits, is correct. The theory of the gene, stating that traits are located on specific places on the chromosome, is also correct. The variability of populations is explained initially by mutation but is enhanced enormously by crossing over and by sexual mixing. This results in a fantastically variable gene pool, and it is on that gene pool that natural selection works—relentlessly and constantly.

In sum: genes provide the variability; natural selection provides the evolution. Darwin lives.

Inside the Chromosome: DNA and RNA

Two elements are needed to form the truth — a fact and an abstraction.

— RÉMY DE GOURMONT

Truth is such a fly-away, such a sly-boots, so untranslatable a commodity, that it is as bad to catch as light.

— RALPH WALDO EMERSON

The atoms of Democritus
And Newton's particles of light
Are sands upon the Red Sea shore
Where Israel's tents do shine so bright.

— WILLIAM BLAKE

Friedrich Miescher: What Are Chromosomes Made Of? Answer: DNA.

If Darwin had lived to see the new evolutionary synthesis perfected, he would have been delighted—and astonished. His delight would have come when he was confronted with the theory of the gene, with the knowledge that an idea he had toyed with nearly a hundred years earlier had proved to have some value. It was not quite as Darwin had visualized it, but the principle was there. In his efforts to understand *how* individuals developed so rigorously into what they were destined to be, and *how* they also developed individual differences, he had proposed the existence of tiny entities that he called "gemmules." For him they served the same purpose that genes would later to a more sophisticated science.

Darwin had nothing to go on but an idea. He saw animals developing according to the most intricate interior plans, but he had not a clue as to how this happened. Somewhere in the body, he reasoned, there had to be particles of influence, so small as to be invisible, which in their own peculiar way told the growing body how to create eyes, livers, legs, and hair. Those particles were his gemmules. Where they came from, he had no idea. How they did their job, he could only guess. So he guessed. He guessed that there was a great variety of gemmules, that each kind of cell had its own kind of gemmule that told it what to do. Gemmules were being created constantly (again he did not know how). At first, he speculated, they were concentrated in the fertilized egg; as the new organism grew they spread out into the body, carried by the bloodstream, and there did what they were supposed to do.

An ingenious idea but a totally unsatisfactory one to a man as devoted to testing and proof as Darwin was. Gemmules lived only in Darwin's mind. He had no evidence whatsoever that they really existed. To repeat, the theory of the gene would have delighted him.

What would have astonished Darwin was that growth was not being directed by little bodies trundling about in the larger body to do their various jobs, but that *every cell in the body contained the instruction for the growth and behavior of every other cell.* Gemmules — or genes, as he would have learned to call them — did not move about. They handed out their instructions neatly and accurately from their stations in each cell, telling the cell not only how to grow but when to stop growing and what to do to keep itself alive and healthy. In addition they supplied a host of bewilderingly complex instructions having to do with the health, the growth, and the function of the huge aggregates of cells that make up organs and, eventually, entire bodies. More astonishing yet, that immense amount of information was somehow crowded into a surprisingly small receptacle inside the cell: the nucleus. Obviously the entities that were dispensing all that information had to number in the hundreds of thousands or millions. Therefore they had to be inconceivably small. Darwin knew nothing of atoms and could not think that small, nor had he any means of calculating the dimensions of such tiny things even if he could have visualized them. But we can do this today. Atoms are measured by angstrom units. One angstrom unit equals 1/250,000,000th of an inch. A single strand of chromosomal material is from ten to twenty angstrom units wide. This thread is actually too thin for the most powerful conventional microscope to see directly, to reveal what it is made of or how it is put together. The thread betrays its existence only as a pucker of refracted light, a blurred wisp of something whose intricacy of function seems incredible when we first learn of it.

To the Morgans and Mullers who were laying the groundwork of modern genetics it must have seemed incredible, too. They, better than most, were able to comprehend the nearly infinite complexity of even a relatively simple creature like a fly. As their knowledge grew, they could not help realizing that the crumpled wings, the odd-colored eyes, and other mutations they were finding were only the visible tip of an evolutionary iceberg. Size, shape, growth, feeling, seeing, digestion, excretion, flight, sexual urges—everything that a fly was and did from conception to death—was under the control of genes.

Contemplating all that structural and behavioral complexity, geneticists had to face the humbling realization that they did not yet know what genes were. They hadn't the faintest idea. They did not know what they were made of, what they were shaped like, or how they functioned. Assuming (and even this could be only an assumption) that they did act as blueprints of some kind, rather than rattling about in the system like Darwin's gemmules, then additional questions flew: How were the instructions in the blueprints handed around? How were they acted on? What on earth did they say? Were evolutionary questions really going to be answered by them?

Here we are again, outside the bicycle factory without a key, unable to see in because the windows are painted over, able to guess what is going on inside only by what we see coming off the assembly line. Those internal activities, we have gradually learned, are chemical and molecular. From this point on, the major thrust of genetics—and of evolution—will be to penetrate the factory, to get down among the molecules, and to try to find out what they are doing.

It is not quite accurate to say that nothing physical was known about genes, only that for a long time the work of geneticists did not take them in the direction of studying that aspect of them. This work would have to be done by biochemists, and it had been started in a desultory way as early as 1869. In that year the first assault on the chemistry of the gene was made.

Biochemistry is an outgrowth of ordinary chemistry, which, in turn, is an outgrowth of alchemy—that primitive precursor of chemistry practiced by inquisitive men who had not yet learned what elements were. Much of their time was wasted trying to turn other things into gold, but in the process they learned a great deal about what other things did turn into. After several hundred years of soaking, boiling, mixing, roasting, drying, pounding, and measuring various materials, the unifying principle of chemistry began to emerge: that the apparently endless variety of things that existed in nature actually consisted of a far smaller number of basic elements that combined in various ways to create everything else. It came as an absolute surprise that water—such a uniform and satisfactory material, different from everything else on earth and regarded for a couple of thousand years as elemental — was really a mixture of two gases. Similarly, one rock, while it was obviously different from another rock and different again from a third, was not different in having its own unique built-in "rockness"; rather each was composed of different combinations of just a few elements like silicon, calcium, iron, potassium — and even a gas or two thrown in. How rocks were put together — and the different proportions of their ingredients — made all the difference in what they looked like and how they behaved: how soft or hard they were, what kinds of acids it took to break them down, and so on.

By the nineteenth century, chemistry had become a highly sophisticated science. The differences between elements and compounds (things made of two or more elements) had been worked out, and the true nature of most inorganic substances was known. It became clear that it took two particles of hydrogen (later understood to be atoms), combined with one atom of oxygen, to make a molecule of water. Always two of hydrogen, never less. The atoms bound themselves together in that proportion, and — miracle — water resulted. It would be many years before the nature of that bond was understood. It is a matter of the number of electrons circling the nucleus of the atom, and of their electrical charge. Hydrogen has one electron, oxygen has two. Thus two hydrogens, and only two, join up with one of oxygen, and the formula for water, known the world over, emerges: H_2O. Another molecule, H_2SO_4, is a little more complicated. Here two hydrogens can combine with four oxygens so long as there is one atom of sulfur (S) to hold them all together. They produce a powerful corrosive liquid: sulfuric acid. Ordinary rust is a mixture of iron and oxygen. Ordinary table salt is a mixture of a metal (sodium) and a gas (chlorine). And so on, right through the list of common and rare substances found throughout the earth. And on out into space, it may be added, for the things that stars are made of are not different from those that exist on this planet. The elements that are now found in the earth, the gold, the silver, the copper, were not made here but in the nuclear furnaces of giant stars that exploded billions of years ago, scattering their contents to drift through space until they were caught up in the process of planet-making that forged our solar system — and presumably countless other systems throughout the Milky Way.

Slowly it was learned that some elements are much heavier than others. A Swedish chemist found a way of calculating what an atom of gold or iron weighs compared to one of oxygen. Precise atomic weights were assigned to all the known elements. In 1871 the elements were arranged on a chart by the Russian chemist Dmitri Mendeleyev in a way that organized them into groups according to certain properties that they have in common and also according to their weights. When it was pointed out that Mendeleyev's organizational chart had some empty spaces in it here and there, he responded that those spaces represented elements that had not yet been discovered because they were rare in nature, but he predicted that they would soon be found. He was right. Three rarities, gallium, scandium, and germanium were quickly identified. In later years all the spaces were filled, as elements even more rare were discovered. Additions had to be made at the bottom of Mendeleyev's table to accommodate extra-heavy radioactive elements that do not exist naturally on earth, but which can be made in the laboratory. Even they fit the logic of Mendeleyev's basic groupings.

So much for chemistry. Biochemistry, the chemistry of living tissue, is far more complex, simply because the molecules of living matter are so much larger and so much more intricately arranged than those of any of the so-called lifeless, or inorganic, materials. It was in trying to find out what *we* are made of, how cells are put together, what tissues consist of, that biochemistry was born. Its practitioners quickly learned that living matter was no different from inert matter so far as its basic components went. We are made of oxygen, carbon, iron, phosphorus, and a number of other elements, just like everything else. The difference is in the arrangement. Our molecules are spectacularly complicated.

In a burst of activity in the middle of the last century a great deal was learned about organic tissue. It proved to be made up of about two-thirds water and one-third various organic compounds, the most numerous being fatty substances and a group of other materials known as proteins. The fats, it turned out, were largely fuel depots. Proteins, large nitrogenous organic molecules, were the all-important building blocks of the body. Muscles, skin, bones, organs, all were composed principally of proteins in various configurations. A particular class of proteins, enzymes, served to break down the food that was taken into the body, turning it into the various sugars and other materials that would not only build more muscles but also repair the body and give it the energy to function. Proteins, as the breakers-down of foodstuffs and the builders-up of tissue, were obviously at the core of life itself.

It is not surprising, with the growing realization of the centrality of proteins in the web of life, that the brightest minds in biochemistry concentrated on the identification and analysis of proteins. This takes us back to 1869 and to a young Swiss biochemist, Friedrich Miescher, who was puttering over some pus cells (dead or dying white blood cells) in an effort to learn more about them. Pus was common stuff in laboratories and hospitals in those days, and he had all he needed to work with. What he learned, after treating some of the cells with a digestive enzyme, was that the enzyme caused the material in the outer cell to disintegrate but left the nucleus untouched. He had inadvertently found a way of separating the nucleus from the rest of the cell!

Now Miescher had a chance to find out what was *in* the nucleus, something that no one else had ever been able to determine. Previous efforts, like Muller's attempts to induce mutations by brute force, had never managed to separate the nucleus from the rest of the cell material. Breaking down one had always smashed the other, leaving an indecipherable mush. Now, with reasonably clean nuclear material, all the other cell debris having been washed away, Miescher proceeded to analyze it, assuming that it would turn out to be a protein. Using the increasingly refined quantitative and qualitative chemical analysis that laboratory techniques were beginning to make available — that is, how much of what was in the sample — he came up with a substance made up of 112 atoms. As protein molecules go, this was a pretty puny molecule. Nevertheless it was a giant compared to the average inorganic molecule. It was composed of five kinds of atoms: carbon, hydrogen, nitrogen, oxygen, and phosphorus, in the following proportions:

$$C_{29}H_{49}N_9O_{22}P_3$$

The next thing Miescher realized was that it wasn't a protein at all! A protein would have contained a great deal more carbon and nitrogen than this nuclear stuff, which emerged as a whitish powder. He named it nuclein.

That powder was not entirely pure, which gave some critics a chance to say that nuclein was contaminated with protein and might actually *be* a protein. But they were wrong. Precise modern analysis has determined the exact formula for nuclein. Allowing for the few impurities in Miescher's sample, his formula was still strikingly close to the true one:

$$C_{29}H_{35}N_{11}O_{18}P_3$$

This correct formula has 96 atoms; Miescher had failed to eliminate only sixteen extraneous ones. His work, for its time, was marvelously precise. Slightly flawed or dead-on, it showed what the hereditary material was made of. Moreover it was an acid. It is now known as nucleic acid.

The discovery that this hitherto unknown acid was the substance of the cell nucleus is something that one might think would draw some attention. It didn't. The role that the nucleus would play in the coding of all life had not yet been guessed. Biochemists were interested in biochemistry—what things were made of, how they were put together. The larger questions, the wonderful gleaming ones, the ones with evolution written all over them, escaped them. They even escaped Miescher, although Weismann's and Boveri's discoveries about chromosomes soon followed. Miescher devoted his next years to refining nuclein, to getting purer and purer samples of it until he had reached his formula. He had discovered that salmon sperm had unusually large nuclei, and since those fish still spawned in the headwaters of the Rhine, he would go down to the river each spring with a net and get a year's supply of the raw material for nuclein. He died in 1895 without ever quite understanding the significance of the crack he had opened in the wall of the closed factory.

Nuclein itself, that enigmatic white powder, sat in its glass laboratory jar, a curiosity—but not much more. For a number of years the only things done with nuclein were chemical things. Other biochemists studied it in an effort to learn how it was put together, and they did learn a little. They discovered that nuclein consisted of several chemical subassemblies. Some of the carbon and hydrogen and oxygen atoms combined to form sugar molecules. Four other kinds of molecules were built around nitrogen. The phosphorus combined with a few leftover hydrogen and oxygen atoms to make phosphoric acid, the force that seemed to bind all the others together. They also learned that those various subassemblies could be strung together in long chains of, not hundreds, but thousands of atoms. This better understanding of the structure of nucleic acid led to its being given a chemically more descriptive name, but one that was a true jawbreaker for the lay person: deoxyribonucleic acid. This has been shortened to DNA. Thanks to modern studies in heredity and evolution, those three letters have become the most potent ever assembled by biology. For it is universally understood today that DNA—Miescher's white powder—makes us what we are.

The only other thing done with DNA for a number of years was to dye it. It was discovered that if it was treated with a certain acid it turned bright reddish purple. This happened not only to the powder in the jar but also to the DNA in living cells. Exposed to a little of the dye, the chromosomes lit up like traffic lights. Nothing else in the cell did—proving that DNA was concentrated in the cell nucleus and nowhere else.

10

George Beadle and Edward L. Tatum: What Does DNA Do? It Hands Out Instructions and Thus Changes Things.

It is the sorrow of science that the more that becomes known about a particular corner of it the harder it becomes for workers in other corners to keep up with developments outside their own fields. This kind of specialization had begun to infect biology by the time DNA had been identified. The men in genetics and in biochemistry were on ever-widening forks in the biological road. The former were still outside the factory, trying to decipher the instructions on the blueprints they had decided must be in there, based on their observation of the products coming out the door. The biochemists had little use for that kind of windy guesswork. By isolating DNA they had managed to drill a hole in the factory wall and had gotten hold of a bit of blueprint. Never mind what the blueprint said. Their concern was: what was it made of? What did it do?

In short, how did genes work? Up to this point nothing whatsoever was known about that. Morgan, for all his brilliant deductions about fruit flies, had been dealing with essentially imaginary things. We hate to keep repeating this, but it is important to bear it in mind. Morgan didn't know what a gene looked like, smelled like, tasted like, how it was shaped or how large or small it was. All he could say was: "I know it is there because I can see what it does."

Since genes were made of DNA, and since DNA was a chemical (it *was* a chemical; there it sat, a white powder in a laboratory bottle), the belief began to grow that the first step in understanding genes was to assume that they expressed themselves chemically.

If, through chemistry, somebody could discover the process by which a gene did something, then a link between blueprint and bicycle could be established. Unfortunately the aforementioned increase in specialization was making that difficult. As one worker in the biochemical field, a young man from Wahoo, Nebraska, named George Beadle, put it: "Geneticists and biochemists were like miners tunneling toward each other in the dark. Would they effect a junction, or were the routes they were traveling going to bypass each other at different levels?" In the 1930s Beadle decided to make a stab at getting the tunnels to meet.

By this time considerably more was known about biochemistry than had been known in 1869. Proteins had been studied assiduously in the more than half a century since Miescher's brilliant isolation of DNA. There was now no uncertainty about the role played by two large classes of proteins. One group made up the structural parts of the organism. The other group—enzymes—were the catalysts that made things happen. Enzymes helped stimulate the turning on and turning off of cellular processes. In thinking about enzymes Beadle asked himself a question: did genes control enzymes? It was a simple question, but it was only after backbreaking work on two continents that he got an answer: they did. At last a link between blueprint and bicycle had been established.

Beadle's start was anything but promising. Like all tunnelers in the dark he wasn't sure where to aim. The direction he chose was judged wildly bizarre when he and a young colleague, Boris Ephrussi, described it to a couple of their superiors in Paris, where they were working. "Don't waste your time on it," they were told. "You're crazy." What Beadle and Ephrussi planned to do was to make microscopic transplants from one fly embryo to another.

In its earliest stages the fly embryo, like every other embryo, does not have recognizable body parts. These appear only gradually, beginning as very small buds, groups of a few cells getting ready to follow a developmental track of their own and grow into distinct organs. It was those minute buds that Beadle and Ephrussi concentrated on. They would slice a bud from one embryo and try to graft it onto another. The embryo was only a tenth of an inch long and the bud far smaller, so all this surgery had to be done under the microscope. For months Beadle and Ephrussi did their slicing and grafting in vain. None of the buds "took" on the new grafted embryo. Then one day they found a fly with three eyes. They had located the particular bud that makes eyes, and in so doing they had hit on a tool for a more sophisticated experiment.

Too often something like this is stumbled over in the laboratory, looked at as a diverting curiosity, and then shelved. It takes a true creativity to figure out what to do with a discovery once it is made. Beadle and Ephrussi were capable of that imaginative jump. Able to create odd flies, they decided to use them by jumping in an unexpected direction. It had to do with some other strange flies that they knew had turned up several years ago in Morgan's Fly Room at Columbia: mutant forms that had both male and female characteristics. A fly would have a male-shaped wing on one side and a female-shaped wing on the other. Another would have one red eye and one white eye. It was as if a bicycle had come off the assembly line with one side painted red and one painted white, in defiance of all proper blueprint instructions.

The Fly Squad had quickly worked out the reason for the gynandromorphs, as those sexually mixed-up flies were called. They were the result of improper division of the sex chromosomes at the time of the egg's fertilization. Morgan and Sturtevant checked that out with a series of controlled crossing experiments with their laboratory flies. Sure enough, the parade of strange-eyed flies could be explained. However, there was one puzzler. When a particular cross was made that should have produced a vermilion eye in the male side, the eye came out dark red. The experiment was repeated. Dark red again. There was no explanation for that. Sturtevant, completely baffled, finally concluded that there was an influence at work somewhere deep in the fly, flowing from the female to the male side, involving a darkening substance that in some unexplained way affected the eye color on the male side of the gynandromorph. Unable to identify this influence (Sturtevant was not a biochemist), he abandoned his inquiry.

Beadle and Ephrussi picked it up. Now that they had eye buds that they could transplant at will, they had a tool for tracing eye color; they also had the chemical skills — they hoped — to run down what made it change. They repeated Sturtevant's experiment and got the same result. Clearly there was an enzymelike substance at work in the fly. Being biochemists, they were able to take the step that Sturtevant had been unable to take; they succeeded in identifying it. It turned out to be an amino acid, tryptophan.

Wildly excited, sure that they had at last found the chemical link between gene and organ (between blueprint and bicycle), they checked their results in a number of further experiments. To their consternation, they didn't work. Tryptophan was there in the fly, they knew that. But it only sometimes did its job. There were certain combinations of eye color in their freak flies (the two men had a number of different ones by this time, with eyes of several colors) in which the tryptophan seemed to be working. That is to say, flies appeared with eye color changes that the men could predict. But in others the tryptophan had no effect whatsoever; the flies' eyes did not change color. And in one totally baffling case, two lighter-colored eyes produced a dark-red one. The only explanation the two bewildered young scientists could think of was that there were other enzymes at work as well as tryptophan and that they had to work in series. If there was a break in the series nothing would happen.

It would be as if, in the bicycle works, a blueprint said at the start of the painting operation, "Put on a priming coat." The next instruction would read "If you see a priming coat, put on red paint." But if the first instruction was missing, the worker with the red paint would not be given the signal to do his thing. Beadle and Ephrussi were left trying to figure out what those earlier enzyme instructions were. They never found them. They too gave up.

But the problem continued to gnaw at Beadle. Back in the United States he did some more experiments with another chemist, Edward L. Tatum. They were as inconclusive as the ones he had done in Paris with Ephrussi. The trouble, he decided, was that the chemistry of a fruit fly was just too complicated. He didn't know nearly enough about it and was not sure he would ever learn. What he needed was a simpler organism whose chemistry *was* known. He asked Tatum about that and Tatum recommended red bread mold *(Neurospora),* a very simple organism whose chemistry he thought he could work out rather quickly.

Molds are better laboratory subjects than flies. Instead of breeding in a couple of weeks, they produce a new generation in a few hours. They have fewer chromosomes and a simple body chemistry that Tatum did work out. He arranged a series of bread mold clusters in test tubes and gave each a full supply of chemicals, vitamins, sugars, and amino acids. Then he began taking things away. He removed a vitamin — the molds flourished. He removed another — they flourished. He took away some amino acids — they still flourished. Eventually he found the border beyond which the molds could not live. They had to have a minimum of a few base chemicals, plus one vitamin: biotin. They *used* other vitamins, but they manufactured them themselves. Biotin they could not make.

This gives rise to some interesting speculation about the nature of the earliest and simplest organisms. They were surely able to extract whatever they needed for survival directly from the dissolved chemicals in the primordial "soup" in which they swam. Anything that today we could call "food" they would have made themselves from those raw ingredients. They would have had to; real "food" did not yet exist. Bread mold remains such an organism, able to survive on chemicals and on one vitamin, biotin. There may even have been a time when it could make biotin also, but if so, it has lost the ability through a later mutation.

We humans, by contrast, are so removed evolutionarily from those early frontiersmen of life that we have long lost the ability to make *any* vitamins. We are the ultimate parasites. We get our vitamins from the plants and animals that we eat. We should be respectful of oranges for their skill in doing something that we can no longer do. Indeed, we are so far from being frontiersmen that we cannot even eat raw chemicals. We must go to the supermarket of life to find those chemicals reprocessed and packaged into "food" — again by those simpler but more self-reliant lower forms.

When Beadle and Tatum had established that bread mold was on that borderland of self-sufficiency, they proceeded to the second stage of their experiment: administering a dose of X rays to see what effect mutations had on the chemical needs of the bread mold. Perhaps a mutation would knock out the mold's ability to make a needed chemical transaction, make an enzyme, make a vitamin.

Having no idea of how frequently mutations would occur in bread mold, the two men agreed to start big, with a thousand mold cultures, hoping to find a mutation in at least one of them. The cultures would be grown in a thousand separate test tubes, each on its bed of minimal nutrients.

Whereas Morgan, in the Fly Room, had been able to detect mutations by observing differences in flies, Beadle and Tatum would have to do it in another way. Bread mold does not have red or white eyes.

In fact it has no distinguishing features whatsoever. The only way that Beadle and Tatum would be able to tell that something had happened would be if one of the mold cultures failed to live. Given its sparse diet, any genetic change might make that diet insufficient —the mold simply would not grow.

Anxiously Beadle and Tatum watched their one thousand X-rayed cultures as the telltale reddish blush began to appear in each, indicating that the mold was growing. But in one test tube, number 299, nothing happened. No red.

Or rather, something ***had*** happened. A mutation had made it impossible for the mold to grow under conditions in which it would normally flourish.

So far, so good. The next question was: what chemical function had the mutation knocked out? Beadle and Tatum answered it by putting individual spores of the mutated mold into a number of separate test tubes and then adding a different ingredient to each. When one of the spores began to grow—and the others didn't—the missing ingredient would be revealed. It turned out to be an amino acid that the mold needed to make the vitamin B_6. Before being X-rayed, the mold could do that by itself out of the raw materials available. Now it had lost the ability.

"So what?" Don said when we were discussing how to write up the Beadle-Tatum experiment. "I don't get this. We know that X rays cause mutations. Muller proved that a long time ago. Now Beadle mutates a bread mold instead of a fly. What's the point? It seems to me that all he's done is to make it harder for himself by experimenting with something on which no peculiar eyes or wings can appear. Now he has to go through a long procedure just to find out if the mold has mutated."

I had done my homework on Beadle and Tatum and was ready for him. "The point is," I said, "not the mutation itself but what the mutation does. They were trying to link a specific chemical reaction to a specific gene. You can't do that with a fruit fly; a fly's too complicated. You can see the result, a red eye, but you don't know how it got there. You *can* do it with a mold. That's what the first part of the experiment was all about, to find an organism simple enough for them to be able to make that chemical connection. Before being X-rayed a mold could produce—all on its own—a specific enzyme for making vitamin B_6. After the X ray it couldn't. This proved that the affected gene directed the production of the enzyme. And don't forget—when Beadle and Tatum started out, they weren't even sure there was that chemical connection between gene and enzyme. When they started getting mutated molds they had a way of finding out exactly what was missing in the molds that didn't grow. Those missing things were the links they were looking for. In time, they and others discovered a whole series of amino acids that mutated molds couldn't make."

"In other words," said Don, "genes direct the making of amino acids."

"Exactly. It had never been demonstrated before. And they discovered another thing. They learned that some of those amino acids worked in series. It took two, sometimes more than two, to get something done in a bread mold. You need A to make B function."

"Well," said Don, "that seems to put us back in the factory with those people painting bicycles."

"Yes. As Beadle had guessed during his first experiment, you need a primer coat before you can put on that final red coat."

The Beadle-Tatum experiment was a classic, and it earned the two men a Nobel Prize in 1958. The biochemists tunneling into the mountain had met up with the geneticists. Both could now see more clearly into the murk. Genes did express themselves chemically; they instructed for the making of amino acids.

But how did they do that?

It seems that every question that is answered by science produces another. This one was exceptionally troubling because it had to face up to the enigma that a fairly simple chemical was making something far more complicated than any chemical. Genes, by the best analysis, were chemical; they were made of DNA. But the amino acids that DNA made were the building blocks of proteins. Proteins were hideously large and hideously complicated molecules that definitely were not chemicals.

Worse, the list of different proteins and enzymes that DNA was making was huge. How could DNA, with its relatively simple formula worked out by Miescher (page 190), direct all that fantastic variety?

Still worse, when more sophisticated analysis of DNA succeeded in breaking down Miescher's formula into subingredients, the ingredients were simpler yet. We mentioned earlier what those subingredients were. Now we should describe them more fully.

DNA consists of three kinds of things:

1) A phosphate group.
   It consists of one atom of phosphorus hooked up with four atoms of oxygen.

2) A sugar molecule.
   It consists of five atoms of carbon and one of oxygen.

3) Bases.
   Bases are simple compounds of nitrogen and carbon arranged in a ring. If there are two rings attached together, the base is known as a purine. If the compound consists of a single ring, it is a pyrimidine. DNA has two of each.

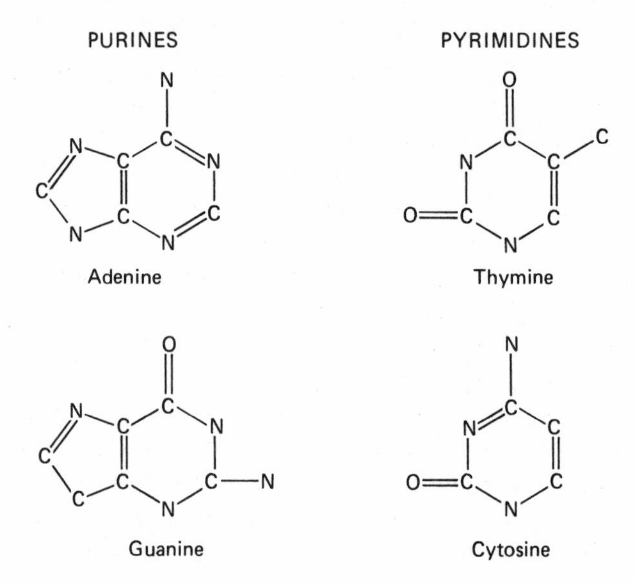

We will be meeting these four again. Generally they are referred to by the letters A, G, C, and T. A phosphate group hooks onto a sugar. The sugar hooks onto a base. The three together make up a nucleotide, one unit in nucleic acid, Miescher's white powder.

That was DNA? That was ALL? True, there were four different bases to be hooked on, to give a little variety, but even so — ! The longer biochemists pondered this simple stuff, the more incredible it seemed that it could direct the endless complexity of all living things. As a matter of plain common sense, it couldn't; there just didn't seem to be any way that this could happen. The suspicion grew, despite the magnificent work done by Morgan, Muller, Beadle, Tatum, and others in pinpointing genes and tracing their effects through mutations, that DNA might not be the source of heredity after all. This idea, ironically, was given strength by increasingly accurate laboratory techniques. The people who were sifting the contents of chromosomes to study DNA discovered that there was always some protein left over when the chromosome was broken down. True, there was none in the DNA itself; it was pure nucleic acid. But there was some protein *somewhere* in the chromosome. Might not that elusive bit of protein have a better connection with the production of all the other protein in the body that DNA had? After all, proteins were complex and varied. DNA was just a dull succession of simple nucleotides arranged in monotonous and apparently meaningless sequences. It was a mere wallpaper pattern and not a very imaginative one at that. If one believed in DNA as the programmer of life, one would have to accept this as the source of it all (see illustration, page 202).

For several years the most sophisticated corners of genetic research were full of rancorous argument: was the secret of life to be found in DNA or in protein? DNA had one advantage in the argument — but it was a purely theoretical one. If it was not connected with heredity, what in the world *was* it for? There it sat, in every cell. It divided and duplicated itself with scrupulous care, with unbelievable accuracy — cell after cell after cell, for thousands, millions, billions of divisions, with hardly an error. The English chemist Francis Crick (whom we will be meeting in a moment) has compared this extraordinary fidelity in copying with what might be the case if it were done by typists copying an encyclopedia. Because of the complexity of chromosomes, Crick, in order to make his com-

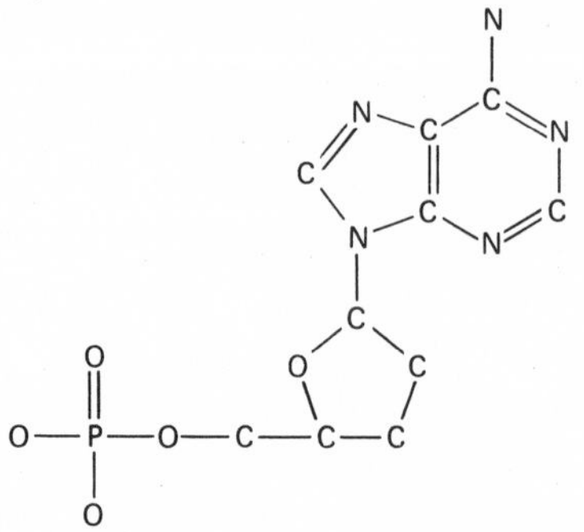

A nucleotide made with adenine, "A"

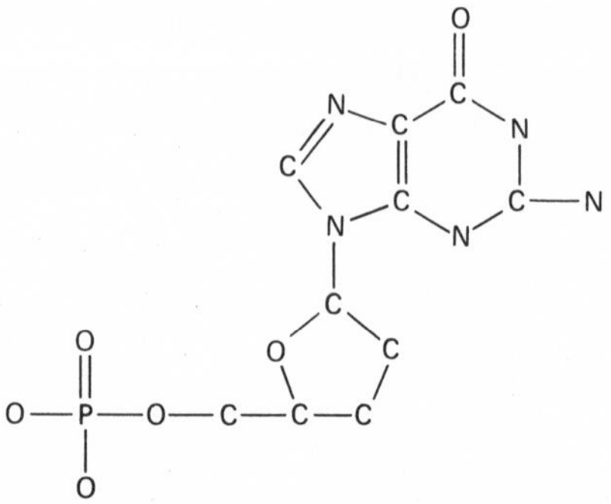

A nucleotide made with guanine, "G"

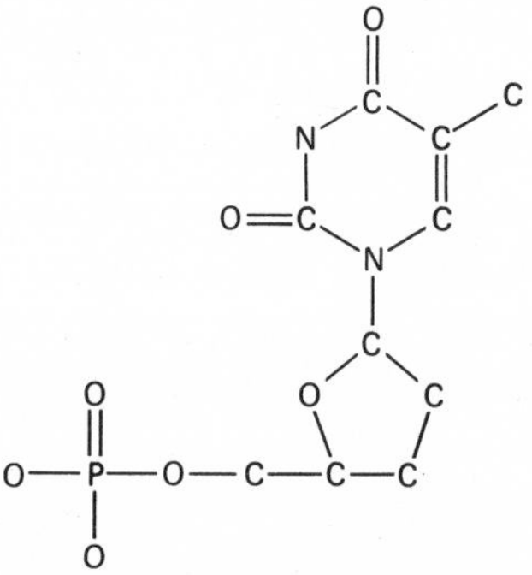

A nucleotide made with thymine, "T"

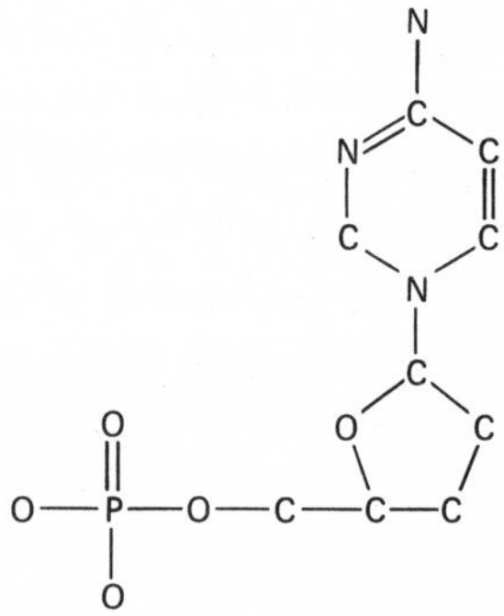

A nucleotide made with cytosine, "C"

parison valid, had to imagine a very large encyclopedia of a thousand volumes, each five hundred pages long, and then imagine twenty typists working to make twenty copies of it. This adds up to a total of ten million pages copied! Even skilled typists make a mistake every five or six pages. But let's assume that our typists are unbelievably super-accurate and make a mistake on the average of once every hundred pages. Even so, they would be ten thousand times as sloppy as the cell is in replicating DNA. It makes the equivalent of a mistake (a mutation) about once every million pages.

Accuracy like that had to have some meaning.

Was DNA the regulator? That was the question that remained unanswered. Beadle and Tatum had figured out what genes did; they made enzymes. But they had not demonstrated whether or not genes were exactly the same as DNA. If genes were made of DNA and nothing else, how could that same white powder make different genes?

In their work with bread mold Beadle and Tatum had made a second and almost inadvertent contribution to biology. They had demonstrated that a very simple organism could be more useful—because of its simplicity—in certain laboratory experiments than a more complicated one. They had struck out with fruit flies but they had struck gold with bread mold. Biologists in general, and geneticists in particular, were grateful for that. Think of the far greater speed in reproduction. Think of the neatness. Think of the small space required. No more messy fruit flies buzzing about, no more rotting bananas. Less need to pass the hat for extra lab funds, for time saved is money saved. Bread mold experiments are much cheaper than fruit fly experiments.

That being so, was there a way to get even cheaper and even faster? Yes, there might be: with bacteria, the simplest organisms known. Ever since their discovery bacteria had been recognized as the lowest, most basic form of life, standing on the bottom rung of the ladder. As for speed, they had it. Whereas it took molds hours to reproduce, bacteria could do it in minutes. So, why not use them? The answer was that bacteria were so unlike other organisms that biologists suspected that conventional evolutionary ideas could not be applied to them.

There was good reason for that suspicion. Bacteria appeared to have no nucleus, no chromosomes, no sexes, and no mating. Then there was the old matter of variability. Unlike a colony of people, a laboratory strain of bacteria was presumed to consist of identical individuals. Start a culture with a single bacterium, and it will grow into a colony of millions of clones in a very short time. It does this by splitting into new cells that are exact duplicates of the original. Those duplicates split and split again. Given enough food, one bacterium can turn into two in twenty minutes, eight in an hour, sixty-four in two hours, two million in eight hours, 2,000,000,000,000,000,000,000 in a day, and into a mass far larger than the earth in a week. Since food supplies are never unlimited, this never happens. Bacteria die as fast as they are created. Thousands, in and on a child's body, will perish while he or she is eating breakfast.

This relentless assembly-line production of clones does not sound like evolution and sounds even less like natural selection. Feed bacteria; clones grow. Don't feed them; they all die. Where is the evolution here?

Better knowledge of bacteria has proved them to be not that simple. As more and more has been learned about them they have become more and more widely used in genetic research. Most of what science knows today about genes has come from research with bacteria.

11

Oswald Avery: Is That Really True? Yes: DNA Is the Transforming Agent.

Bacteria may have seemed so unlike other organisms as to have prevented their use in laboratory experiments for a long time, but they got into the laboratory eventually. They forced their way in—not through the efforts of geneticists but through medicine, through the work of ordinary doctors. This step was inevitable after the French scientist Louis Pasteur had discovered in 1877 that bacteria caused diseases. It compelled physicians for the first time to focus their attention on those tiny organisms. Only by understanding them and how they affected people would medicine ever be able to learn how to deal with some of the most lethal ailments of the Western world.

Today the killers of the nineteenth century, diphtheria, tuberculosis, typhoid, scarlet fever, pneumonia, and so forth, are no longer the scourges they once were, because medicine has learned how the body's immune system works and has developed appropriate sera and antibiotics to deal with invading bacteria. At the turn of the century none of this was known, and those infectious diseases were carrying off people by the hundreds of thousands, with medicine helpless to do anything about it. It has been said of that time: "Physicians put drugs of which they knew little into human bodies of which they knew nothing."

It is not easy to envision the state of medical art in 1900, or how abysmally ignorant all doctors were about a great many medical matters that we take for granted today. The very idea of medical research in disease control was a new one. In 1901, when John D. Rockefeller, Sr., lost a grandchild to scarlet fever and learned to his consternation that nothing at all was known about it, he was persuaded to establish an institute for investigating the causes of infectious disease. At the time there were only two other medical research institutions in the country. One was at Johns Hopkins in Baltimore. The other was a small establishment in Brooklyn, the Hoagland Laboratory. An elflike little physician who had been working at Hoagland was recruited by the Rockefeller Institute to do bacteriological work there. He stayed thirty-five years. When he left he had demonstrated what nobody, anywhere, had been able to show before: that the regulator of life, the thing that makes organisms what they are, that transforms them, is DNA. His name was Oswald Avery.

T. H. Morgan, the "inventor" of the gene, invites instant contrast with Avery, the man who proved its connection with heredity and change. The two could not have been more different. Morgan was a tousled, uninhibited, outgoing, somewhat careless man, a great talker, mover and shaker, lecturer, and political force within the world of science, one who made his mark early. He expressed his opinions flatly and often wrongly, scattering mistakes as he went, confident that the debris would be swept away by the great discoveries that would follow, so long as he hammered away at what he did not know—always looking, always discarding, always asking questions.

Avery was a questioner too, but otherwise he was comically unlike Morgan. Lewis Thomas has described him as a "small, vanishingly thin man with a constantly startled expression and a very large and agile brain." Small puts it mildly. Avery was tiny. He never weighed more than a hundred pounds. He was an extremely quiet man, fussy, fastidious, private, cautious. His entire life was devoted to the work he did at the Rockefeller Institute. He spent summers in Maine and loved to go sailing, but otherwise his world was bounded by his laboratory and by nearby bachelor quarters he shared with another doctor. When that man went out to parties or the opera, Avery stayed at home and read. His friends were almost all found among his scientific colleagues. He opened up to them, particularly to young workers at the Institute. They called him Fess, short for Professor, and worshiped him.

Avery would sit for hours in a cell-like laboratory almost as small as he was. Everything was stripped down, spare, scrupulously neat. He did small, careful experiments with meticulous precision, often using ridiculously primitive apparatus. He thought long and deeply about them and embarked on them only when he felt sure he knew where he was going. He got his scientific start rather late in life; fame came to him even later. He did not seek it. In due course he received honors and medals from such distant places as England, Germany, and Sweden. He never traveled to accept a single one of them. The work he did earned a Nobel Prize—not for him, but for a team of men who, with other experiments, would later prove what Avery had already proved. He was nominated for a Nobel but was passed over for various reasons. One was that his great discovery was "before its time"; science didn't really take in its importance. Another reason, given later and when his work could no longer be disparaged or ignored, was that he himself had not been sufficiently positive about it; he had lingering doubts about whether he had eliminated every other possibility in his identification of DNA as the long-sought director of shaping and change. As an added irony, it was also suggested that by the time he had completed his work he might have become a little too old for the prize. At any rate, it went, not to Avery but to those other men, Max Delbruck, Salvador Luria, and Alfred Hershey, who were fashionable and flashy scientists (brilliant ones also, we must emphasize) working in a trendier field, genetics, than plain, dull, arduous biochemistry—which was where Avery had been laboring.

Avery was born in Nova Scotia, the son of a somewhat mystical Baptist minister who followed a call to set up a mission church for derelicts in the Bowery in New York. Avery grew up there, on the edge of the city's worst slum. As a boy he learned to play the cornet and on Sunday mornings would perform on the steps of his father's church to attract passersby. At college and medical school he was given the nickname "Babe" because he was so small and looked so childish.

After graduation from medical school he practiced medicine for a few years but decided to give it up for research. Unable to help his patients in the ways he thought he should, since medical care was all too often useless at the time, he felt that he might become more useful investigating the causes of disease. He accepted a research post at Hoagland and worked there for six years, during which he became a competent laboratory technician. He joined the staff of the Rockefeller Institute as an assistant in 1913.

Quite early in his career at Rockefeller, Avery began to specialize in respiratory diseases and soon became involved in what would turn out to be a nearly lifelong battle with some of the enigmas of pneumonia. In his years of studying the structure and chemical functions of the pneumococcus that caused the disease, he learned an enormous amount about it. As a consequence, he was as startled as others in the field were when he learned of some extremely peculiar experimental results obtained in England in 1928 by a Britisher, Frederick Griffith. It is inaccurate to say that Griffith's strange experiment led Avery directly to his DNA breakthrough. It is more likely that he stored it in the back of his mind along with a growing mass of other information about the pneumococcus, much of it derived from his own pioneering experiments. The breakthrough would not come for two decades. But when it did, it answered a stunningly important question that Griffith's work had posed.

Griffith had been studying lobar pneumonia, the disease's most lethal type. He had learned that the pneumococcus that caused lobar pneumonia came in two forms: a so-called smooth form and a rough form. The smooth form was the deadly one, to laboratory animals as well as humans. It was called smooth because it was covered with a coating, or capsule, of fine hairlike sugar molecules. These helped protect the invading bacteria from the defense antibodies of the host animal. Without the coat, the antibodies would attack the bacteria, and the host would not get pneumonia.

This kind of warfare goes on in animals all the time. Invasion by potentially dangerous agents triggers off the production of defensive materials that destroy the invaders. If the body cannot produce the right ones, or cannot produce them fast enough, it gets sick. Recovery from an infectious illness represents the eventual conquering of the invaders by the body's own immune system. What made the smooth pneumococcus dangerous was that the host's antibodies could not get at it.

The rough form of the pneumococcus looks bumpy under the microscope because it lacks the hairlike coat. Its failure to grow one is now known to be the result of a mutation, a bad one from the point of view of the pneumococcus because, without the coat, it cannot protect itself against the immune system of the host that it hopes to invade.

What Griffith stumbled over in his studies of the two forms of the pneumococcus was something never encountered before in the history of biology. He discovered that one form could be transformed into the other. He learned this during a series of routine experiments on laboratory mice.

Griffith's first step was to inject his mice with a dose of the rough form of the pneumococcus. As he expected, nothing happened; the rough form was unable to infect the mice, and they all thrived. Next he added a dose of the smooth form. Knowing it to be lethal, he took the precaution of heating the sample beforehand to kill all the bacteria and render them harmless. With this apparently innocuous mix of inoffensive and dead bacteria in them, he expected that those mice, too, would thrive. To his astonishment they did not. They began keeling over in the last throes of pneumonia. He examined them and found them to be swarming with live specimens of the deadly smooth type. Where had they come from?

Griffith's first reaction was that he had bungled his test. He decided that he had not heated the smooth bacteria sufficiently and that some of them must still have been alive when he injected them into the mice. To test that, he shot some more of the same heated sample into another group of mice. They were unaffected. Clearly the bacteria in the heated sample were all dead.

So he repeated the original experiment: a mixture of live rough bacteria and dead smooth bacteria. He repeated it a number of times. It always came out the same; the mice died. He published his findings: an unknown substance in a dead bacterium could cause a permanent change in a live one. It was tantamount to saying that a glass of medicine could change a blue-eyed man into a brown-eyed one. More than that, the brown-eyed man's descendants would have brown eyes.

The reaction to Griffith's paper was one of mingled doubt and dismay. The doubters dismissed Griffith as an unknown and probably unreliable man. He did little to help that image. He was an even more private man than Avery. Pathologically shy, he could not even bring himself to read his scientific papers before audiences. Once, hounded into reporting on some important work he had done, he read his paper in a whisper, and nobody heard a word he said.

The dismayed were fellow workers who knew Griffith to be an exceptionally careful and reliable man. No matter how self-effacing he might be, his scientific work was rock solid. Crazy as it might sound, it would have to be respected. His experiment was repeated by others and confirmed. It crossed the Atlantic and astounded Avery, whose deep researches into the complexities of immunology and the chemistry of the coatings of bacteria told him that it was impossible for such a change to take place. Nevertheless he forced himself to accept the possibility of the impossible, and in due course had workers in his laboratory fussing with the problem. They, too, repeated the Griffith experiment and confirmed it.

The question now became: what exactly was it in the dead bacteria, what part, what chemical, that was causing the change? A young researcher of Avery's, J. L. Alloway, tried to find out. He smashed up some smooth cells, filtered out all the broken cellular parts, and came up with a syrupy white substance that could do the job. The syrup was by no means pure. It was so full of different things that no one could guess which was the effective one. Sometimes it worked, sometimes it didn't. But when it did work, it worked with a vengeance. Rough cells turned to smooth as if some kind of wand had been waved over them. Mice were not needed. The transformation could take place in a test tube. People in the lab called the syrup the transforming principle. Avery decided to refine the syrup and learn what the active part of it was. It took him ten years to do so.

"Disappointment was my daily bread," Avery reported more than once during those years. He enlisted the help of two assistants, Colin MacLeod and Maclyn McCarty. Improved filters and centrifuges were employed, more refined purification techniques. In an elaborate series of steps, one extraneous contaminant after another was squeezed out of the syrup. This was the old game of throwing things away one at a time, pausing after each extraction to see if the mixture still worked. If it did, more filtration, more cleansing with chemical baths would follow. Finally there emerged a liquid in which only one ingredient was thought to remain suspended. As a last step, that ingredient was precipitated out of the solution by the delicate addition of alcohol, one drop at a time, while the solution was gently stirred. As the technicians watched, a cloud of minute white threads appeared, clinging to the stirring rod. That was it. There was nothing left in the solution. The white threads were DNA.

Whereas Beadle and Tatum had connected two tunnels in a mountain, Avery had leveled the mountain itself. With MacLeod and McCarty he published what would come to be recognized as one of the landmark biological papers of the century. Biochemists and geneticists everywhere were fixated by it, although Nobelist judges later would not be. Indeed the three future Nobel Prize winners previously mentioned were sufficiently stimulated by Avery's work to decide to try to verify it in another way. Their way was to get rid of all the cellular "junk" by using a virus or "phage" that invaded a bacterium and killed it. That virus acted like a hypodermic, jabbing the bacterium and squirting its DNA inside—and leaving everything else behind. Thus it had to be DNA that was the activating force. Results from the two approaches were the same: pure DNA had been obtained and identified as the transforming agent; all the other material had been eliminated. As noted above, when the Nobel Committee met to consider its honorees, elegant genetics was deemed more worthy than trudging biochemistry.

One question remains: if Avery *had* won a Nobel Prize, would he have gone to Stockholm to claim it?

The "phage group"—Delbruck, Luria, and Hershey—may have come along a little late in the game to deserve all the Nobel spoils, but they did some enormous things on their own. By confirming Avery's finding—and by doing it in an entirely different way—they removed any lingering suspicion that proteins might be involved in DNA's transforming process. That possibility had nagged workers from the beginning. Avery himself was never certain that he had eliminated the last trace of protein from his "pure" final sample. To his death he worried that a stray molecule or two might still be there. He, better than most, knew how contaminating some of those molecular residues could be, how potent their reactions, how little of a substance was needed to flaw an experiment. His own pure transforming substance, for example, could do its work in a solution of one part in a hundred million.

Now, with protein effectively eliminated from the equation, science could turn to the next great riddle of life: how did DNA work? How did it do what seemed to be a double job? That duality, inexplicable at first, was a logical necessity. If DNA was to persist, clearly it would have to be able to make more of itself. If it was a transforming agent, it had to be able to make other things. As the phage group, working with viruses, got deeper and deeper into their studies of those minute entities, it began to be increasingly clear that DNA was a two-job substance.

At first, viruses seemed totally mysterious. They had been detected about forty years earlier but virtually nothing was known about them. They were too small to be seen with ordinary microscopes; they passed right through the finest filters, filters that had no difficulty in trapping the smallest bacteria. Their very existence was impossible to verify except indirectly—they invaded and destroyed bacteria. They were in a class with genes, speculative entities, known to exist only by what they did. Their name "phage," which first identified them, comes from the term *bacteriophage,* which means destroyer of bacteria.

Greater familiarity with viruses gradually forced their acceptance as real things. At first they were believed to be ultrasmall bacteria. Later they were thought not to be alive at all. That idea arose when some tobacco mosaic virus, one that attacked tobacco leaves, was refined and dried in a laboratory. It turned out to be a chemical, a bunch of crystals. But when the crystals were dissolved in a liquid and exposed to tobacco plants, the disease spread again.

Ordinary crystals "grow," it is true—that is, they make more crystals. Put them in the proper chemical medium, add the ingredients needed to get the crystallizing process going, and one crystal after another will form, indefinitely, on and on as long as the crystal-supplying material is not used up. The crystals are identical in structure, rigid, incapable of doing anything. Clearly they are not alive. No one regards a crystal as having life any more than needles of ice forming on the surface of a freezing puddle are alive. They are simply inert minerals, organizing themselves into predictable cubes and other neat solid shapes according to well-understood laws of chemistry.

So, when refined tobacco mosaic virus—commonly known as TMV—turned out to be a crystal, a legitimate question arose: could it be a living organism? Few people thought so until the phage group began learning some peculiarities about viruses. If they were not entirely "alive," they certainly clung to the edge of life, because they were able to reproduce themselves in more elaborate ways than ordinary crystals. They could, however, do this only inside other cells, which seemed to classify them as some kind of semi-alive crystalline bacterial parasite, able to conduct their own reproductive transactions—whatever they were—only by using the currency of proteins, sugars, fats, and nucleic acids that they stole from the vault within the bacterial bank.

When viruses were isolated and their chemical properties refined, they were found to consist of two kinds of material: a protein skin and an internal supply of DNA* — and scarcely anything else. What they seemed most like was primitive DNA-making machines. When the electron microscope was invented in 1939 and its monster eye focused on viruses, they became visible for the first time. They *were* discrete animate objects. They varied greatly in size; the largest had six or eight thousand times the bulk of the smallest. They came in different configurations. Some were round, some were rod shaped, some looked like minute hypodermic needles, with a head at one end and a sharp point at the other. This was totally unexpected and totally astonishing. As one elderly biologist, who had studied viruses for half a century and was at last shown a photograph of the hypodermic-shaped type, exclaimed, "Mein Gott, they've got tails!"

Tails they had. And it was those tails that not only looked like but were used like hypodermic needles to pierce the skins of bacteria. Having pierced, the DNA in the head was squirted into the cell, leaving the empty shell of the virus sticking to the bacterial casing like a hollow porcupine quill. Meanwhile the deadly work of making more little hypodermics out of the bacterial cell's own resources was conducted ruthlessly inside at breakneck speed, rupturing the cell in less than half an hour and spewing out a swarm of tiny virus replicas to seek and consume other bacteria cells.

Definitely alive. What ordinary crystal could do that — make more crystals and at the same time make a protein shell to enclose the crystals? Here is an extraordinary situation: only the DNA of the virus gets into the cell. And yet, once there, it is capable of making not only protein for its own skin but also more DNA, constructing both from the supplies it finds in the cell. Here is the evidence for the claim of duality. DNA is heterocatalytic (it makes protein) and also autocatalytic (it makes itself).

Rather suddenly DNA was everywhere. It could make itself, it could make other things, it could alter other DNA, it could even do this when the organism that contained it was dead, as with Griffith's pneumococcus. The experiments of the 1930s and 1940s were drenched in DNA. Every path that led to mutation, to heredity, eventually — and this is important to our main story — to evolution,

---

* Some viruses contain a different nucleic acid: RNA. See page 250.

also led to DNA. DNA might be a dull, stupid, repetitious molecule, but by 1950 its central role as the moderator of all life processes was becoming overwhelmingly clear.

As to how DNA did its two jobs, no one had a clue. Everyone knew what it was made of: four bases, a phosphate group, and a sugar (page 200). They also knew that those simple ingredients were strung together in long chains of molecular units, not dozens but tens of thousands of them. But how the chains were put together and what they did was an utter mystery. Here we can compare a man looking at DNA to a man looking at his wristwatch. He knows what the watch is made of: steel, brass, glass, a few small jewels. Those are the watch's elementary materials and they correspond to the sugars and phosphates and bases of DNA. He also knows that his watch does something — it tells time, just as DNA does something. But in order to find out *how* it tells time, the man must learn how his watch is put together. Similarly with DNA. It was becoming increasingly clear to all who had anything to do with it that if it were ever to be understood, its architecture would have to be learned.

12

James Watson and Francis Crick: How Is DNA Put Together?

We must learn about the architecture of DNA—that is what a young man named James Dewey Watson told himself.

Watson had grown up in Chicago. He was a precocious youngster, a member of an infuriatingly bright team of Quiz Kids on a famous radio show that made older, stupider listeners alternately blink with astonishment or grumble with suspicion that the Quiz Kids had been coached up beforehand to answer the knotty scientific questions asked them. But the suspicion was unjustified. All were bonafide little protogeniuses. Watson, already fascinated by biology, entered the University of Chicago when he was only fifteen and from there drifted under the influence of the phage group.

Watson studied under Luria and learned a good deal about phages. He had a suspicion that they might be naked genes. That was a weird idea, but Watson seemed a weird young man. He had protruding pale blue eyes, a skinny neck, a burr haircut, strange clothes, and a manner that veered erratically from shyness to self-mockery to outright arrogance. But Luria knew he was bright and helped him get a small postdoctoral fellowship to study chemistry abroad.

As a fringe member of the phage group, Watson had been trained mostly in genetics. That discipline is more structural than chemical. That is, it concerns itself less with processes than with the spatial relationships of molecules, with the positioning of genes on chromosomes, with the structure of chromosomes themselves. Watson, trained to think like a phage man, knew a great deal about genes but was not so strong in biochemistry, in the riotous activity that took place inside the cell. He was advised to brush up on his biochemistry under the eye of Herman Kalckar in Copenhagen.

Watson found both biochemistry and Kalckar excruciatingly dull. Bored, restless, underemployed, he loafed about Europe, attending scientific congresses here and there. In Naples he blundered into a lecture about how the structure of crystals could be determined—not by prying the crystals apart, but by looking at photographs of them taken by an X-ray machine. The technique is known as X-ray diffraction photography. A young English crystallographer, Maurice Wilkins, was giving the lecture. During it he showed a diffraction photograph of a sample of DNA. That picture, admittedly, was a poor one. It was blurred and said very little that was comprehensible about the structure of DNA. The reason: DNA, being an organic molecule, albeit a crystalline one, is far more complex than the molecules of simple inorganic crystals. The latter can be made to reveal their structure with remarkable clarity in the pattern of dots and lines on the photograph caused by the precise and repetitive superposition of actual atoms in the crystal. The technique used is formidably difficult and delicate and requires considerable skills of interpretation. Nevertheless great progress had been made in it over the years at the Cavendish Laboratory at Cambridge University. The head of the laboratory, Sir Lawrence Bragg, was perhaps the world's most experienced X-ray crystallographer. He had invented the technique and for it had been awarded a Nobel Prize in 1915. He was only twenty-five years old when he got it; no younger man has ever been so honored.

Maurice Wilkins did not work at the Cavendish, but at the laboratory of King's College in London. He was a much younger and considerably less experienced man than Bragg, but he had taken on a more difficult job: trying to learn something about organic molecules through crystallography. At the Naples lecture he explained the technique and its limitations and expressed the hope that, with better equipment and better samples, better pictures might some day be obtained—and then flashed his DNA picture on the lecture screen. Smudged as it was, it shot a shaft of light through Watson.

The inertia and goallessness that had infected Watson since his arrival in Europe vanished. He had never seen a diffraction photograph before, didn't know such things existed. Now it occurred to him that the mystery of the gene, something that had been itching him since he was a teenager, might be elucidated through X-ray diffraction. He knew, as did everybody in the field, that DNA in its pure form was a crystalline powder. He decided to give up his drab and reluctant liaison with biochemistry for a more thrilling mistress. Somehow or other he would attach himself to the Cavendish Laboratory at Cambridge, the acknowledged center of crystallographic work.

Watson made that decision at one of those breathless moments in science, not unlike that of more than a hundred years earlier when a new idea almost too large to be contained had begun to flood through Darwin's consciousness. Now another enormous idea was slowly surfacing. The foundation stones of life were chemical. What transformed them from dead chemicals to pulsing vitality seemed to be locked in the structure of a particular chemical, DNA. Unlock that structure and one might cross the threshold between nonliving and living matter—and discover the process that changed one into the other.

That idea—that hope—was far from being exclusive with Watson. Members of the phage group had been wrestling with molecular structures for a number of years. So had Linus Pauling, one of the great scientists of this century. Pauling had started life as an atomic physicist, then moved over to apply what he knew of atoms to the structures of molecules. In his laboratory at the California Institute of Technology, he invented a way of making molecular models by designing a kit of different-colored balls and knobs representing different elements, with joints and flat surfaces that corresponded to the chemical links between elements. With that kit he labored for years to construct ever more complicated molecules. "I was simply entranced by chemical phenomena," he once wrote, "by the reactions in which substances disappear and other substances, often with strikingly different properties, appear; and I hoped to learn more and more about this aspect of the world."

One of Pauling's greatest assets was the ability to think in three dimensions. That is why he had his knobs and balls built, to help him visualize, by putting them together, the various configurations that the laws of chemistry require. Organic molecules are not flat things. Actually they are as three-dimensional as a roast turkey, their atomic connections sticking out at precise angles like drumsticks. Those chemical angles and atomic distances are exquisitely exact. For Pauling's Tinkertoy set to be useful, the flat surfaces on the balls and their angular relationships to each other were machined to a fine tolerance of one to one two-hundred-fiftieth of an inch. They had to be; they were built to a scale of one inch = one angstrom unit and were duplicating atomic angles and distances of a few billionths of an inch. If they were to fit together properly, they would have to do so according to what he had learned about those angles and distances. He was playing with atoms, and he would have to play by their rules.

To help him in his reconstructions of molecules, Pauling learned crystallography as a way of looking at them and slowly promoted himself from simple structures to more complex ones. He worked out six rules for determining crystalline structures. To this day they are known as Pauling's rules and are still observed. Along the way he also picked up an interesting fact from an English crystallographer, William Astbury. Astbury, as it happened, was working with proteins, not DNA, but what he was learning about their structure would later turn out to be a key to the identification of the structure of DNA also.

The interesting fact that Astbury had stumbled over had to do with sheep wool. He learned that the diffraction pictures he got from wool that was stretched were quite different from pictures of unstretched wool. With unstretched wool Astbury got what he called the alpha structure; with stretched wool he got a different diffraction picture: the beta structure. The two were consistently different and easily recognizable. Not much else could be derived from Astbury's rather primitive pictures except a fuzzy suggestion that the atoms of sheep wool were arranged in some sort of coil and that the coil repeated itself every 5.1 angstrom units—approximately twenty billionths of an inch.

That was the first hint that protein chains, or any other organic molecular chains, might exist in the form of coils, or helices. A helix—it should be explained here—is not a spiral. A spiral is like a snail shell, a curve winding on itself and getting larger with every revolution. What is commonly but mistakenly called a "spiral" staircase, whose every coil from top to bottom is the same diameter, is a true helix.

Pauling, with his acute three-dimensional sense, realized that what Astbury's photographs suggested was congruent with the rest of his knowledge about protein molecules, that in fact the way to organize the various parts of a long protein chain so that its chemical and electrical bonds would come together at the proper angles and distances would probably require a coil. As early as 1937 he spent considerable time trying to produce a protein coil that would satisfy the numerical dimension—5.1 angstrom units per revolution—suggested by Astbury's diffraction photographs of sheep wool, or, more properly, keratin, the protein from which sheep wool is made. He failed.

A decade later Pauling took up the problem again, having made innumerable tests and calculations to check and refine his atomic angles and distances. By this time he *knew* that those infinitesimal figures were correct. Somehow or other, a long protein—or polypeptide chain, as it was called—would have to yield to that knowledge. He was in bed one day with the flu, playing with a piece of paper, when the answer came to him. He drew a polypeptide chain on the paper, then began twisting it in varying degrees of tightness until he got a configuration that suggested that the various chemical bonds could all be made to fit. He confirmed it with a Tinkertoy construction and succeeded in producing a working model for a protein, a helix—the world's first. The only trouble with it was that its coiling period was not 5.1 angstrom units, as Astbury's keratin diffraction picture said it should be, but 5.4 angstrom units. Three-tenths of an angstrom unit is scarcely one-billionth of an inch, not a very large distance, but enough to have thrown Pauling off for twelve years. The anomaly later turned out to be a misreading of Astbury's photograph.

Pauling named his coiled protein structure the alpha helix. It made a profound impression on all the crystallographers, geneticists, molecular biologists, and biochemists who heard about it. It particularly impressed Max Perutz, a Viennese who was doing crystallographic work at the Cavendish Laboratory. His subject was not sheep wool but a different and more intractable protein: hemoglobin. Hemoglobin also comes in two forms, two shapes, depending on the amount of oxygen in it. Since it is the molecule in the red blood cells that supplies oxygen to the body, it has to be something that can absorb oxygen quickly and efficiently from the air in the lungs and also release it generously to the body tissues that need it. By Perutz's own estimate the structure of hemoglobin was on the order of a thousand times more complicated than that of DNA. He said that a number of years *after* the structure of DNA had been worked out, but he said it reflectively and without malice—no doubt thinking back over thirty or forty years of his own life spent on hemoglobin and the ferociously complex models that he built, culminating in his winning a Nobel Prize.

THE FIRST MODEL A DISASTER

WATSON has been called "Lucky Jim" for having had the good fortune to arrive at the Cavendish Laboratory in Cambridge at the moment he did. DNA was in the air, but nobody was paying much attention to it. There was a vacuum—more or less—for him to fill. Others there were preoccupied with proteins. Even Maurice Wilkins, at King's College in London, was only fiddling with DNA in a desultory way, partly because he was having trouble deciphering the rather poor X-ray diffraction pictures he was getting, partly because he could not stand a young woman colleague who was, he thought, supposed to be supplying him with better ones. She was under the impression that she had been brought to King's to work independently on DNA. She resented Wilkins's assumption that she was subordinate to him and would not speak to him. As a result, progress on the structure of DNA was not being made either in London or Cambridge.

Watson may have been lucky, but he forced his luck. His presence at the Cavendish was engineered by him in an elaborate maneuver. He had to persuade the people in the United States responsible for funding his fellowship that it was perfectly proper for him to abandon the work he had been sent to Copenhagen to do and take up something else (crystallography) about which he knew almost nothing and whose relevance to his supposed field of study was not clear. He burned his bridges, leaving Copenhagen and showing up at the Cavendish for an interview with Sir Lawrence Bragg before his grant switch had been approved. Boldness sometimes pays off. He was taken in; his grant followed.

Where Watson *was* lucky was in almost immediately running into a man at the Cavendish named Francis Crick. Crick had been doing some thinking about DNA but had not gotten anywhere with it. He was, by any measurement, an extraordinary man. The first sentence in chapter one of Watson's book, *The Double Helix,* published eighteen years later, reads: "I have never seen Francis Crick in a modest mood."

Not many others had either. Crick had a racing mind, a manic personality, unbounded energy, a booming voice. He never stopped talking. He poked his nose into the work of all his colleagues, giving them unsought (but unnervingly perceptive) advice about their problems. He drove his boss, Sir Lawrence Bragg, straight up the wall with his noise, his ubiquity, and his apparent inability to settle down to steady and productive work on hemoglobin, the subject assigned him. When Watson met him he was thirty-five years old, still only a graduate student. He had dillied and dallied about getting his Ph.D. and was conspicuously over age in grade: a brilliant man, ferociously intelligent, ferociously impatient, exploding in all directions and threatening to blow himself up for good. Horace Judson has written of him: "Restless, inquisitive, looking for simplifications rather than the relentless detail of crystallography with large molecules, for more than two years Crick moved from problem to problem, doing other men's crosswords but not finishing his own."

Crick had grown up in physics and had spent the war years designing magnetic mines for the British Admiralty, but then switched to biology. He had become possessed with a fever to penetrate the borderland between the living and the nonliving, to swim in the submicroscopic world of the phage group. He wanted to know more about viruses and genes. He particularly wanted to know how the message written on a gene got out and created a protein, the totally different stuff that organisms were made of. He wanted, in short, to jimmy his way into the central office of the bicycle factory and find out what happened when a blueprint was handed to a worker. He wanted to know who the worker was, where he came from, how he read the blueprint, and what he did when he had read it. There is a consuming mystery there. What is it in a genetic blueprint that makes a worker reach up to a shelf for a wheel and then turn and attach it to a bicycle? Just as mysterious: how does the wheel get on the shelf?

Crick realized that answers to such questions would come only through knowledge of the structure of DNA. He went to the Cavendish and got a job as a graduate student working on hemoglobin. He studied crystallography, impatiently and in savage bursts of effort.

When he first met Watson and discovered that this awkward young American—who had suddenly splashed down into the Cavendish like an adolescent heron into a marsh—knew a great deal about genes, he grabbed him. Watson, for his part, was wholly seduced by Crick, by the swoops of his mind, by what he knew of chemistry that Watson didn't, but mostly by his realization that both he and Crick were really interested in the same thing: solving the riddle of life by figuring out the structure of DNA. They began talking. They continued talking. Soon they were assigned a room together at the Cavendish where they could talk the day down without distracting everybody else. Most of their talking was about DNA.

The trouble was that neither of them was supposed to be working on DNA. In a rather clubby fashion, and without anything having actually been written down, DNA had been assigned to King's College. It was Wilkins's private property. For people at the Cavendish to play with it was like poaching grouse on someone else's moor; a gentleman just didn't do that. Furthermore, as Watson quickly learned, Crick and Wilkins were friends; that made the situation even more awkward. And even if Wilkins could be persuaded to cooperate, it was not going to be easy to get good diffraction pictures from him. He had rather unwisely given all his best samples of DNA to his female associate, thinking that she was working for him. With those better samples she was beginning to get better pictures than Wilkins—but, still thinking she was working independently, she was still not speaking to him.

That woman's name was Rosalind Franklin. She was a determined, rather stubborn, frighteningly dedicated person, direct and abrupt in her manner, made defensive and more than normally contentious by the atmosphere in which she was forced to work. Quite apart from there being a legitimate question about her relationship with Wilkins (was she merely his assistant or was she an independent scientist answerable directly to the head of the laboratory in which both worked?), she was a woman forced to make her way in a peculiarly British old-school-tie, masculine, exclusive, scientific fraternity. The warmth and effectiveness of that system, for those inside it, lay in its casual and cozy atmosphere. There was a great amount of cross-pollination as crystallographers, chemists, mathematicians, and geneticists picked each others' brains in each others' offices and spent innumerable off-hours in freewheeling discussions in neighboring pubs, at lunch or dinner in groups. Rosalind Franklin was totally excluded from that. She was not even permitted to use the dining room at King's; it was barred to women.

Watson's picture of Rosalind Franklin in *The Double Helix* is a disturbing one. It turns out to have been somewhat of a hatchet job. He was young and irreverent, with a tendency to be contemptuous of people who were not certified geniuses. He was also — in a somewhat unfocused way — contemptuous of women in general. He seemed to regard them as playthings and earned a reputation while at Cambridge as an indefatigable woman chaser. Looking at Rosalind from that corner of the ring, and through the eyes of Wilkins, who detested her, it is not hard to see why Watson's picture of Rosalind Franklin, as it emerged in his book, was a biased one. That it was biased is attested to by Anne Sayre, in her own book *Rosalind Franklin and DNA,* and by the testimony of Franklin's assistant Raymond Gosling and of her scientific executor Aaron Klug. All pictured her as a far warmer, more approachable, less tunnel-visioned person than Watson did. They conceded that her exterior was rather formidable but claimed that it had been forced on her by circumstance. She knew she would either have to fight or recede into a mousy role, her work directed by someone else. Since she was a fighter by nature, she chose to fight.

At any rate, Watson chose to call her Rosy, a kind of patronizing nickname which, according to Anne Sayre, was not commonly used by her scientific associates and was never used by her family or friends. Watson pictured her as a frumpish, aggressively unfeminine woman. "I wondered how she would look if she took off her glasses and did something novel with her hair," he wrote. He went on to characterize her as extremely rigid in her thinking: "It was downright obvious to her that the only way to approach the DNA structure was by pure crystallographic approaches. . . . Model building did not appeal to her. . . . The idea of using Tinkertoy-like models to solve biological structures was clearly a last resort."

Added to all this, never brought into the discussion but lurking in the background, was the fact that Rosalind Franklin was Jewish. Through both her parents she was connected to distinguished banker-merchant families with outstanding records in politics, the law, and public works. Although she chose to live a Spartan life on her own small earnings, she was wealthier and better connected than most of the impoverished young scientists with whom she rubbed elbows. Certainly she was better situated than Watson, who had nothing at all.

That was the climate in which Rosalind Franklin lived. Being a woman and Jewish, she found herself unable to breathe the scientific air she so badly needed. Except for one graduate student assigned to her, she worked alone.

For Watson, chafing to get started on DNA, the prospects were not good. Somehow or other he would have to harness the energies—and the knowledge and cooperation—of Crick, whose enthusiasm ran up and down like a yo-yo and whose attention it was not always easy to get; of Wilkins, a distant and secretive man who worked slowly and, it often seemed, without any real hope of accomplishing anything; and of the steadily enraged and uncooperative Rosy.

Watson's first small breakthrough was in getting to know Wilkins. They had met in Naples, but Wilkins had more or less snubbed him. Now, attached to the Cavendish, Watson was allowed to attend a colloquium at which both Wilkins and Franklin spoke. The two men met again and had dinner together. This time Wilkins was much friendlier. He warmed to someone brought up in genetics who seemed to see the relevance of his arcane and, so far, unproductive labors in the cool universe of crystals. He spent a good deal of the evening complaining about Rosalind Franklin.

Rosalind's talk at the colloquium had been a straightforward one. In the preceding months she had discovered that DNA came in two forms. One was the so-called A form, definitely crystalline. It produced X-ray diffraction pictures of a certain type; that was what she and Wilkins and others working on DNA had been photographing. Now Franklin had learned that DNA also came in a B form, a more elongated shape that resulted from the addition of a number of water molecules to it. In lay terms, "soaked" DNA looked different in diffraction photos from unsoaked DNA. Franklin had taken some soaked B-form pictures, but they were fuzzy rather than sharp, reflecting the difference between a highly crystalline substance and one that had been somewhat altered by water and had become more fibrous. The latter had a great many ultrasmall crystals facing every which way, which did not make for clear pictures. Those pictures tended to be smudged or smeared in appearance. As a result, it was difficult to derive precise estimates of molecular distances and relationships from the X-ray patterns that those minute crystals produced.

And yet precise measurements were what Watson and Crick needed. Watson listened to Franklin's presentation but took no notes. He was unable to tell Crick the next day how much water was in the B form, nor could he remember some of her other figures. Sometimes Watson seemed almost to relish playing the role of Young Dope, and here he played it to the hilt.

Just what did Watson and Crick know about DNA? They knew its chemical composition. They also knew that crystallography had revealed it to have a regular pattern. That meant an orderly arrangement of its constituent parts. Chemistry told them how the phosphates and sugars could join together to form a backbone for the molecule, with the four bases tacked on somehow:

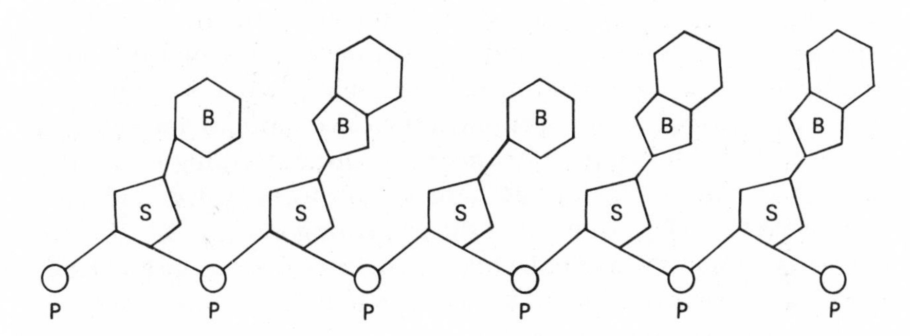

P = phosphorus
S = sugar
B = base

Crick had done some mathematical work that excluded a great many impossible models. This simple, basic, repetitive chain could only come together in a relatively few plausible constructs. Crick and Watson began fiddling and fitting and eventually came up with a model for the B form that consisted of three intertwined helices with the backbones in the center and the bases sticking out in all directions like bristles.

The result fitted what the two men knew of the existing crystallographic evidence. It did not contradict any of the chemical input that Crick was able to supply. Greatly excited, they invited Wilkins to come and have a look at it. To their surprise, Rosalind Franklin came too. That morning was a huge embarrassment to Watson and Crick. Franklin was abrasively critical of the model. "To her mind," Watson wrote, "there was not a shred of evidence that DNA was helical." With pinpoint disdain she also explained that they had underestimated by a factor of ten the amount of water that the molecule should have.

The model builders were devastated. The party adjourned for lunch, during which, according to Watson, "Though Francis could not help dominating the luncheon conversation, his mood was no longer that of a confident master lecturing hapless colonial children."

The upshot of the model debacle was a suggestion by Wilkins's boss, John Randall, to Crick's boss, Lawrence Bragg, that the two young troublemakers cease and desist from working on DNA. This idea was particularly appealing to Bragg, who could not endure the prospect of Crick's hanging about at the Cavendish for another two or three years, shooting off his mouth about everything under the sun, while still neglecting the hemoglobin work that would earn him his Ph.D. and get him out of the place for good. The word came down to Crick and Watson that henceforth they were to abstain from work on the structure of DNA and leave that to Wilkins—and, by inference, to Rosalind Franklin.

THE SECOND MODEL A TRIUMPH. THE DOUBLE HELIX

THE collapse of their model, with the subsequent order from Bragg to stop poaching, was a crusher to Watson and Crick. It removed them directly from the DNA scene for more than a year, but it did not keep them from a kind of indirect, submarine connection. They continued to talk about it and to do some work that they hoped would be productive in the end. Watson, by this time, had become convinced that the cracking of the DNA structure would provide a first-class ticket to a Nobel Prize and that a three-way race was shaping up between the Cavendish and King's and the looming shadow of Linus Pauling, six thousand miles away in California. Pauling, Watson knew, had by far the deepest and most agile mind of any who might snag the prize. He also knew more about the chemical behavior and bonding of atoms than any other man alive. He had written a classic work, *The Nature of the Chemical Bond,* which Watson read and reread that year, hoping to learn from it, while at the same time shuddering at the menace of its brilliance. Between bouts of reading Pauling and worrying about him, Watson did some crystallographic work with tobacco mosaic virus. He learned something of the technique and got some pictures that suggested to him that its nucleic acid structure might be helical. Meanwhile Crick did some thinking.

What Crick thought about most was the four bases that are present in DNA. The reader should now turn back to page 202 and take another look at the diagrams there: of adenine, guanine, thymine, and cytosine—or A, G, T, and C, as they will be called hereafter. What nagged at him was the fact that they were the only parts of DNA that had any variety at all. If DNA was to prove to be the blueprint that coded for the manufacture of other things, then that coding would have to be supplied by some variation in those bases.

There was no other way. In whatever fashion the four bases might be found to fit into a hypothetical helical structure, it would have to be differences in *them* that held the key to coding, because they were the only variables. On well-established chemical and molecular grounds it was becoming more and more certain that the other elements in DNA (the phosphate group and the sugar) were entirely regular. They fell together in the continuous chain shown on page 225, monotonously, one unit after another, thousands and thousands of them. They constituted the backbone of the structure. They were uniform throughout the molecule. Therefore, programming or coding could not possibly be done by them. It had to be done by the bases.

Pondering the remarkable accuracy with which chromosomes replicate themselves during division, Crick wondered if that might not be explained by a pairing of the bases. Could A be bonded to A, and G to G? He discussed this with a young mathematician colleague, John Griffith, and was told that some preliminary calculations that Griffith had made on his own suggested that A did not bond with A but with T, and C with G. There was no likely pairing of like with like, said Griffith, but more likely a pairing of like with unlike. Crick stowed this information in the back of his head.

He hauled it out again a few weeks later during a visit from America by the biochemist Erwin Chargaff. Chargaff had devoted years to the study of the four bases and had discovered that the amount of A in a sample of DNA always matched the amount of T. Similarly, the amount of C always matched that of G. Chargaff had then gone on to show that though there was this unexplained equalization among pairs, it did not follow that there was an equal amount of all four bases in a sample of DNA. In other words, there could be more A and T in a given sample than C and G, or there could be less. Each organism whose DNA Chargaff had examined differed in that respect. Since he was a biochemist interested in processes and precise amounts and not a geneticist interested in structures, Chargaff did not pursue the implications of the ratios he had discovered. He merely published papers about them. Watson had read them, Crick had not. When Chargaff met the two young men in England he was astonished, even irritated, by their ignorance—and by their brashness. Here they were, trying to construct molecular models without even being able to remember which base matched which (Crick had rather embarrassingly mislaid that information in talking about it to Chargaff), and without having taken into account his one-to-one rules when they built their first ill-starred model.

This was vacuum-cleaner time for Watson and Crick. They were forced to go about sweeping up what crumbs of knowledge they could. The Chargaff crumb would turn out to be a boulder of significance a few months later.

Some other things they learned: DNA probably did have a helical form. All the crystallographic evidence they could get their hands on pointed that way; so did the few pictures that Watson himself had taken of tobacco mosaic virus. However, none of it was conclusive. The phosphorus and sugar backbone was probably on the outside of the helix, with the bases sticking inward. This they had belatedly gathered in a roundabout way from Rosalind Franklin and from calculations made by Crick. Again, nothing sure, but it was the opposite of what they had thought when they constructed their first, and flawed, model. It had had the backbone on the inside.

This was not much to go on. Watson became increasingly fidgety. He suddenly became desperate when he learned that Pauling had focused the giant searchlight of his mind on DNA and had written a paper describing its possible structure. For Watson, once he got hold of a copy and read it, it was like walking away unscratched from a head-on automobile collision, for Pauling had made some elementary mistakes in chemistry. His model was a three-chain one with the backbone inside—not all that different from Watson's and Crick's first abortive try. "Though the odds still appeared against us," Watson wrote, "Linus had not yet won his Nobel."

Unable to keep still in the face of the great surge of relief that went through him, Watson let the news fly around Cambridge that Pauling had blundered, then went off to London to tell Wilkins and Franklin. There he had a strange encounter. Poking his nose into Franklin's laboratory unannounced, he met a frosty presence, impatient with his talk about models. When he persisted in discussing helices she erupted. "Suddenly Rosy came from behind the bench that separated us and began moving toward me. Fearing that in her hot anger she might strike me, I grabbed up the Pauling manuscript and hastily retreated."

This account is a bizarre one. Franklin was a rather small woman. For her to have attacked Watson seems highly unlikely. But for her to have been angry is not in the least surprising. She had been working with desperate concentration on the structure of DNA, trying to tease out some sense of it from her crystallographic evidence. It eluded her. She must have been sorely frustrated at that particular moment, past all patience with the continued exclusions and condescensions of men, and especially with this particular nosy and undisciplined man, this would-be model builder. She knew far more about DNA than he did. She knew the diameter of the molecule. She knew that it occurred in two forms. She knew the length of the repeat pattern in the B form: 34 angstrom units, which was exactly ten times the 3.4 unit size for the single nucleotides. She had driven herself nearly crazy trying to decide whether the A form was helical or not and had finally decided that it was not. But she had a very powerful hunch that the B form was; her notes say so.

It is irony of the acutest kind that the outburst at Watson should have opened an all-important door for him. It propelled him from her office directly into the arms of Wilkins in the hall. Wilkins gathered him up, took him into his office, said that Rosy had often been on the verge of punching him—***and then showed him Franklin's best picture of the B form, the one from which she had derived her information.***

Its pattern hit Watson like a hammer. It bellowed "helix!" For more than a year he had talked, thought, and dreamed DNA. He had attended Franklin's colloquium in which she described the B form. Until that instant he had not taken in its significance. Now, at one stroke, he knew as much as she did—or almost as much. He hustled back to Cambridge. Armed with the absolute certainty that DNA was helical, he dared bring up the matter with Bragg, pointing out to him that it was only a matter of a short time before King's would have the secret worked out, but—more terrible—Pauling might do it even sooner. Bragg,who felt extremely competitive with Pauling already, since Pauling had derived his rules for crystalline structures almost under Bragg's nose, now reversed himself. Watson would be permitted to resume model building of DNA. He would even be permitted to get the necessary Tinkertoy parts made in the Cavendish machine shop.

Watson gave this walloping piece of news to Crick, who, at a lunch a few days later, began urging Wilkins in his customary windmill argumentative style to start building models immediately, or Pauling would get there first. Wilkins demurred. He wanted to wait until Rosalind Franklin was gone—he had gotten his way; she was leaving; she had become utterly frustrated with her life and work at King's. She was going to move to another laboratory at Birkbeck and continue with crystallography, but not DNA.

Well, Crick said, if Wilkins wasn't going to make DNA models immediately, was there any objection to his and Watson's trying it? Wilkins thought a moment and said no. They were back in business.

Watson was so anxious to get going that he could scarcely wait for the model pieces to come from the machine shop. After several false starts he opted for a two-helix structure, on Crick's advice, and put the backbones of the two twisting threads on the outside of the molecule. With a brilliant deduction from Rosalind Franklin's research, which was now available in a report, Crick suddenly realized something from her figures about angles and twists that she had not. The only way to explain them properly was to have the two helices running in opposite directions, one up, one down. He patiently explained to Watson that the structure had to have a pair of twisted backbones, both turning on the same central axis, but one of them a "down-staircase," and the other an "up-staircase." The chemical reasoning that went into that idea is beyond the scope of this book.

Watson accepted it and turned to the only matter that crystallography could not deal with: how the four bases were packed on the inside of the helix. The patterns of X-ray diffraction photographs said nothing about that.

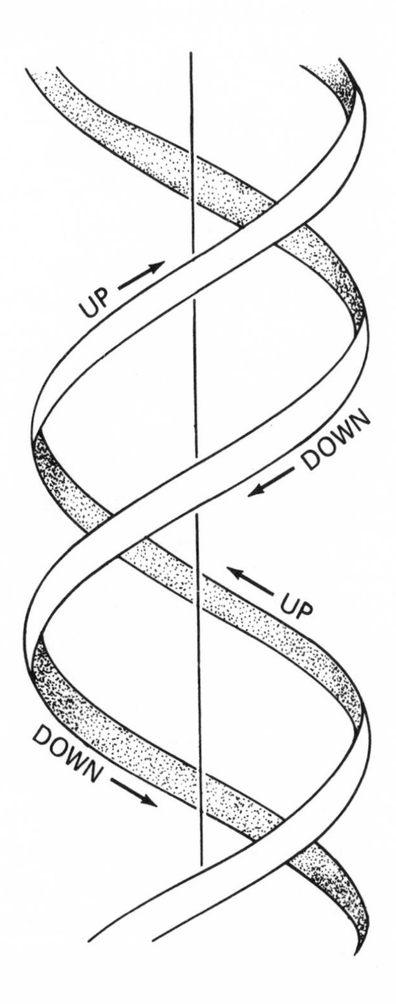

Watson wrestled with the matter for several days. Again, chemical evidence told him that the bases would have to be fastened at regular intervals to the twisting backbone, and at right angles to its vertical plane. Since the backbone was now on the outside of the molecule, the bases would have to jut inward like horizontal flanges toward the central core. But a terrible problem arose in how to fasten them together: that is, how to join up a base that was attached to the down-helix with one attached to the up-helix. If the two twisting backbones are visualized as the outer frames—the actual structure—of a turning staircase, then the bases become its steps, extending from one side across to the other, joining in the middle and holding the whole thing firmly together.

The problem was that the four bases were not the same size. Another glance at the diagram on page 200 will show that two of them, the purines, are considerably larger than the two others, the pyrimidines. Fitting purine to purine and pyrimidine to pyrimidine just didn't work, because one pair would be conspicuously larger than the other. The result would be a bumpy helix, too thick in some places, too thin in others. X-ray photography denied the possibility of bumpiness.

It was then that Watson remembered Chargaff's rule: the amount of A always equals the amount of T, the amount of C always equals the amount of G. *Could that mean that A JOINED with T?* Together they measure exactly 20 angstrom units, the known width of the helix. He tried it. The whole thing came together very quickly. It satisfied all Crick's chemical requirements. It satisfied Chargaff's rule. It confirmed all the predications of the X-ray diffraction photographs. Against all odds and by a masterful patching together of bits and pieces of information from a dozen sources, they had done it.

They had used Franklin's B-form picture, which Wilkins probably should not have shown Watson without Franklin's permission. They had used all Franklin's measurements, derived solely by her after many months of lonely and excruciatingly difficult work. This information again came to them without her knowledge; they had dug it out of a report she had written to her superiors summarizing her progress over the previous year. They were in no sense stealing that knowledge; it was part of a larger report to which people at the Cavendish also had to furnish information and was thus available to Crick's superiors. Nevertheless the work had been Franklin's and she received too little credit for it. She had failed to clear the last hurdle. She had been unable to visualize Crick's masterful deduction: one helix ran up, the other down. That made all the difference. Watson and Crick went on to win a Nobel Prize in 1962—just as Watson had predicted—for cracking the structure of DNA. Wilkins also won one for his X-ray diffraction pictures. Franklin, although she had taken the best picture and understood the molecule best, got nothing. She was dead, having succumbed to cancer in 1958 at the age of thirty-seven. She deserved better.

The model was an instant success. Everyone who saw it was impressed by it. Wilkins, although he must have felt grievously scooped, examined it carefully and approved. So did Rosalind Franklin. In fact she was genuinely friendly, her scientific nature warming to the brilliance of its achievement. She did not realize that she was the one who had made it possible. She was under the impression that Watson and Crick had worked it all out, not from her data but from calculations they had made themselves.

Even Sir Lawrence Bragg, the head of the Cavendish Laboratory, was pleased beyond measure. His seasoned crystallographic eye could find no flaw in the construction. His bad boy, Crick, had, against his worst forebodings, produced something worthwhile—better, a work of heroic stature, something on the order of Michelangelo's *David.* The Cavendish had ended up a winner in a race that Watson had more or less promoted himself. In his *The Double Helix,* published some years later, Watson stressed throughout it his craving to win a Nobel Prize, and that the others involved were driven by the same craving. That was heatedly disputed by Wilkins and Crick, both of whom declared that they were driven only by curiosity, by scientific interest in solving an important problem, and that the goal of a Nobel was not in their thoughts at all. Both were enraged by the book and scathingly contemptuous of it. They felt that it was self-serving, inaccurate, and betrayed confidences. Crick was so inflamed that he wrote a long and sulfurous letter to Watson, even sending a copy to the president of Harvard, where Watson was teaching, in an attempt to get the book suppressed. He failed. The book was published to great public acclaim and a certain amount of professional snarling. It had—of all things—a graceful foreword by Bragg, who added: "Those who figure in the book must read it in a very forgiving spirit."

The Double Helix is an extremely lively and interesting book. It became a best-seller, and is still in print two decades later. It made the public aware of DNA in a way it might not otherwise have become. While not everyone who read it may have quite understood the size of the door that it opened into the cell nucleus, or visualized the torrent of workers who would rush in with pick and crowbar, few can have missed its message that science is ruthlessly competitive today. In the last analysis, it is a catfight.

Don said: "Well, you certainly made a big production of the cracking of DNA. Is it really worth all that?"

"Everybody else thought so. The Nobel Selection Committee did. They gave out three prizes for it, because you have to count Wilkins's as part of the overall one."

"Even so—"

"Pauling himself thought so. He was very generous about it. He was at work on a faulty model when he got the news that Watson and Crick had won the race. He said of the structure: 'I think that the formulation . . . may turn out to be the greatest development in the field of molecular genetics in recent years.' He was right. It did."

"On second thought," Don went on, "Pauling's right. Knowing the structure puts us in the bicycle factory, doesn't it? Right in the central office. At last we have the blueprints in our hand. We can see what they say."

13

Questions for Crick: What Does the Code Say? How Is It Read? Is RNA Involved?

What Watson and Crick had done was far more than put together a three-dimensional jigsaw puzzle. They had opened the way to an examination of the "how" question: how did genes do what everybody thought they did? Watson and Crick realized this. In fact, while they were working on their model, one of the conditions they felt they had to satisfy was that the model provide a mechanism for copying: for replication. It would have to show, or at least suggest, how a gene could duplicate itself accurately, and also how it could program the making of other things. In short: autocatalysis and heterocatalysis.

The model does that. And the two men signaled to the world that they were well aware of the larger implication of their discovery when they published their first paper on it in April 1953. They wrote: "It has not escaped our notice that the specific pairing we have postulated immediately suggests a possible copying mechanism for the genetic material."

There is a world of meaning packed into that modest little statement. To extract it, it is worth taking another look at the helix, this time with the base pairs drawn in (page 236).

Two important things immediately pop out from an examination of this picture. The first is that ***the base pairs can be arranged in any sequence.*** An A–T pair can be followed by another A–T pair, or

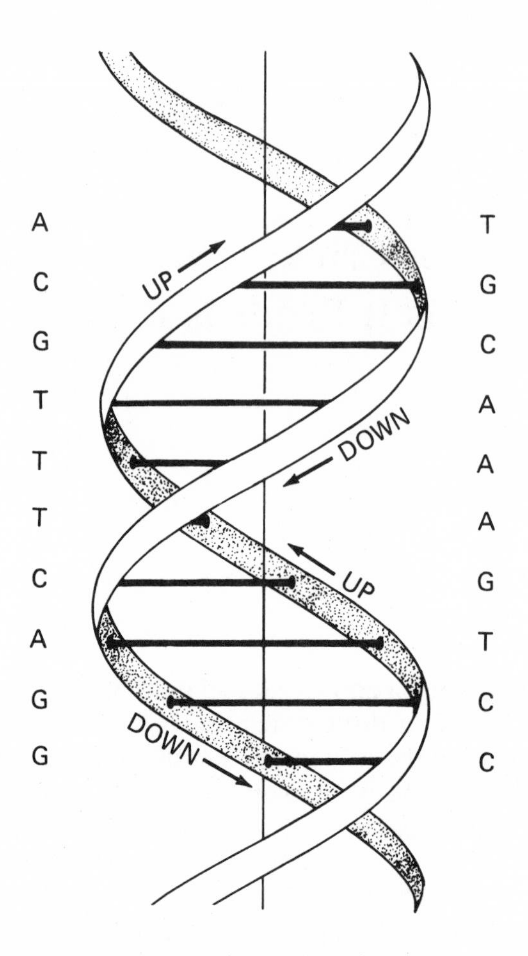

a dozen of them. Or they can be interspersed with T–A, C–G, or G–C. Structurally it makes no difference. Genetically it makes all the difference in the world, because it is that very order of pairs that determines what the gene will do. For heredity the order is absolutely critical. It has been established in the laboratory that a change of one pair can change the genetic instruction.

"Just *one* pair?" said Don. "How are we going to get our readers to swallow that?"

"They'll have to take it on faith," I said. "Informed people in the field don't question it."

"But just one pair? That's incredible."

"It is. But that's not to say that every change is critical to what the gene says to do. Knocking out one base pair, or reversing it, or substituting another, might do nothing. It might take several pairs to have any meaningful effect on what the gene is instructing for. But it could be done with just one. That is what a mutation is. That's the message we want to get across here. A mutation is nothing more than a change in the order of base pairings in a little strip of DNA."

"Wait a minute. You're telling me something that's hard to accept. How could such a tiny change, only half a dozen atoms, have such a big effect? They're too small, too simple."

"You mustn't think of them as effecting the change directly. All they do is instruct. The letters S-T-E-P written on a sheet of physical exercises are simply an order. They don't constitute any movement in themselves. Something else is going to have to do the moving."

"The person who's reading the instructions and doing the exercises?"

"Exactly. And you could change one of those letters so that the exercise sheet said S-T-O-P. That would be the equivalent of changing one base pair. S-T-O-P means something quite different from S-T-E-P. But both are still only instructions. All the base pairs do is code for the production of a particular amino acid. It's the amino acids, the way they build into proteins and enzymes, that actually show up as something tangible in an organism. Each enzyme is coded for by a precise pattern of base pairs. Change the pattern and you get a different enzyme. Base pairs may be little in themselves, but they cause big results.

"That may be. But the spiral staircase you've shown has only two kinds of steps, A–T and C–G."

"Four kinds. A–T is not the same as T–A."

"Okay, four kinds. But how are you going to make the whopping number of differences that you find in the human body with only four kinds of things? I read somewhere that there are a trillion cells in the human body."

"Somewhere between a billion and a quadrillion. But do you know how much DNA there is in the human body?"

"Well, not too much. After all, it's all in the cell nucleus, and the cell nucleus is pretty small."

"You're wrong. Granted that the cell nucleus is small, remember also that DNA is extremely thin. Also, it's twisted into a helix. And that helix is twisted on itself in a larger coil and that, perhaps, in an even larger coil. It's like a piece of rope or telephone cable: coils on coils on coils. That makes DNA very compact. It's crammed into the nucleus. A single strand of DNA may be only a few hundred-millionths of an inch wide, but if you were able to straighten out that gossamer thread, how long do you think a human chromosome would be?"

"I haven't a clue."

"About two feet. So, take all forty-six human chromosomes from all the trillion cells in the human body, straighten out their DNA, and set it end to end—how far do you think it would stretch?"

"Quite a way, I expect."

"That's understating it. It would go to the sun and back more than a hundred times. Actually, it's closer to two hundred."

"Wait a minute!"

"I won't wait. I'm just getting started. We're talking about little clusters of atoms. They're so much smaller than you can imagine that it's possible to pack more of them in a small space than you can imagine. Stretch them out, and that's what you get. The important thing, though, is not how far all the DNA in the human body will stretch, but how many different combinations of those four bases can be fitted on one chromosome. That's a really large figure."

"How large?"

"Well, if I were to write it down, it would start like this: 10,000,000,000,000,000,000,000,000,000,000,000,000,000,000,000,000,000,000,000,000,000,000,000,000,000,000,000,000,000,000,000,000,000,000,000,000,000,000,000,000,000,000,000,000,000,000,000,000,000,000,000,000,000,000,000,000,000,000,000,000,000,000,000,000,000,000,000,000,000,000—"

"Why are you stopping?"

"Because I'm getting tired. To finish that figure I'd have to write a good many million pages of zeros. If I wrote down zeros at the rate of one a second, starting now, and not taking any time off to eat or sleep or go to the bathroom, I'd be finished in about thirty or forty years."

"You make your point."

"I want to bang it down hard," I said. "That figure is so much greater than all the atoms in the universe that the two don't bear comparison. For all intents and purposes, the theoretical variability of the human genetic system is infinite. So, don't worry about there being only four bases. When Watson and Crick figured out that they could occur in any combination on a very, very long chromosome, they made the present population of human beings on earth—

about five billion and each of them different — look like peanuts. As far as human variability is concerned, we haven't scratched the surface. We never will. That's one of the mind-bending things about the double helix."

"And the other?"

"The other is *DNA's duality, its ability to direct the production of more of itself and more of other things.*"

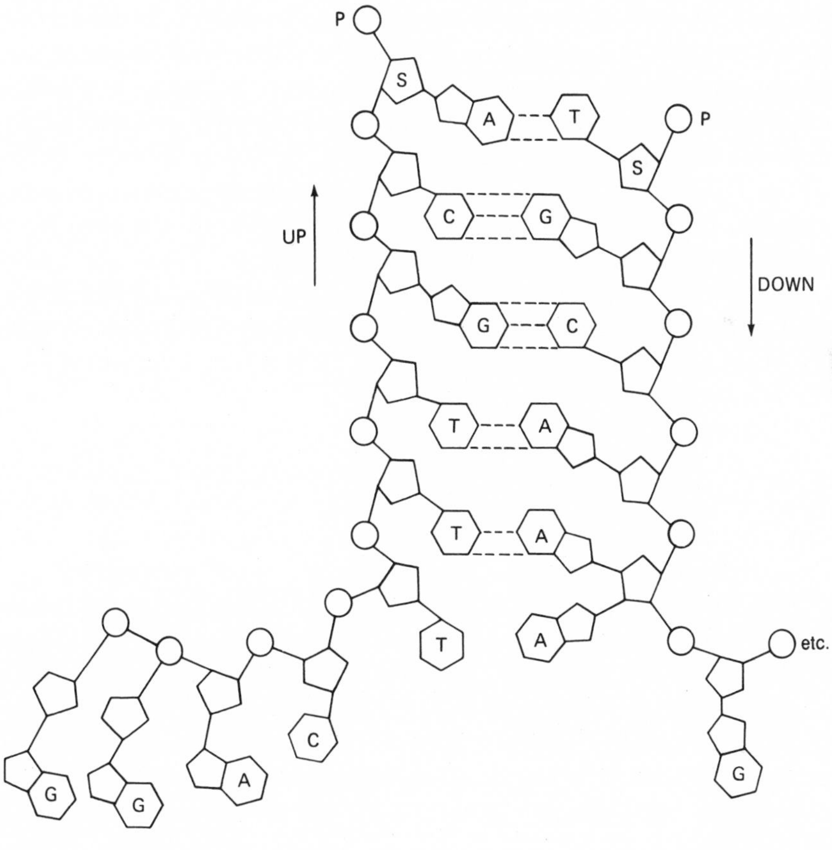

The model suggested to Crick a way of DNA's doing that — subsequently proven right. If, as Crick suggested, the helix were to unwind and then come apart like a zipper by the breaking of the chemical bonds that hold the base pairs together, there would be two strands floating loose in the cell. Then, suppose there were little free units, single bits of A and T and C and G, also floating around. If there were some way they could attach themselves to the single chains, they would have exactly duplicated the original chains. That would explain how DNA, in cell division, is able to replicate itself so exactly.

In the diagram on page 239 the helix has now been uncoiled, flattened out in two-dimensional form, and is beginning to break apart at the bottom. The bits of phosphorus (P) and sugar (S) are now clearly visible as a regular, repeating backbone on the outside of the structure. The bases stick inward. They do this irregularly because of their differences in size. But when they are matched up according to Chargaff's rule, all pairs are exactly the same size, permitting a constant width for the entire structure. The bases are held together by hydrogen bonds *(dotted lines),* and when those bonds break, the bases dangle free at the bottom—without partners, *inviting* partners. At the left we see some G's, an A, a C, and a T waiting to join up with any stray matching bases that may be in the neighborhood. Similarly, at the right, there are a loose G and A also looking for partners. As they find them, two new helices will begin to form, one base pair at a time, whereas there was only one helix before. When the two reassume their coiled shape, they will be identical with the single helix on which they patterned themselves.

Or so Crick thought. He had yet to prove it.

To see this a little more clearly, let us use symbols—as most texts do—for the four bases. Their shapes and sizes show how they fit together:

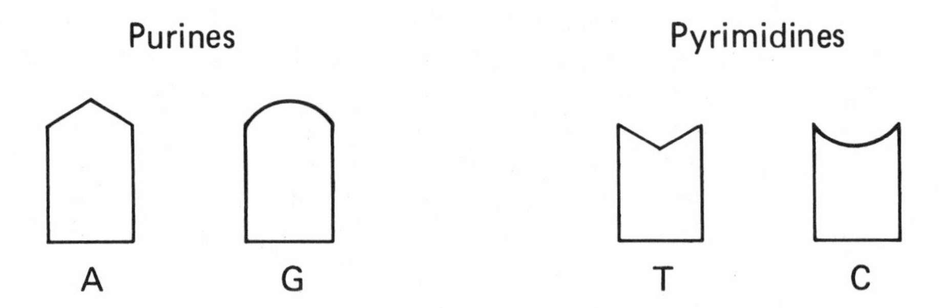

These symbols have no reality in themselves. A is not actually pointed at the top and G is not round. This is simply a way of showing that the base pairs are complementary in their chemical structure and that they do fit: A with T and C with G—and not otherwise. A simplified diagram of the helix appears on page 242.

Take a good look at that apparently simple diagram. In it, believe it or not, are hidden—in the endless permutations and combinations of base pairs—the coded instructions that, according to Crick, allow DNA to replicate itself and also to direct the organization of all the other material that goes into every cell of every living organism. From the beginning Crick had insisted that the model serve his belief that DNA played that dual role: it made itself, it made proteins. Somehow it would have to be shown capable of each function. It was a brilliant model; Crick's idea was brilliant. But for some time they were only that—an unproved model and an unproved idea. It would take Crick and half a hundred other scientists the better part of a decade to arrive at the working principle of how the double task was accomplished. That principle, as majestic, essentially as simple, as astonishing as DNA itself, is now well understood and its details fairly well explored for bacteria. For higher organisms, with cell nuclei, more and larger genes, and infinitely more complex structures and behavior, it is not all that simple. We still know very little about how the game is played in octopi and humans, although we have the rule book.

That rule, it would slowly be learned, after endless slogging through perplexing, contradictory experiments and wrong turns, involved two matters that Crick already had given considerable thought to:

1) What did the code of the four bases say?
2) How were its instructions carried out?

We are now inside the locked office of the bicycle factory, looking over the shoulders of the designers as they send out their blueprints and watching the workers as they move about and do things in response to instructions on the blueprints. We examine the blueprints hopefully. But, to our dismay, we don't understand them. We cannot relate what is printed on them to what a worker does—why

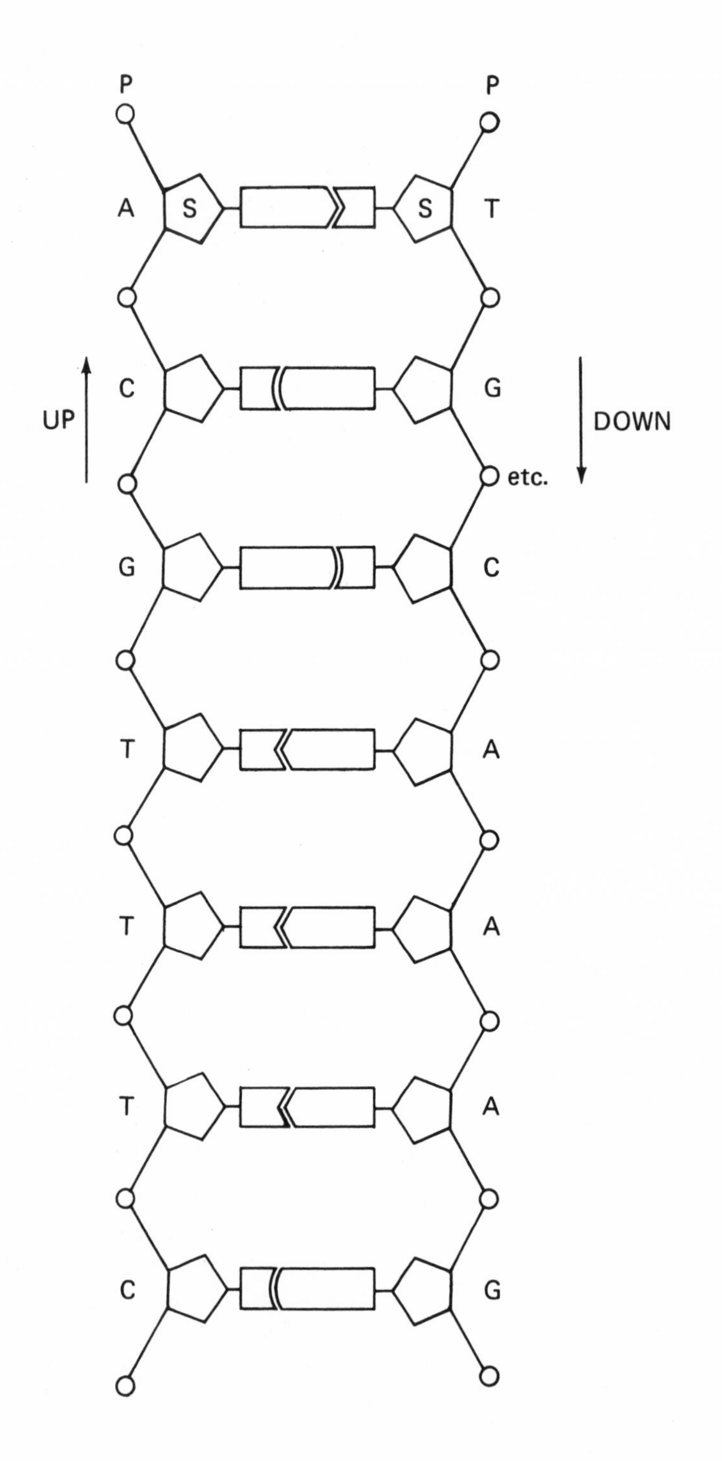
P
P
A
S
S
T
C
G
UP
DOWN
etc.
G
C
T
A
T
A
T
A
C
G

he goes to a particular spot, why he picks up a particular part, why he turns a particular knob, what gives him the initial shove.

Crick understood none of those things. He was made painfully aware of how great was the gulf between the model and knowing how it functioned by the remark of a fellow scientist, who reminded Crick that, though it seemed overwhelmingly logical that DNA coded for the construction of proteins, nobody had yet demonstrated it. It could be done by showing that a mutation in a specific gene caused a change in a protein sequence, but nobody *had* done it. It was still only an idea.

But what bedeviled Crick far more than that lack of proof was that he could not crack the code. At first he could make no sense of it whatsoever. Those four letters A-T-C-G, in their enigmatic combinations, stared at him without betraying the slightest clue as to what they meant. It was a time of vast frustration. He had the blueprint in his hand, but he could not read it.

Crick has always been a daring speculator. As recently as 1982 he was speculating that all life on earth came here in the form of microbes from outer space and did its evolving subsequent to that arrival. He has maintained that good theoretical thinking—productive speculation—can be most effective if it is not hemmed in by too much evidence, for evidence sometimes narrows the range of the mind as it tries to accommodate to that evidence. As a result, it may miss a possibly larger and correct idea that could float into it if the evidence were not there to trip it up. Evidence, while "true" in the sense that it is a bundle of observed data, is not always understood properly the first time around, when one is still blundering toward a principle. If the evidence should happen to be misinterpreted, the principle may never be thought of.

In his contemplation of the riddle of the double helix code Crick had scarcely any evidence at all. There was no direct proof that DNA coded for proteins, although he was virtually certain that it did—ultimately. What it probably coded for directly was amino acids, the smaller units from which proteins are built. If you know how to make bricks you can probably learn to make a brick chimney. It was on this supposition that Crick began his thinking about the coding problem. He would have to know what coded for bricks.

Amino acids are remarkable things. They are chemically simple structures and there are not many of them, and yet they combine to make incredibly complex things. Study of proteins had already revealed something that Crick knew well: all proteins are made from the same basic group of approximately twenty amino acids (there are a couple of exceptions that need not concern us). But there is a far greater number of different kinds of proteins in the living world — many thousands of them. Therefore, Crick reasoned, it would have to be the *sequence* of amino acids in a protein that determined what it was. Amino acids were like a twenty-letter alphabet; arrange them in different ways and you could get a huge number of different words out of them. And, just as in the spelling of words, where the substitution of one different letter (S-T-E-P into S-T-O-P, for example), or the rearrangement of the same letters (S-T-E-A-L, S-T-A-L-E, L-E-A-S-T, T-A-L-E-S, T-E-A-L-S, S-L-A-T-E) changed the meaning completely, so it was with amino acids. Mix them ever so slightly, and you got a different protein. Indeed, the extreme specificity of that arrangement became clear as the amino-acid sequences of certain proteins were unraveled in the laboratory. Intense work over many years revealed that the hemoglobin molecule, for example, contained four protein chains, two of them 141 amino acids long, and two of them 146 amino acids long. Investigation of the blood disease, sickle-cell anemia — in which the red blood cell itself is misshapen and as a result causes serious illness in the sufferer — revealed that a single amino-acid switch from glutamic acid to valine on one of the longer chains caused the disease.

That is almost inconceivable — one tiny switch in a chain of 146 amino acids causing illness and death — particularly when we learn that there is a lot of other glutamic acid and valine scattered through that long sequence. It is not as if it were missing; it's *there,* in abundance. It's just in the wrong place. The precise location of an amino acid is as critical as its presence. In the case of sickle-cell anemia one difference makes all the difference. There may be a dozen identical small screws in a watch, but for the watch to work, each must be where it is supposed to be. Put one in a hole too large for it and it will fall out; the plate it is supposed to hold in place will move, a wheel will fly loose, and the entire mechanism will cease to function properly.

So, asked Crick, if twenty amino acids can combine as neatly and precisely as they do to make any and all proteins, what does it take to make an amino acid? For an answer to that question he was forced to turn to the four bases in the DNA helix. He had to look there; they were the only variables in it. Furthermore, it was clear enough that one base pair did not code for one amino acid. If it did, there could be only four kinds of amino acids, for that is the maximum number of combinations that four single base pairs can produce. If, however, he assumed that it took two base pairs working together to do the coding, those, in their various combinations, could produce sixteen different amino acids. But that was still not enough; it fell short by four. Crick was forced, by logic and without a shred of evidence, to conclude that three base pairs, a triplet, was the fundamental instructional unit on the helix. That allowed for sixty-four combinations, more than enough. He considered, but discarded, four base pairs as the unit. This would have produced 256 different combinations; it was unreasonable to expect a parsimonious natural system to be so lavishly over-generous.

"You're going much too fast here, Mait," said Don. "I'm not sure people are going to get this."

"What won't they get?"

"Where you're heading. They'll hear what you're saying, but they'll lose track of how it connects. You're supposed to be talking about how DNA makes proteins. How does it do that?"

"It doesn't—directly."

"Well, amino acids, then."

"DNA doesn't make them either. It only instructs for them. I keep repeating: DNA is the blueprint, not the worker."

"Then who is the worker?"

"Now you've asked a really important question. At that point Crick didn't know. Nobody did. But he realized that if he was ever going to find out, he had better understand what the DNA code said. If he couldn't read that, he was never going to get anywhere. Maybe we should go over this again to make it a little clearer."

"Okay."

"I'll start by asking you to take a couple of things on faith."

"Fair enough."

"First, DNA *is* the information center."

"Okay, okay, you needn't belabor that one. Cells can't function without it. Change it, and you change things; you alter flies, you alter sea urchins. You've been blabbering about that for chapters."

"Second, DNA is a nucleic acid. Most of everything else is a protein. What we are made of, what builds us, what holds us together, what makes us work—tissues, hormones, enzymes: all proteins. A nucleic acid is not a protein. It is an instruction sheet for the making of proteins."

"Yes, yes, yes—"

"Those instructions have to be read. Therefore, they have to be in the form of a code or an alphabet of some kind. That's all I'm saying. Since the only differences within DNA are in the arrangement of the four base pairs, A-G-T-C, they carry the code. They *are* the code. Crick knew that. He said so. In a second paper that he and Watson published they wrote: "The precise sequence of bases is the code which carries the genetic information."

"Still with you."

"Now take another thing on faith. There are twenty amino acids that form long chains to make thousands of different proteins. That's another way of saying you need only twenty kinds of bricks to make any kind of building you can think of. You just put them together in different ways."

"So—"

"So, where do you get those twenty bricks?"

"You just said—out of the arrangement of the bases."

"Yes, but how do you read that arrangement?"

"Why, you—uh—"

"Exactly. That's what Crick said: "Uh—." When he looked at the DNA helix, he couldn't read that apparently random sequence of bases at all. The only thing he could be sure of was that it took more than one base pair to make an amino acid. Using an A or a G or a T or a C as a mold, he could get only four kinds of bricks. He needed twenty. If he used two base pairs he could get only sixteen combinations—I'm repeating exactly what I said on the previous page."

"Yes, but you're making it a little clearer."

"Two base pairs would be the fundamental unit in a so-called doublet code: AA or AG or GC, say. But, as I just said, doublets produce only sixteen combinations, not enough to make twenty amino acids. That is why Crick went to triplets. That gave him sixty-four combinations, enough and some to spare. He could imagine things like AAG or GAC, for example, and assume that each coded for a specific amino acid. That way, all the seemingly random bases on the helix could be matched to the equally random amino acids in a protein."

"Like how?"

"Like this. You can't predict how the triplets are going to repeat themselves in the helix. The DNA code is apparently totally haphazard. But it's the haphazardness that makes original and unusual patterns, things you might notice. Take the triplet AAG. If you had a regular pattern:

1	2	3	4	5	6	7	8	9	10	11	12
AAG	GAC	AAG	GAC	AAG	GAC	AAG	GAC	AAG	GAC	AAG	GAC
x		x		x		x		x		x	

you wouldn't learn much, since this pattern doesn't suggest a match with any of the irregular patterns that show up in amino acids and proteins—in the stuff of the real world. But the helix isn't regular like that. You'll find AAG scattered around in a lot of places, perhaps like this:

1	2	3	4	5	6	7	8	9	10	11	12
AAG	CAG	GAA	GAC	GAA	AAG	AAG	CGA	AGG	GGC	GGC	AAG
x					x	x					x

That 1-6-7-12 pattern is a very distinctive one. You would recognize it wherever you saw it. If a certain amino acid showed up in exactly the same pattern in a protein chain you would spot it. The two would match. That would be a pretty strong indication that AAG was the code for that amino acid.

"So that's how they did it."

"Approximately. It's a matter of learning how to sequence a strip of DNA and a strip of protein and compare the two for matching patterns. At least, that is how you *could* have done it if you knew enough about proteins. But when Crick began thinking about codes, it was not possible to sequence DNA, and protein sequencing was just getting under way. No protein had yet been completely sequenced for its amino acids. The first one was bovine insulin—the insulin protein derived from cows. That took one of Crick's colleagues at the Cavendish Laboratory, Frederick Sanger, six years, working steadily between 1948 and 1955. It was an extraordinary achievement."

"What was so hard about it?"

"Anything that involves sorting out and arranging individual parts of protein molecules is hard. Their amino-acid structure is irregular, their scale is incredibly minute, and your tools are blunt. Getting back to bricks as a metaphor for amino acids—remember we have twenty different kinds—suppose we want to copy a row of brick houses in a city street. Unfortunately, the man who built those houses is gone. He took his plans with him; we can't find them. If we're going to build duplicates of those houses, we'll have to do it by examining the original houses. That involves breaking them apart. If we smash them to rubble we learn nothing; all we have is a pile of assorted bricks and no way of knowing their former arrangement. Instead we take a wrecking ball—it's a crude tool, but the best we have—and tap the house as gently as we can to knock out a small piece here and there. We are, in effect, taking a protein apart piece by piece, one amino acid at a time. By comparing the pieces, a little hunk here overlapping another hunk there, we may eventually be able to reconstruct how the entire house row went together from the proper arrangement of its twenty kinds of bricks—or how a protein is put together from the proper arrangement of twenty kinds of amino acids. Sanger had to break up thousands of houses in order to learn how one was put together. Finally he knew which brick went where, from foundation to chimney top.

"Once you know that, you can look at the blueprint that instructs for the arrangement of bricks, and compare. Today the functions of nearly all the triplets in the DNA code are known. That is to say, the amino acids they code for have been identified. *Why* the amino acids do what they do, once they are arranged on a protein, is another question. It's a Chinese puzzle inside a Chinese puzzle."

"What kind of wrecking ball could Sanger possibly have used on a protein?"

"He used an enzyme. He had learned that certain enzymes can break up proteins by the way they operate on specific amino acids. It was a remarkably accurate wrecking ball. He could aim it between two windows on the second floor of the third house from the corner and knock off just one brick."

"Well, I have to admit, that's very neat. And, I suppose, as other proteins began to be sequenced, it may have helped Crick figure out how the code worked. But that doesn't solve the big problem: how you make a house. What I mean is that just because you can read something doesn't mean that you can *do* anything with it. Suppose you know what AAG codes for—by the way, what does it code for?"

"Lysine."

"Fine. AAG codes for lysine, an amino acid, right? You could just as well call it a hubcap, right? I mean, there's this blueprint sitting there, telling you it's a hubcap. Right there in the factory office you are always talking about. But if that hubcap is going to be installed, somebody is going to have to do it. The blueprint can't."

"That's what I've been telling you."

"A messenger or somebody is going to have to hand it to somebody else. That messenger's going to have to read it—okay, Crick has explained how he can do that—but then he has to *do* something himself. How does he do that? And, by the way, who the hell *is* that messenger? Did Crick know any of that?"

"No."

"My point."

"Your word 'messenger' is a good one," I said. "Crick did imagine a messenger service of some kind. But as to how it worked, nobody knew. It was one of the most difficult and frustrating times in the history of modern biology. People worked at the problem from two directions. Some worked from proteins toward DNA. Others worked from DNA toward proteins. The closer they got to each other, the more they seemed to be stumbling over another nucleic acid I haven't mentioned yet, but I should now. It's called RNA. Also I'll have to mention some enzymes."

"Just a minute. We're not signing up our readers for a course in chemistry. They're here to learn about evolution."

"Then they're going to have to learn a little about RNA. At the heart of modern evolution theory is the gene, the sequence of chemicals in a strip of DNA that gives instructions for the creation of living organisms. RNA is the bridge that helps translate those instructions into bricks and then helps organize the bricks into buildings."

RNA is short for ribonucleic acid. It will be remembered that the chemical name for DNA is deoxyribonucleic acid (*deoxy* = unoxygenated); it has one less oxygen atom in it than RNA. Also one of the bases in RNA is slightly different, this time a change in one atom of carbon. The new base is called uracil—here we'll call it U—and it replaces the T in DNA. So, when we talk about RNA we will be dealing with A-G-U-C instead of A-G-T-C. A now pairs with U in the helix. To repeat, the molecular differences between the two are extremely small, a couple of atoms. But it should be clear by this time that the organization of life is being examined at the atomic

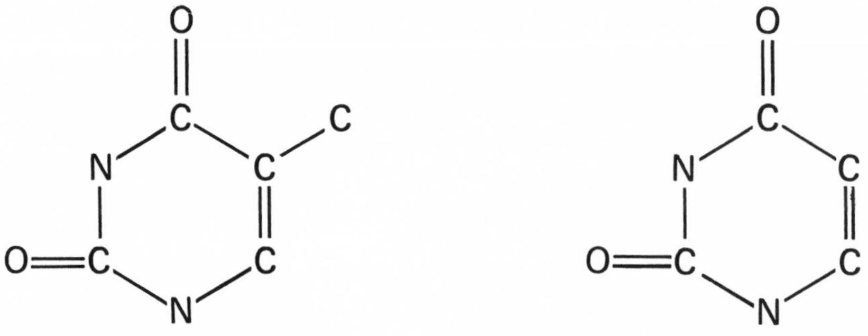

Thymine, or T

Uracil, or U

Only difference: the carbon atom attached to T.

level; at that level a small change of an atom or two at the beginning can lead to great differences later on.

The existence of RNA had been recognized for a good many years before Watson and Crick went to work on DNA, but not much was known about it, other than the slight atomic differences between the two. RNA seemed to come and go in the cell. After cell division it apparently disappeared. All sorts of theories about DNA transforming itself into RNA and back again were batted about. They didn't make good sense; they were discarded. What *was* known was that while DNA sat coiled in the cell nucleus, RNA was found (all or mostly all of it) outside the nucleus, floating about in the cell. Also, when protein was actively being manufactured in the cell, there always seemed to be a good deal of RNA in the vicinity. Other than that, what RNA was or did was pretty much an enigma. It was not even clear for some time whether RNA had a single or a double helix, or how big it was. It broke apart easily—so easily that it took a good many years for anybody to realize that it might be a very long structure. That is—sometimes a very long structure. Sometimes it seemed to be a very short structure. Maybe there was more than one kind of RNA. Mischievous, mysterious stuff.

14

The Dual Nature of DNA

MATTHEW MESELSON AND FRANKLIN STAHL: DNA MAKES ITSELF.

Did RNA truly fill the gap between DNA and the manufacture of protein? That was a hot question, and it was getting hotter. But before it reached the boiling point, some other matters had to be disposed of.

One was the validity of the Watson-Crick helix. While its structure had a magisterial logic, fitting all the stereochemical knowledge available, conforming to all the agreed-on physical measurements, confirmed by the best X-ray crystallographic evidence, it had some dreadful practical things wrong with it. Watson quickly found that out when he returned to the United States in 1953. His model was intellectually compelling, but it offered no explanation of its implied deeper task: how it performed physically. It was a double coil, held together by hydrogen bonds, and it was immensely long—hundreds of thousands of nucleotides long. How could such a chain unwrap in minutes and catch, in *exactly the right order,* matching bits of *exactly the right stuff,* floating by at *exactly the right instant it was needed,* to make two identical chains where one had existed before? And that unwrapping helix would be only one of forty-six in a single cell. The body had billions of cells. Could such a torrential molecular scramble, involving many thousands of chemical transactions every second, possibly occur without causing a mixup of such monumental proportions that all sensible coding would turn into gibberish? Many scientists thought that exact transcription on such a massive scale was just plain silly.

The second problem that the helix encountered involved the awkward necessity of proving colinearity: a meaningful match between the order of amino acids in a protein and the order of nucleotides in DNA. With these two problems, here were autocatalysis and heterocatalysis again: both necessary, neither proven.

Of the two, autocatalysis had priority; it was the basic task. If, during cell division, a helix could not unwrap itself, find the necessary ingredients to replicate itself and wrap up again, one might as well forget about unwrapping to code for something quite different in another part of the cell.

Autocatalysis: unwrapping and rewrapping with a new matching piece identical to the original match. The way to demonstrate that would be (it seems obvious now) to find some way of distinguishing between the original strand and the new matching strand. An experiment was devised by two young men working in Linus Pauling's laboratory at the California Institute of Technology: Matthew Meselson and Franklin Stahl. They decided that the best way to detect differences between the two strands would be by differences in their weight. They would put into the diet of some bacteria a form of nitrogen, N^{15}, that was heavier by one neutron than ordinary nitrogen, N^{14}. These slightly different forms of elements are known as isotopes, and they are common. Forcing their bacteria to dine on only heavy nitrogen for many generations, Meselson and Stahl would wind up with a population whose DNA contained only the heavy isotope. They would then feed those heavy bacteria a few light lunches containing only N^{14}. If their hunch was right, the next generation of bacteria, responding to that lighter diet, would themselves be lighter.

It may seem foolish to try to distinguish between two bacteria whose weight difference depends on the difference between the isotopes N^{15} and N^{14}. That difference is only one atomic mass unit, or AMU, the weight of a single neutron. It is a small weight. It takes approximately 2,724,000,000,000,000,000,000,000,000 AMUs to equal a pound, a figure too large to make much sense of unless it is compared to something. Charles Price, a chemist from the University of Pennsylvania, compared it to the number of stars in our galaxy (about a hundred billion), and then to the estimated number of galaxies in the universe (another hundred billion). He found that all the stars in all the galaxies in the universe would fall short of the number of AMUs in a pound — far short. It would take about twenty-seven thousand universes to make up that figure.

It is not hard to see that there is no balancing scale delicate enough to detect a difference like that. If Meselson and Stahl expected to weigh their bacteria, they would have to do it indirectly. The instrument they chose was a centrifuge, a machine that whirls liquids around and around at great speed, gradually causing the various ingredients of the liquid to separate, the heaviest ones moving to the outermost (or bottom) part of the container and the lightest ones remaining near the top.

Meselson and Stahl put their heavy bacteria in a test tube full of water and then set the tube to spinning like a spoke in a wheel. They had previously put some salt in the water to make it heavier. When the centrifuge started turning, the salt molecules in the water were driven outward toward the end of the test tube. This was a gradual thing, the result being a steady increase of saltiness from one end of the test tube to the other. Somewhere in the middle there was a point at which the weight of the salt exactly matched the weight of the heavy bacteria that were also being spun around in the centrifuge. After several hours of spinning, all the bacteria were concentrated in a narrow band at that point, just below the middle of the test tube. The two scientists marked the spot and removed the sample.

The heavy bacteria are driven by the centrifuge toward the bottom of the test tube.

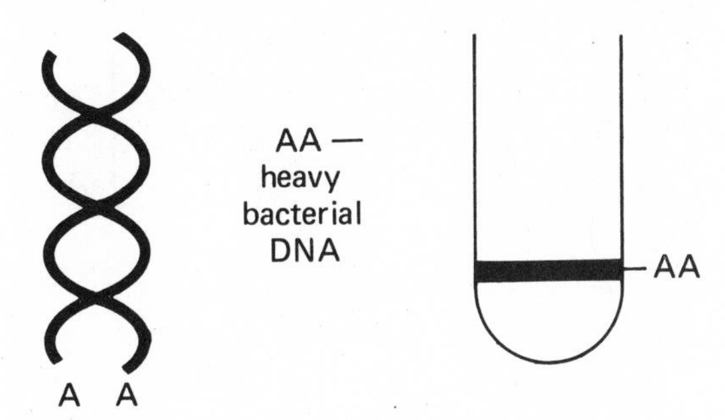

They then did the same thing with a culture of light (N^{14}) bacteria and found that after centrifuging, these formed a band higher up in the tube.

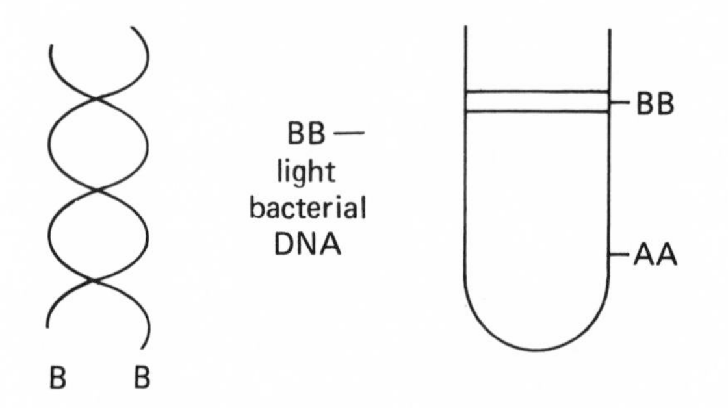

The light bacteria form a band near the top of the test tube.

Now came the critical part of the test. They switched the diet of some heavy bacteria to light nitrogen, leaving them to feed on it just long enough to divide once. Then they put that sample in the centrifuge. They reasoned that a new cell, made during that first cell division and with only light nitrogen to draw on, would end up, after replication, with a DNA helix consisting of one old heavy strand and one new light one. After many hours of spinning, the sample concentrated itself into a band about halfway between the other two, confirming their prediction beautifully.

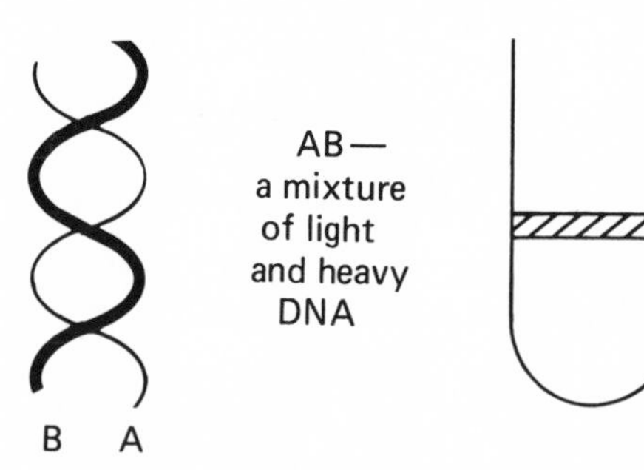

The mixture settles halfway between the other two, proving that a new strand of DNA has been made and that DNA does replicate itself.

So, DNA makes itself. Autocatalysis occurs. But how? Like an uncertain homing pigeon, that question circles and circles, but will not land.

ARTHUR KORNBERG: HETEROCATALYSIS—FINALLY A DIRECT LINK BETWEEN DNA AND PROTEINS.

PART of the answer was supplied by an American biochemist, Arthur Kornberg, in 1956, not long after the Watson-Crick helix had been unveiled and while Crick was still bedeviled by the coding problem. Ruth Moore quotes Kornberg: "I got tired of feeding things into one end of an animal experiment and watching what comes out of the other, without understanding what goes on in the middle." He was a brilliant youngster, on an even faster track than the Quiz Kid Watson; he graduated from college when he was only nineteen.

To find out "what goes on in the middle" Kornberg began by delicately prying open a bacterium and inserting some radioactive nucleotides. He used those free bits of "hot" A-G-T-C because he knew he could follow them wherever they went in the cell during the moment-to-moment chemical churning going on in there. If any of them were used in making DNA he would spot that when he killed the cell and examined the DNA.

This happened; he did get some small amounts of DNA with "hot" nucleotides in it. Then he addressed himself to the more difficult part of his exercise: finding out what caused those nucleotides to move into the DNA. He did it by using the same method that Oswald Avery had used more than a decade earlier (page 210), by eliminating. He prepared some cultures of bacteria and then began taking various bits of their cellular material away from them, one by one. Anything that was not involved in the making of DNA was removed. At every step along the way he stopped to grow new bacterial cultures to make sure that radioactive nucleotides were still being used.

If you had twenty men working on an assembly line, and it was one man's job to paint the final decorative stripe on a bicycle, and you couldn't get into the factory to see which man did that job, the only way to find out who did it would be to yank one man at a time off the line to see what happened. When you got a bicycle that did not have a decorative stripe on the day that a man named Rudolf was absent from the job, you would know that Rudolf was the stripe maker. In a similar manner Kornberg narrowed his available cell substances down to one whose only function seemed to be to push loose nucleotides onto helices. It was an enzyme, polymerase. Like all enzymes, it was a protein.

With this new protein Kornberg had the secret of life in his hand, or so newspapers and magazines trumpeted. They were riveted by what Kornberg had done. A white-coated Prometheus had put some polymerase and an assortment of nucleotides, not in a cell, but *in a test tube!* The polymerase had turned them into DNA. It had taken a bagful of chemical nuts and bolts and transformed them into a nucleic acid. If that wasn't creating life, what was?

A closer look at the kinds of nucleic acid that Kornberg got makes it clear that he was not exactly creating life. Kornberg's DNA helices looked like this:

A–T
A–T
A–T
A–T
A–T
A–T

This was not very useful DNA. It coded for no living substance, no amino acid, no protein. It could only make itself. These monotonous repetitions of bases could not be regarded as "real" life because they could not instruct for it. They resembled a factory blueprint that said "attach, attach, attach" over and over again, or "cotter pin, cotter pin, cotter pin," without giving a clue as to what was to be attached or how a cotter pin might be used. Still, Kornberg had brought organic and inorganic matters unimaginably close together. He had identified an enzyme—a "sort of" organic thing that could turn stray nucleotides—not quite organic things—into a DNA sequence, a blueprint for life. As the Nobelist George Beadle has said: "He knocked the final prop out from under the comfortable assumption of mankind that 'life' is inherently different from 'nonlife.'"

When we get that close, right on the edge between life and nonlife, we find that Beadle was right. There is no edge, only a gray area, a continuum. The slide over from the nonliving to the living is not sudden. It is plaguingly gradual, and it is noticed in chemistry by the coming together of certain elements that are labeled—for definition's sake—organic molecules, although they are far from being alive. However they are the precursors. Life appears to depend, not on some magic elixir, but on the organization of those chemicals in new ways, in slightly more complex ways in which atoms would not ordinarily glue themselves together. They do so when they have picked up bits of energy to hold them together in those unusual ways — and even attract to them other units. In that sense, a nucleotide, the building block of DNA, might be said to be the nearest to a living thing that exists among the chemical compounds on earth. It has the energy source (the sugar and phosphate bits that are attached to the base), and it uses hydrogen bonds to bind itself to a matching base. Here, at the lowest level, in DNA or in a single amino acid, is the beginning, a glimmer only, of the increasingly intricate processes that end in protein chains, cell walls, blood, living organisms. It is not possible to see — looking at them only — what the destiny of those simple things is. Only through the hindsight of what they have done can we see. But once life is seen as an artificial holding together of matter that otherwise would not be so held, then the nature of death becomes easier to comprehend. It is broken bonds. An organism is reduced again to its inert chemical parts when those bonds slacken and snap. In that sense an amoeba is a clenched thing, a rock a relaxed thing. Life, it may be said, starts with odd chemical assemblages and is kept in business by supplies of raw materials and energy to hold those assemblages together.

When the sense of continuity between nonlife and life becomes an acceptable idea, then it becomes easier to think about evolution. Darwin's theory, although it could not, through ignorance, address the matter of where and how life originated, did, nevertheless, raise the question in the mind of any alert reader of the *Origin*. Life had to start somewhere. If it did not arrive on heavenly wings or from another galaxy, then, by logic, it had to create itself out of inorganic things right here on earth. That matter will be explored in chapters 16 and 17. Here it is important only to emphasize that Kornberg made thinking about evolution and the origin of life much easier by demonstrating how close the interrelationship of organic and inorganic things is at their lowest levels.

Of more immediate significance, he demonstrated a direct linkage between DNA and something that was not DNA, a protein. The importance of that linkage cannot be overestimated. It throws a shaft of light directly on the second of DNA's presumed jobs: heterocatalysis, the making of something else. Meselson and Stahl had demonstrated autocatalysis: DNA makes itself. Kornberg's work was the first step in determining that DNA did indeed make other things as well. But how—? That question would not be answered until more had been learned about what real-life DNA actually said, until the code that held its message had been cracked. "Not until you can read me," DNA was stubbornly muttering, "will you have that answer."

Fortunately a couple of able code crackers were lurking right around the corner.

MARSHALL NIRENBERG AND JOHANN MATTHAEI: THE DNA CODE IS CRACKED.

ENERGY. Obviously it is needed to do any job. Flour, sugar, and eggs do not mix themselves to make a waffle. They need two things: a hand to bring them together in the right proportions and energy to measure and stir and cook. The bringing-together hand had to be RNA. More and more scientists were beginning to suspect it. They kept finding it lurking here and there in their experiments. By the 1950s identification of the materials in the cell nucleus had reached a point of remarkable exactitude. Aside from a certain amount of protein—always present but not involved in the instructional activity of the nucleic acid—the overwhelming mass of the stuff in the cell's central office was DNA. But there was also a small yet persistent amount of RNA. Outside the nucleus RNA was common stuff—rare inside, much more abundant outside. A battery of experiments had revealed that the more abundant cellular RNA seemed to have at least two forms. Both of those seemed to be closely associated with some structures in the cell called ribosomes.

One of the great astonishments of modern biology was the ribosome, first brought to light when the electron microscope was focused on the cell. It will be remembered that the feebler instruments of a century before had succeeded only in locating the cell nucleus in what otherwise seemed to be a wholly transparent container. Later came the identification of chromosomes within the nucleus, and still later a growing suspicion that the apparently clear liquid that comprised the rest of the cell's contents—the cytoplasm—might be more than just a liquid. It was known to be rich in dissolved materials; might it not also contain very small *things* of some kind?

Although they were prepared to find something, few biologists were ready for the elaborate plumbing that suddenly sprang into view under the scrutiny of the electron microscope. This marvelous instrument operates on an entirely different principle from a conventional light microscope. The latter's ability to see, since it depends on light waves, does not allow the detection of anything shorter than a wavelength of visible light. Its actual ability is considerably less. Design problems, the nature of the materials used in lenses, the behavior of light as it passes through them, all combine to limit the conventional microscope's ability to enlarge. The smallest things it can see are hundreds of angstrom units in diameter. Anything smaller slips through the cracks in the floor, as it were, and is unseen. As far as the microscope can tell, it isn't there; it doesn't exist.

The electron microscope does not depend on light waves but on beams of electrons that are far closer together. The best electron microscope today can pick up objects that are only a few angstrom units across. It notices their size, their shape, their structure. Individual bases that are only 3.4 angstrom units apart on the DNA helix can be detected. They are just within the instrument's range of visibility, although it cannot determine the actual atomic structure of those bases. That must wait for the completion of an improved instrument now on the boards that promises to distinguish between things that are only half an angstrom unit apart. Since atoms are spaced about one angstrom unit apart, the ability to view—other than theoretically—the atomic structure of a molecule may be just around the corner. Think of it: atoms at long last seen as hypothesized—or maybe seen entirely differently?

Getting back to the cell: its interior, viewed by the electron microscope, overnight became not an empty aquarium but one teeming with strange-shaped objects. There were mitochondria, an endoplasmic reticulum, Golgi bodies, plasmids, ribosomes, and other things. Here we will concern ourselves only with ribosomes, for the phage group and others had begun to be very interested in them.

Over the years the work of cellular biologists had become increasingly complex. Phage workers had taken to using radioactive tracers that enabled them to follow the material of a virus—nucleic acid—as it traveled into a bacterial cell, to try to determine what was happening and when. In the course of this work a number of observers had noticed tiny specks of radioactive nucleic acid scattered through the cell. Whether the specks were *all* nucleic acid or consisted of other materials as well, they could not say. The particles were extremely small, considerably smaller than viruses. The best size estimates gave them a diameter of about 150 angstrom units. They were barely detectable by the most powerful light microscopes, so far out on the edge of visibility that they were not really visible in the accepted meaning of the word. They managed only to disturb light rays slightly without revealing anything about their structure or their function. It was suspected that they were involved in some way with protein building and that they might have short lives, being made and unmade as needed.

Under the electron microscope those specks clarified themselves dramatically. They emerged as roundish objects swarming in the cell by the thousands. Their structure was complex, consisting of about fifty different kinds of proteins and some RNA. They seemed to come in two parts, like a fat hamburger bun sliced a little above the middle so that the top part of the bun was considerably smaller than the bottom part. Those remarkable objects occurred sometimes as whole buns and sometimes as separate pieces. Either way, they did not appear and disappear as had at first been suspected. Rather, they were durable, regular inhabitants of the cytoplasm, long-lease tenants. In 1958 Richard Roberts of the Carnegie Institute in Washington suggested that they be named ribosomes. The name stuck.

Did ribosomes make proteins? Experiments showed that they did not. So, what were they for? And how *did* protein, apparently being manufactured on the surface of the ribosome, get made?

The thought grew that a ribosome was like a workbench, a handy surface where items could be gathered and then fastened together. If it were an unusually sophisticated workbench, it might even help with the work, join in the sorting process, hold things in place, move them along. But an empty bench can do nothing. Clearly it needs the help of something else. And that something else had to be RNA because RNA was always there when protein was, clustering there as predictably as children cluster around an ice-cream truck on a hot day. Logically, RNA had to be in the picture.

By this time three kinds of RNA had been identified. There was some inside the ribosome; it could be ignored, as its function turned out to be connected with the structure of the ribosome itself. Then there was messenger RNA, conveniently called mRNA. This was the stuff that seemed to gang up on the surface of the ribosome at protein-building time. Finally there was soluble RNA, tiny bits that seemed to float free in the cytoplasm—maybe in solution, maybe not—identifiable only as minute pricks of radioactivity when someone succeeded in irradiating them.

The obvious candidate as protein-maker: mRNA.

Back to autocatalysis and heterocatalysis for a moment, and another trip into the bicycle factory. If bicycles are to continue to be built according to blueprint, then the designers in the central office are going to have to make blueprint copies to hand to the workers. They must not hand out the originals. Whoever they are, however they do it, they must be able to *guarantee* that there is always a master blueprint in the office file. Lose the master—no copies, ultimately no anything. Therefore autocatalysis: the ability of DNA to make copies of itself and preserve itself as the master. That an original strand of DNA does endure—and in that sense is immortal—and is thus able to issue reliable duplicates of itself once, twice, or a hundred billion times was demonstrated by Meselson and Stahl in the experiment described on pages 253 and 254.

How does DNA do that? As Meselson and Stahl showed, its helix unwinds and matches fresh bases to the old original ones that are dangling free. Then it wraps up again. The result is two identical helices where one existed before.

But DNA cannot just make itself over and over again. It has another duty, to see that protein is made: heterocatalysis. This forces the conclusion that sometimes when it unwinds it matches up with something other than DNA. We know that this must happen; the matching sequences of bases in DNA and amino acids in protein demonstrate it. Change a base, and an amino acid changes. It never fails; there has got to be a relationship. Therefore the only really tough remaining question is how DNA does this second job. The answer is that it does use something else. It uses RNA as a messenger to carry information from the helix to the ribosomes, where the work of assembling proteins is done.

As to how this happens, let us suppose that the blueprints in the central office are in the form of long strips of paper, and that the instructions consist of individual sentences, separate little hunks of information running down the strip. Suppose further that they are written in a wet or sticky ink. Would not the easiest way of getting that information out of the office be to press another piece of paper (or a series of RNA bases) onto that wet-ink master (DNA) to get an impression and carry that to the assembly line?

That, it was eventually learned, is exactly what happens. Some loose RNA bases that are floating about in the cell come along and stick themselves to the unwrapped single strand of DNA. They do that in the proper order by matching base to base. The result is a strand of RNA that corresponds exactly to the master blueprint.

An example:	T-C-C-T-T-T-G-C-A	DNA sequence
Snap on some matching RNA bases, one after another, to make a complementary chain:	A-G-G-A-A-A-C-G-U	RNA sequence

Note the U at the right end. Remember that RNA has uracil, not thymine, and that U pairs with A in RNA. Otherwise the two base sequences are a perfect match. But that one difference is profoundly important. RNA, with U in it, moves about, changes form. DNA, with T in it, stays put in the nucleus and does not change. That is how the master blueprint is preserved. Unless altered by a mutation, it is there — theoretically forever — to preserve and distribute the message encoded on it. The RNA is free to travel out into the cell and see that something is done with that message. Its name, messenger RNA, is an apt one. It does indeed carry DNA's message.

The implications of this simple difference are shattering. One carbon atom in one base enables messages from the otherwise rigid, self-perpetuating DNA to be released into the cell and be acted upon.

"Yes, it's shattering," said Don. "But you've left something out. The RNA strip is not just like the DNA strip. It's a mirror image of it; it's backward. If, as you say, position of bases is the all-important thing, then there will have to be some way of getting back to the original sequence again. Is there such a thing as a 'backward protein' or an 'upside-down protein'? That's what you'd get from a mirror image piece of RNA."

"I don't know about backward proteins," I said. "But you are right about the mirror image. There is a further stage in which the RNA does, in effect, reverse itself. We'll get to that in a minute. The thing we must talk about first is the division of RNA into triplets. We must break that strip of A-G-G-A-A-A-C-G-U into ACG, AAA, and CGU."

"Why triplets?"

"I've already told you," I said. "It's because of the coding problem. Francis Crick figured that out. If you're using twenty amino acids to make proteins, and you have only four bases, you *must* have triplets, three bases that go together to make up what's called a codon. Codons provide the variety you need. That way you can get eight different combinations with A and G alone: AAG, AGA, GAG, and so on. A single A or a single G won't give you any variety at all. With a doublet, all you can get is AG or GA. It's not enough."

"Okay, you need triplets."

"Yes. And given this coding unit you can begin trying to figure out which triplet will code for which amino acid. Do that, and you're really in business. That is, if the *concept* of triplets is correct. I have to keep repeating that all these experiments proceeded from ideas that needed to be proved in some way. There was still no direct evidence that a given triplet coded for a given amino acid. That breakthrough was not made until 1961, by an American, Marshall Nirenberg and a German, Johann Matthaei. And they weren't even thinking about triplets. They were thinking about how to crack the coding problem by simplifying it. They hit on a most ingenious method."

Nirenberg and Matthaei did their experiment with a new laboratory technique that was just coming in; it has since facilitated immense contributions to medicine and genetics. It was the so-called "cell-free" system, a way of doing things in test tubes instead of within actual cells. Whole cells, they thought, were far too complicated to use; instead they decided to use only certain ingredients of cells: selected parts, separated, purified, and extracted. Particular bits of DNA and RNA, specific amino acids, ribosomes, other radioactively labeled particles—all these and more could be measured out like ingredients for a cake. They could then be mixed to order in a test tube by cordon-bleu cellular chefs to see if the amino acids were really stringing themselves into proteins, and if so, how they did it, and what sorts of other things were needed to make them do it—and, equally important, make them stop doing it. It was during the course of this work that the role of RNA as messenger was discovered, as was also the fact that ribosomes were involved.

The great advantage of the cell-free system was that extremely simple things could be used. Real RNA, as the example on page 262 shows, is an irritatingly irregular molecule. Its reactions are hard to monitor. Something simpler was needed, and something simpler was at hand: synthetic RNA, recently developed in the laboratory. There was U-U-U-U-U-U, for example, an endless chain of nothing but uracil. It was called poly-U.

What Nirenberg and Matthaei proposed was to put poly-U into the cell-free system instead of the more complicated real RNA, and see what came out. Logic told them that with one base going in one amino acid would come out.

Which is exactly what they got. They didn't get it right away. The experiment was dauntingly difficult and took many months. But what finally did come out of the test tube was a single amino acid: phenylalanine. And that amino acid made a protein composed of a continuous strip of nothing but phenylalanine. It wasn't really a protein at all. It didn't occur in nature. It made nothing, did nothing. But this made sense since it was produced by another unreal thing: synthetic U-U-U-U-U-U. Nirenberg and Matthaei had made their point, one of stunning importance; the triplet UUU instructed for phenylalanine. A connection had at last been made between the message in the nucleus and the operating material in the living organism.

"What you're saying, of course," said Don, "is that the UUU in RNA is actually the messenger for AAA in the DNA of the cell nucleus."

"Of course."

"I mean, it's the AAA message that gets carried. UUU is just the delivery truck."

"In that sense, yes."

"Well, let's nail that down. This is how you get the connection between DNA and a protein, how the message gets out."

"Yes, that's the important thing. That link is now forged."

Nirenberg and Matthaei wrote a paper that Nirenberg presented at a monster scientific congress held later that year in Moscow. He was completely unknown to the phage group, to the Cavendish people, their friends, and their friends' friends, all of whom were in Moscow, gossiping, exchanging views, reading papers of their own. He appeared on a long program at the end of a long day in a long week. Everyone was exhausted by the blizzard of good papers, bad papers, irrelevant papers. When it was Nirenberg's turn nearly everyone had left the hall, and he found himself speaking to a small and drowsy group. But one who did stay was Matthew Meselson. He was electrified. He realized instantly that the code had been cracked. The first word on a blueprint could now be read.

Nirenberg did not remain obscure long. Meselson immediately told Crick of the sensational kernel in the obscure scientist's ill-attended talk. Crick arranged for Nirenberg to give it again, this time to a much larger and more alert audience. He returned to the United States a marked man in the world of those interested in the ultra-small. It was as if he carried a radioactive tracer of his own. Back in his laboratory at the National Institutes of Health in Bethesda, Maryland, he refined his techniques and went on to identify the links between other amino acids and their RNA complements. He found himself in a scorching race for further identification with Severo Ochoa, a biochemist at New York University. Ochoa, incidentally, was the one who had discovered how to make poly-U and had received a Nobel Prize for that in 1959. Nirenberg got his a decade later for breaking the code. His partner Matthaei, apparently not comfortable in a climate of competitive cut and thrust, went back to Germany. Like Rosalind Franklin, he got only the respect of colleagues who knew his work.

As for the code itself, Nirenberg's work forced the decision that it was indeed in triplets. Although his breakthrough experiment had been with poly-U—and there was no way of telling whether only U or UUU was involved, since the entire thing was uniform from one end to the other—his later matchups were not of uniform bases, and they could be explained only by triplets. Triplets, in fact, began to spring up everywhere in the matching process. For example, it is now known that AUA makes isoleucine, but AUG makes something entirely different, methionine. As Nirenberg clearly showed, it takes three to tango.

The gap in the bridge between DNA and proteins—between blueprint and bicycle—was, as we hope we have made clear, a terrible impediment to an understanding of the genetic process, and, ultimately, to an understanding of the process of evolution. Filling that gap has required a long chapter that may seem to have wandered off in several directions. But each apparent digression has led finally to an answer, to a linkup.

The distinction between autocatalysis and heterocatalysis—DNA makes itself, it makes something else—is crucial. Scientists had to recognize this distinction, and then show that each process happened. In their elegant experiment with light and heavy strands of DNA, Meselson and Stahl proved that autocatalysis happened; DNA did make itself. Then, with his "hot" nucleotides, Arthur Kornberg demonstrated heterocatalysis, that there was a link between DNA and something outside, a protein. Finally, Nirenberg and Matthaei took Kornberg a step further. They broke the code by showing that a particular bit of DNA made a particular amino acid. Throughout these experiments the role of RNA became increasingly clear; it was the messenger that carried instructions from the central office to the workers in the bicycle factory. What an extraordinary story that is!

"That, and more," said Don.

"More?"

"Yes. We now know that DNA makes proteins, but how does it do that?"

"It does it on the ribosomes," I said. "We'll get to it."

15

The Triplet Code and the Ribosome. Crick Enunciates the Central Dogma.

The code was triplets. More and more research confirmed it. It also confirmed another idea that Crick had had. He had asked himself: "If we need only twenty amino acids to make all proteins, what are the other forty-four combinations for?" He hazarded that the code was "degenerate," that is, several triplets might code for the same amino acid. That turned out to be true. Here is the entire code as it was ultimately worked out:

UUU phenylalanine	UCU serine	UAU tyrosine	UGU cysteine
UUC phenylalanine	UCC serine	UAC tyrosine	UGC cysteine
UUA leucine	UCA serine	UAA nonsense	UGA nonsense
UUG leucine	UCG serine	UAG nonsense	UGG tryptophan
CUU leucine	CCU proline	CAU histidine	CGU arginine
CUC leucine	CCC proline	CAC histidine	CGC arginine
CUA leucine	CCA proline	CAA glutamine	CGA arginine
CUG leucine	CCG proline	CAG glutamine	CGG arginine
AUU isoleucine	ACU threonine	AAU asparagine	AGU serine
AUC isoleucine	ACC threonine	AAC asparagine	AGC serine
AUA isoleucine	ACA threonine	AAA lysine	AGA arginine
AUG methionine	ACG threonine	AAG lysine	AGG arginine
GUU valine	GCU alanine	GAU aspartic acid	GGU glycine
GUC valine	GCC alanine	GAC aspartic acid	GGC glycine
GUA valine	GCA alanine	GAA glutamic acid	GGA glycine
GUG valine	GCG alanine	GAG glutamic acid	GGG glycine

It will be noticed that some of the amino acids occur more frequently than others. Leucine, arginine, and serine occur six times; methionine and tryptophan only once. It can be assumed that the former are used more often in making proteins; that, too, turns out to be so.

Three triplets, or "codons," are labeled "nonsense"; they code for no amino acid. It turns out they are punctuation marks. They signify "stop," just as a period would in a sentence. After all, ordinary writing needs punctuation. Without it we would find things like this:

Don't go there is something the matter.

Those words can be punctuated in two ways with entirely different meanings:

Don't go. There is something the matter.
or
Don't go there. Is something the matter?

Also, where a message starts is critical:

SAM ASK DAN AND MEG AND ANN FOR THE BIG TOY BOX

Suppose there is a mistake in transcription and the triplets are all moved over by one space:

S AMA SKD ANA NDM EGA NDA NNF ORT HEB IGT OYB OX

Such a rearrangement makes no sense at all. A corresponding rearrangement of amino acids, that in their correct position make a useful protein, may in this different position, make nothing. The ribosome takes care of that problem. And it has a lot of other things to do; no wonder it is a complicated structure. First, it must recognize the RNA strip being assembled on the DNA master blueprint and attach itself to it. Second, it must make sure that it has hold of the "start" end of the message and has not got it wrong end to. That is not easy; a strip of mRNA may be a thousand or more nucleotides long. It may be twisted, folded back on itself, bunched in a tangle. It is the ribosome's job to see that the strip is straightened out and that the mRNA arrives and passes through, like a tape going past the magnetic heads of a tape recorder, in exactly the way that it is intended to. The larger part of the hamburger bun handles all that.

Meanwhile the smaller part of the bun is just as busy. It attaches itself to the larger part, bringing with it one of those tiny specks of soluble free RNA mentioned above and matches it to the mRNA strand that has been positioned on the bottom part of the ribosome. This third RNA is called transfer RNA, or tRNA.

Now for the real surprise. ***Each tRNA fragment brings with it an amino acid.*** The tRNA molecules are very short. Each specifies for a particular amino acid. Therefore there must be at least twenty different kinds of tRNA cruising about in the cell, each looking for its proper amino acid partner. As soon as it finds it, it is put in its proper place by the smaller half of the hamburger bun.

"This could be where the mirror-image problem gets straightened out," said Don.

"Yes."

"Let's go back to the example on page 262. There we had a DNA sequence of T-C-C-T-T-T-G-C-A. The mRNA mirror image of that was A-G-G-A-A-A-C-G-U. That's backward, right?"

"According to the original DNA code, yes."

"Now, if you make a mirror image of *that* with those bits of tRNA, you'll be back to T-C-C-T-T-T-G-C-A again. You'd have the original DNA pattern."

Don is correct. That is how the original blueprint instruction gets on the ribosome, with each codon in its proper order. That final reversing step may be seen taking place in a series of diagrams of the ribosome at work (pages 270 and 271).

When the second tRNA molecule has been brought in and positioned, the first releases its amino acid to the second. These two constitute the first two building blocks of a protein chain—*and they are in the correct order.* It sounds complicated. It is. The diagrams may make it easier to follow:

That ever-lengthening little column of squares coming out of the top of the hamburger bun is what it is all about. It is the payoff, the making of a protein. Despite the apparent complexity of the diagrams (and they would be more complex yet if the enzyme activity

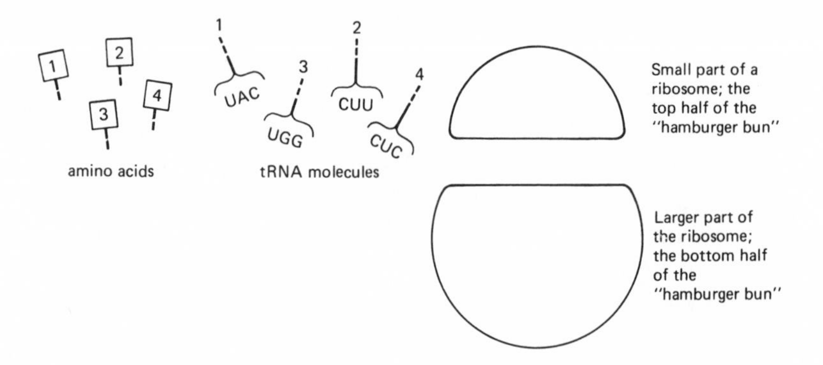

Stage I: All the elements of the process are loose in the cell.

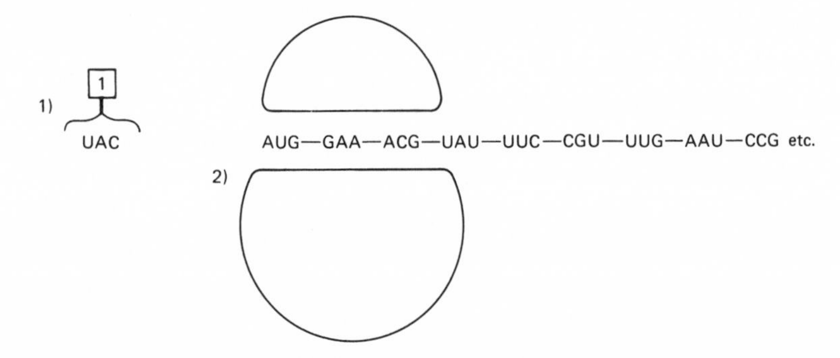

Stage II: 1) A tRNA molecule has succeeded in picking up its proper nucleic acid. 2) At the same time, a strand of mRNA, bearing information from the DNA, is positioned on the bottom half of the bun.

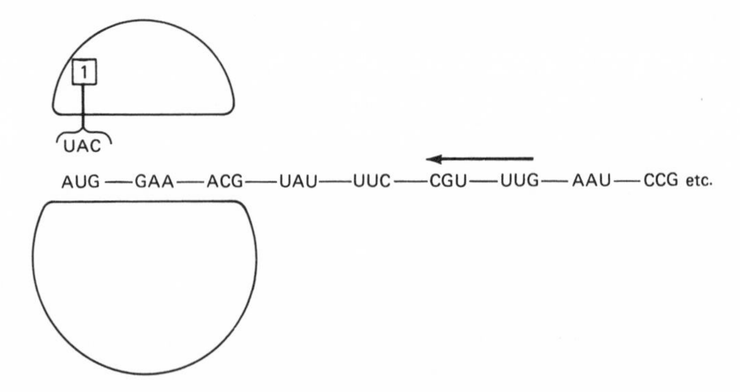

Stage III: The small part of the hamburger bun has joined the larger part, bringing the tRNA (and its associated amino acid) with it and matching it to the first triplet on the mRNA strip.

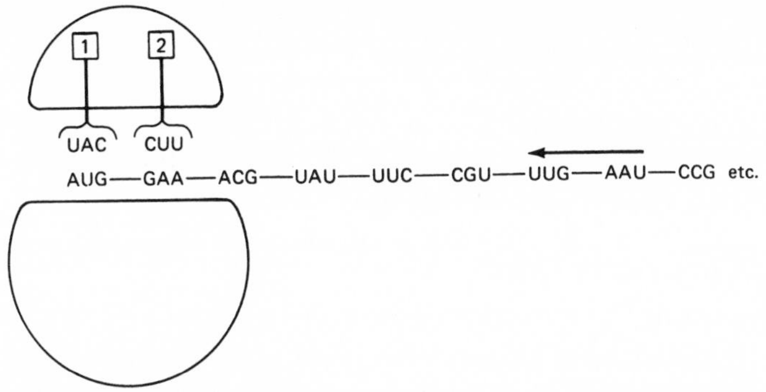

Stage IV: A second tRNA molecule, with its captured amino acid, is brought in. It matches *its* mRNA triplet.

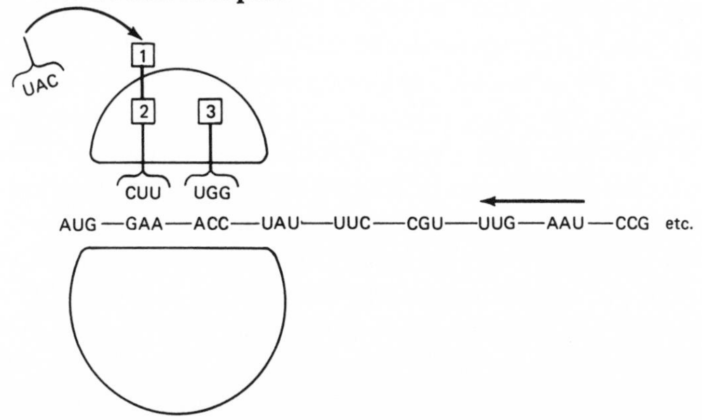

Stage V: The mRNA strip now moves one step to the left on the ribosome. The first tRNA molecule breaks loose and releases its amino acid, which attaches to the second amino acid. Meanwhile, a third tRNA and amino acid have moved into place.

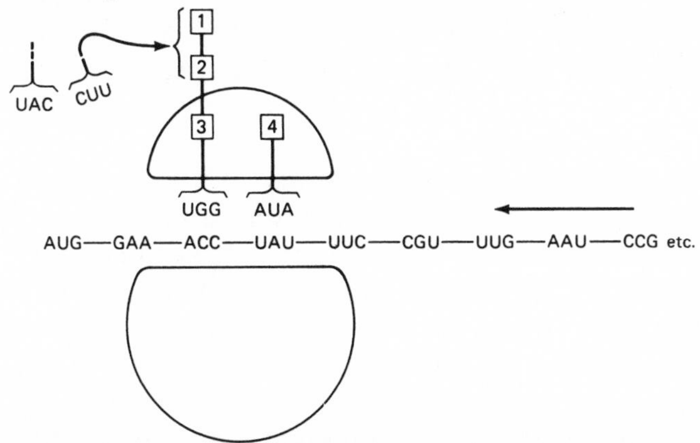

Stage VI: The mRNA strand moves another step to the left, and a fourth tRNA unit joins up. The second tRNA (CUU) now releases the *two* amino acids attached to it. They travel to the third. The released tRNA units (UAC and CUU) are now both free to seek out matching amino acids again and rejoin the chain at appropriate spots farther down the line. This process continues until the protein chain, growing longer with every amino acid that joins it, is completed.

that *causes* those things to happen were to be included), the process is essentially simple. It involves only two steps:

Transcription: The copying—or transcribing—of the nuclear message onto a strip of RNA.

Translation: The transformation of that message, via another strip of RNA, into a series of amino acids—and ultimately into a protein.

That is all. But it is a lot. It is a process that took the best minds in the biological sciences a decade to unravel. It is a truly overwhelming thing to contemplate: a long slender strand of crystalline particles, arranged in a certain order in the center of a cell, can transmit a message to a similar strand that carries that message to a part-workbench, part-power plant, elsewhere in the cell, where still a third strand of that generally crystalline material brings in *other substances* in the exact order necessary to create a bit of material that any biologist would recognize as being alive.

There lies one of the great questions of the universe: what to call life. Is it in the atoms themselves? It can't be. They can do nothing except attach and separate according to the unfailing laws of chemistry and electricity. Is it in atoms joined in larger, more complex configurations? Well—that depends. Is a piece of quartz alive? Again, it can't be. It can do nothing but sit there as an unchanging, unyielding amalgam of a couple of elements joining in regular crystalline form, molecule after molecule, for as long as the materials and the conditions for joining exist. Quartz cannot do anything else. It cannot add other things to itself. It cannot turn corners. All it can do is break apart. If the forces for breaking are present, its elements drift away to be constituted into other similarly "unlifelike" combinations somewhere else.

DNA is crystalline, too. That is to say, it has a regular pattern of atoms that repeat themselves. That they are arranged in a coil makes the crystal a rather complex one; that the bases occur in a random sequence makes it more so. But it is crystalline. If it were not, crystallographers could not make X-ray diffraction photographs of it. The X-ray beams would produce an overall gray blur signifying nothing.

So what makes this crystal different from quartz? Biochemists would say that the difference is that it contains carbon. Anything containing carbon, according to them, is an "organic molecule." That is not to say that it is alive. After all, carbon dioxide, the gas we exhale, is an organic molecule, and it is certainly not alive by any of the standards of life we have long come to accept. But carbon, whether with or without oxygen, attaches itself very easily to a wide range of other elements, notably hydrogen and nitrogen. Chemists call it the friendly atom. And when those larger associations—with carbon—begin to come into being, then the first stirrings of life may be detected. Other things drift by, bits of phosphorus that act as energy sources to help bind those larger organizations together, metals like magnesium that speed up the ability to grab or discard. Bit by bit, as if by magic, raw chemicals begin to behave as raw chemicals should not. Repeating what we said a few pages back, that is what life must be.

From the breakthrough that established the relationship between DNA, RNA, and protein came two matters of profound importance to evolution.

First was what Francis Crick enunciated as the "Central Dogma." It stated that genetic information traveled in one direction only: DNA → RNA → protein. It could not reverse itself. There was no way that protein, working its way back up through ribosomes and RNA, could have any effect on DNA. The central factory office was indeed closed, bolted, as impregnable as a bank vault. It issued blueprints; it got nothing back. Molecular biology had at last made that plain and had at last driven a stake through the heart of Jean Baptiste Lamarck.

It is interesting that Lamarckism lingered on in small pockets of dispute right up into the 1960s. It was flourishing in Russia at the time of the great biological conference in 1961. Nirenberg's paper, which for the first time established a clear connection between an RNA base and an amino acid, must have seemed like an avenging ball of flame almost too bright to be stared at directly by many of the assembled Soviet scientists. They had dutifully gone along with the official party doctrine that conventional Darwinian theory was bad. It had to be. It ran counter to the Marxist-Leninist-Stalinist theory that a true socialist state's population could be molded—that the giraffe's neck could be permanently lengthened through inherited effort, and not through inheritance of a different kind.

That idea had been given some semblance of scientific credibility by a Soviet plant geneticist, Trofim Lysenko, who had managed to win the confidence of Stalin by falsely representing some experimental work he had done in developing strains of cold-hardy wheat.* He convinced Stalin that Soviet genetics was superior to Mendelian genetics, that genes and chromosomes and all the experimental work done by people like Morgan to establish the existence of those things were false. Since Russia had its share of able scientists doing good work, Lysenko found himself in dispute with them. Things came to a head in 1948 at a large Party meeting at which Lysenko, with the approval of Stalin, forced the Soviet scientific community to buy Lysenkoism—or else. The "or else" was at best professional extinction, at worst imprisonment and death. In 1961 Lysenko was still the most politically influential scientific man in the Soviet Union. He was president of the All-Union Academy of Science, and in that year approved this statement from another highly placed party-line scientist: "The assertion that there are in an organism some minute particles, genes, responsible for the transmission of hereditary traits is pure fantasy without any basis in science."

Lysenko's iron thumb brought the respectable study of genetics to a standstill in the Soviet Union for decades. His fraudulent experiments with wheat that would somehow evolve wonderfully and independently of Mendelian principles had an effect on Russian crop yields that is still being felt in the huge stocks of grain that the Soviet Union is obliged to buy from the United States. Lysenko fell finally with the fall of Khrushchev in 1964. The full bill for his mischief will never be properly rendered.

But before we look too contemptuously at those doctrine-bound Russians and at the follies of Marxist-Leninist-Stalinist science, let us remember that here in this country there is a fundamentalist "Creation" science that is just as scientifically untenable and—in the long term—just as destructive. It is worth saying again that the only proper science is scientific science. Marxist science and Creation science have about as much validity as Bulgarian science, or April science, or Forty-Second Street Science.† Russian scientists today

* What he had actually done was to freeze some seeds, thereby promoting more rapid germination in the spring. This process is called vernalization. It has nothing to do with the evolution of wheat.

† This point had already been made, ironically by a Russian, Anton Chekhov, who wrote: "There is no national science, just as there is no national multiplication table. What is national is no longer a science."

acknowledge that. They, too, have buried Lamarck. The Central Dogma is generally accepted worldwide—except by Creationists.

The second fallout from the solution to the DNA → RNA → protein puzzle was the realization that this was merely the first chapter in a longer story. Almost all the work that resulted in breaking the code and establishing the one-way nature of the Central Dogma had been done in cell-free experiments or with bacteria. It soon became obvious that the genetic workings of higher life forms were much less easily understandable. A maple tree or a gorilla has an infinitely more elaborate structure and life cycle than a one-celled *E. coli* bacterium—the one that inhabits the human intestine and which has served as the guinea pig for genetic experiments for several decades. A list of the ingredients, large and small, that make up a maple tree—from root hairs to bark to hormones of various kinds to seed coats to elaborate sexual organs, each kindled by its own genetic instruction, each represented by its own specific protein or complex of many proteins—would fill pages of this book. For a gorilla, a similar list would fill a volume.

To take care of that order of complexity, a corresponding complexity in chromosomes should be expected. And that is exactly what we find. A gorilla has forty-eight chromosomes, compared to *E. coli*'s one. And they are much larger, permitting the storage of a far greater amount of information: hundreds of billions of times as much. For a gorilla is not merely a collection of different kinds of tissues. Those tissues are assembled into elaborate organs like eyes and kidneys. Not only must they be made of the right kind of stuff but they must also work. They must be the right shape, the right size, organized into the right patterns, and come on-stream at the right time. Each must start growing and functioning at the right moment in the development of the gorilla embryo. There is a numbingly intricate series of meshed steps that must take place in precisely the right order in any higher organism if it is to be born at all.

That orchestration does not stop at birth. A gorilla must continue to grow according to its own master blueprint, and not any old which way. Fingers must be three inches long, not twenty-three inches. A leg, with its truly marvelous assemblage of bones, muscles, tendons, nerves, blood vessels, flesh, and skin, must become a leg and not an arm. And it must be a gorilla leg, not a human leg, with gorilla toes, and on the toes, gorilla toenails. When that leg is formed, it must remain a leg, growing in concord with the rest of the body, neither too little nor too much. When the time comes for growth to stop, it must stop.

All the while the gorilla must survive. It must breathe; its heart must beat; it must eat and sleep. Many of its life activities are taken care of automatically. Others it must slowly learn—with a brain that must be programmed to grow also and to store the information that the gorilla's life experiences pour into it. Every one of these things must ultimately be traced to the gorilla's genes. Do we know how this flood of genetic information is read, which genes say "lung" and which say "breathe"? We do not. About those things we know virtually nothing.

But we do know, at least, what the directing process is. From that, and from a couple of decades of intensive analysis, we have learned a great deal about *E. coli.* Its single chromosome has been mapped almost completely as to the location and function of its different genes.*

A genetic map made today for a gorilla or a human would show a Sahara of emptiness: thousands of genes, millions of triplets whose meaning is utterly incomprehensible. Human chromosomes have long repetitive sequences of base pairs that remind one of stuck keys on a teleprinter; they seem to make no sense at all. They are called "junk" genes. The question: are they really junk or is their function not understood?

Some other questions from this spaghetti bowl of ignorance for which answers are only just beginning to emerge: where does a specific gene start? Where does it stop? How many base pairs does a gene consist of? What is its relationship to genes next to it? Do they affect each other? If so, how? What about genes a little farther apart, or even very far apart? Does it take one gene to accomplish something or does it take dozens? What turns a gene on? What turns it off? Do genes regulate genes? Are those regulator genes regulated by yet other genes?

All those questions become relevant as the crushing complexity of higher organisms becomes apparent. It takes a mass of instructions to assemble the materials for as simple a thing as a single cotter pin, to make the stamping machine that will shape it, to position the

* A more complex task, determining the exact nucleotide sequences for the entire bacterium, is just now getting under way.

August Weismann. *In his speculations about how a fish always grew into a fish and never a frog, a hand into a hand and never a foot, Weismann came to the conclusion that the chromosomes were responsible; they were the regulators of every part of every living organism.*

Oswald Avery, *an elfin biologist from the Rockefeller Institute in New York, devoted years to breaking down the chromosome to discover what material in it was responsible for heredity and change. He wound up with DNA; it proved to be the controlling agent.*

***T. H. Morgan,** shown here in his cluttered laboratory, the Fly Room at Columbia University, proved, despite his own earlier doubts, that genes existed. He did it by finding mutations in fruit flies and then tracing their descent through innumerable crossbreeding experiments.*

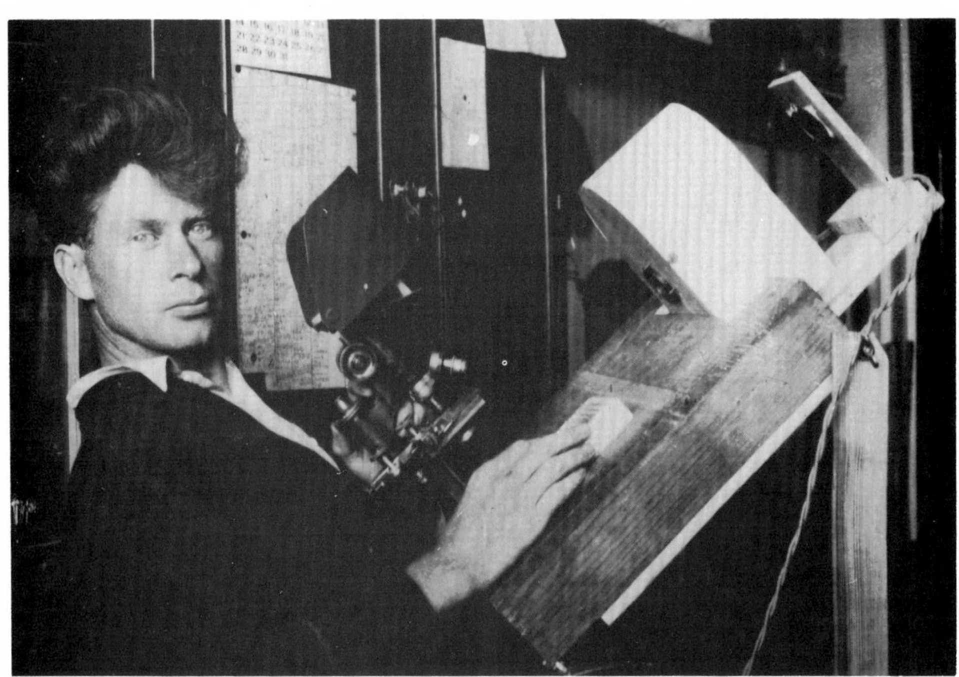

***Calvin Bridges** at his work station in the Fly Room. He was the first to find a mutation in a fruit fly. "Labeled" flies could then be followed in later generations and their effects traced to specific short sections (genes) on the much longer chromosomes.*

James Watson, *who with Francis Crick worked out the double-helix form in which DNA occurs in the cell nucleus, is shown here with a model of it. That shape made possible their contention that DNA can not only make itself, it can also make proteins.*

***Stanley Miller**, using this elaborate array of laboratory equipment, was the first to demonstrate that a mixture of gases such as were believed present in the early earth atmosphere, could, when energized by electricity, produce organic molecules.*

***Carl Woese** has been analyzing RNA in bacteria to learn their evolutionary history and what their long-vanished ancestors may have been like. On the way he has learned that bacteria themselves come in two forms, as different from each other as from all higher life.*

Robert Broom, *an itinerant South African physician and fossil expert, took up the cause of the Taung Child when it was mistakenly considered to be a chimpanzee, and helped establish australopithecines as upright-walking hominids by finding other specimens himself.*

Louis Leakey *gave the australopithecine cause another huge boost when he discovered his celebrated "Zinj" skull* (right) *in the Olduvai Gorge in Tanzania. New dating methods made it possible to give Zinj an age of nearly 2 million years, making hominids far older than previously thought.*

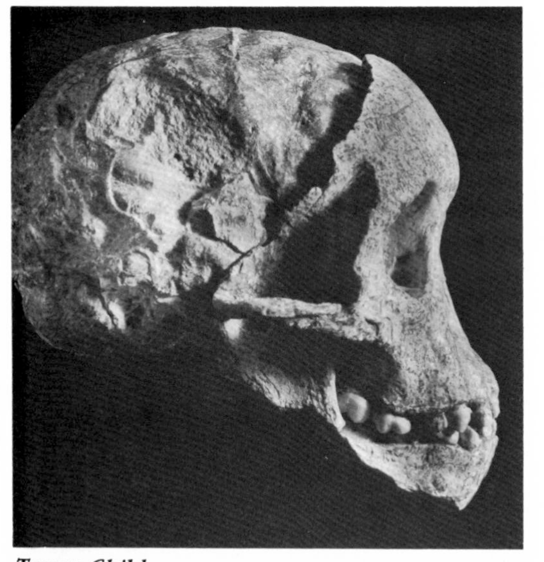

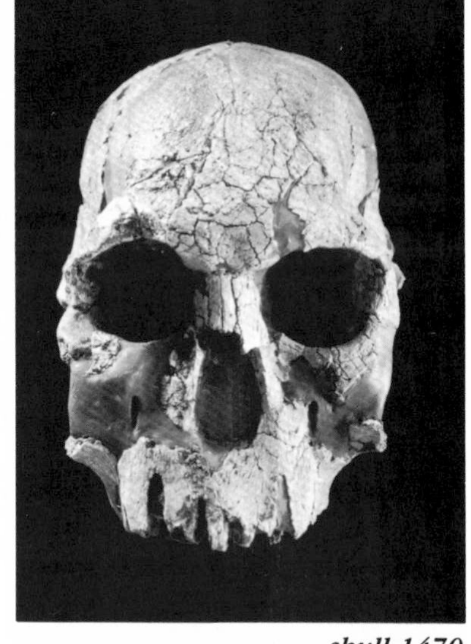

Taung Child

skull 1470

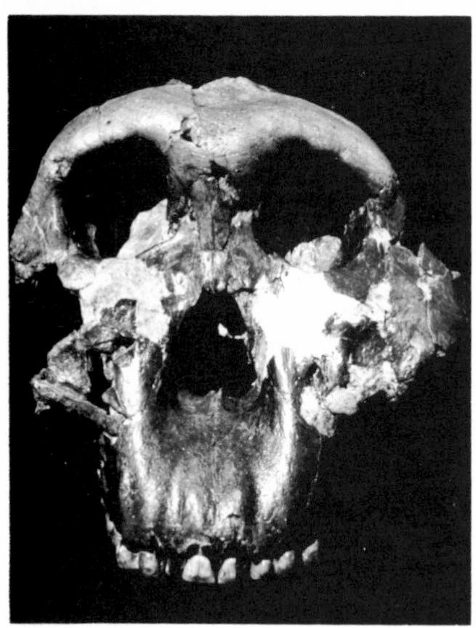

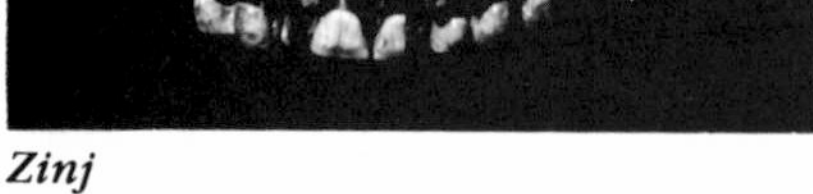

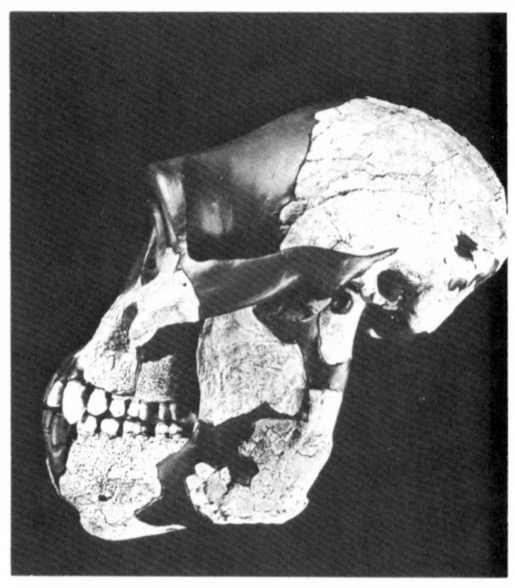

Zinj

Lucy type

Four famous early hominids. *Taung Child* (Australopithecus africanus), *2+ million years; skull 1470 (a human ancestor,* Homo habilis*), 1.7 million years (preliminary reconstruction); Zinj* (A. boisei), *1.8 million years; Lucy type (the oldest known hominid,* A. afarensis*), about 3.5 million years.*

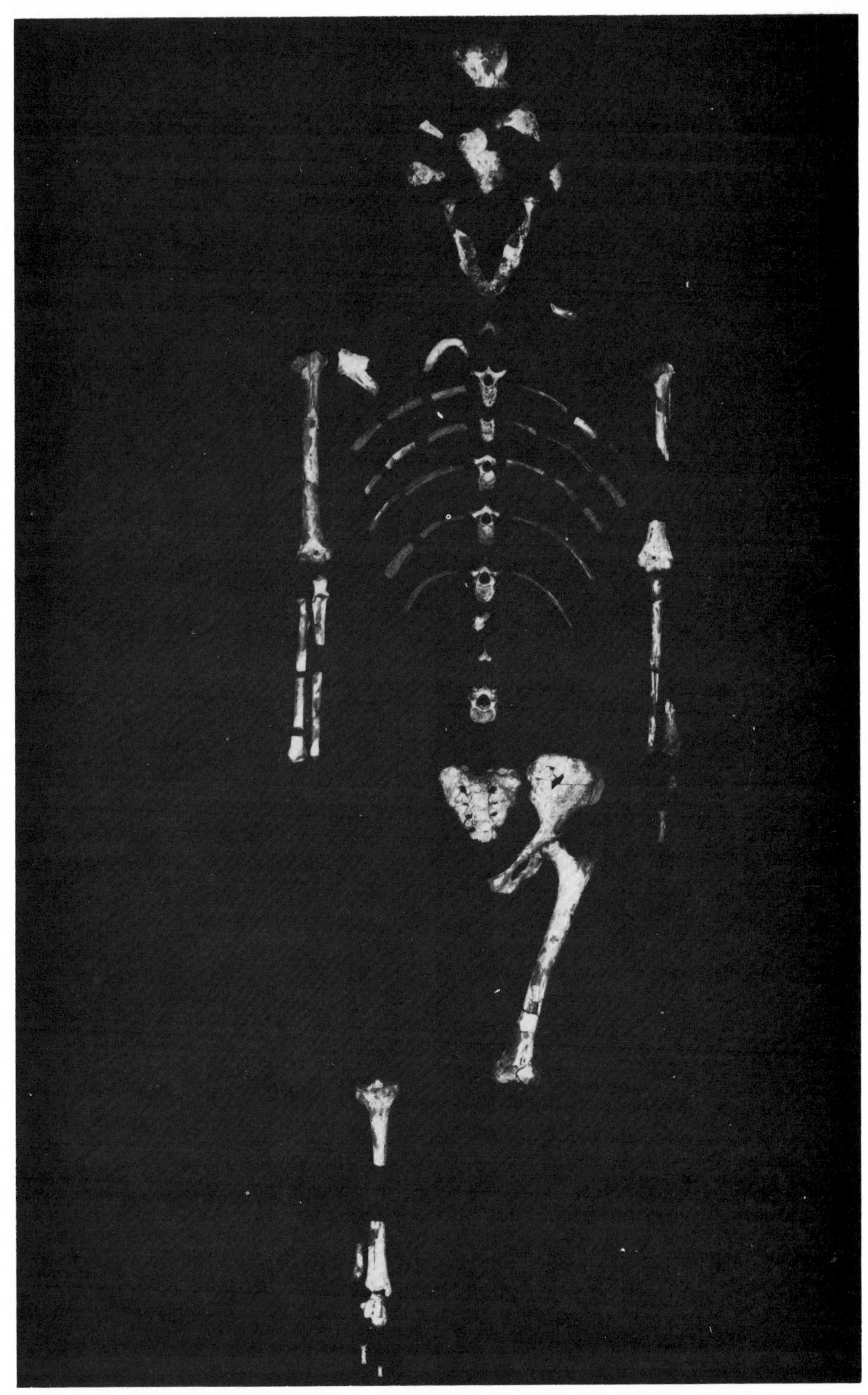

***Lucy**, unearthed by Don Johanson in the Afar desert of Ethiopia in 1974, is the most nearly complete specimen of our oldest known hominid ancestor. At death she was about twenty years old, and 3.5 feet tall; she lived about 3 million years ago.*

***Tim White** helped Don Johanson figure out that the Hadar fossils belonged to a new and ancestral australopithecine species. He is seen here at the famous Laetoli site, having persuaded Mary Leakey that the footprints there were hominid and worth preserving.*

***Vincent Sarich** works in this California-Berkeley laboratory, surrounded by the equipment that he used to compare the blood proteins of ape and human, proving to skeptical fossil experts that the twc had a common ancestor at 5–6 million, and not 15–20 million years.*

metal, to stamp out the pin, to store it on a shelf, and then take it off the shelf just as the axle comes down the assembly line and the wheel is put on. We could say that the juxtaposition of wheel and axle is the trigger that stimulates the presentation of the cotter pin at just the moment that it is needed. We could say that a cotter-pin-delivery gene has been activated, which triggers off an insertion-in-the-hole gene — the hole itself, of course, having been bored in the axle on instructions of still another gene that is part of a whole axle-producing complex of instructions being carried out in another part of the factory. Two entire instruction manuals must coordinate exactly. As a final fillip, when the cotter pin is actually in the hole bored for it, there must be an instruction to bend its ends so that it won't fall out and the wheel drop off.

In these terms the human factory becomes not one assembly line but thousands of them, each sensitive to the needs of others, each programmed to turn on at the proper signal and turn off again at another signal. Tracing this bewildering interaction of genes to construct a complete blueprint for the human genetic program may seem like a hopeless task. We do not yet know how a leg becomes a leg or an ear an ear. But there is no reason to say that some day we will not know, for we have the method. The way *E. coli* uses its simple blueprint is no different from the way human beings use their complicated ones.

Therein lies the second potent endorsement of evolution theory that the DNA → RNA → protein breakthrough provides. The men who made the breakthrough were not thinking particularly about Darwin as they did their work, but he could not help but come to mind as the interior of the cell revealed a wonderful unity in all living things; all use the same four bases to make the same twenty amino acids. The triplet code is the same. The ribosomal devices for assembling amino acids into proteins are essentially the same. The only thing that is different is the message itself. It is the message that makes a bacterium or a man; the process in each case is identical.

That unity speaks with a thundering voice for the validity of evolution theory. We all appear to come from the same source and in the same way. Indeed, a surprising amount of all DNA is the same, suggesting that it has the same remote origin and represents truly basic life-organizing principles that are common to all living things. More than that, it is interchangeable. A bit of bacterial DNA, if it were found to be useful, could be put into a human being to do the work that the human's own DNA — perhaps because of a bad mutation — could not do. *E. coli,* meet *H. sapiens.* You both carry within you many of each other's genes. With the right needles and threads they can even be swapped.

Looking at life from the molecular level, a common ancestor to bacterium and man no longer seems bizarre. In fact, it is quite easy to see ourselves as descended from the earliest known life forms, those that existed on earth more than three billion years ago. What comes as a smashing surprise is the realization that parts of those early organisms may continue to live in us. Our DNA may have changed a great deal, it may have grown immensely longer and more complicated, undergone countless mutations, but some of it has not changed. Thus it has never disappeared. It is here in us right now. We are so accustomed to thinking of our bodies as discrete hunks of protoplasm that begin in conception and end in death that we tend to forget the small coil at the center of each reproductive cell that does not die. It connects one generation to another as long as life is life. Somewhere in all of us, there may lurk bits of original DNA that actually are more than three billion years old.

VI

The Origin of Life

In the physical world one cannot increase the size or quantity of anything without changing the quality.

—PAUL VALÉRY

It is often said that all the conditions for the first production of a living organism are now present, which could ever have been present. But if (and oh what a big if) we could conceive in some warm little pond, with all sorts of ammonia and phosphoric salts, light, heat, electricity, etc., present, that a protein compound was chemically formed ready to undergo still more complex changes, at the present day such matter would be instantly devoured, or absorbed, which would not have been the case before living creatures were formed.

—CHARLES DARWIN

Science has proof without certainty. Creationism has certainty without proof.

—ASHLEY MONTAGU

16

Stanley L. Miller and Manfred Eigen: A Look from the Bottom Up.

In a book about evolution one is apt to get sidetracked. The subject is so large and extends now to so many different and apparently unrelated scientific disciplines that it is sometimes hard to keep one's eye on the main path. Thus, for a good many chapters, this book has had to wander off into what may appear to be a complicated—but what is actually a painfully simplified—examination of the cell and the function of its most intimate parts. That has been necessary, as we hope we have made clear. Modern evolution theory turns on what goes on inside cells. Elucidation of that has had to come, step by step, out of the work of more than a hundred years. Now we know something about how cells replicate. Now we have the truly startling insight that all cells do it in very much the same way. Now we can no longer avoid the question: how did they begin doing it?

We said at the end of the previous chapter that, with our present knowledge of the DNA → RNA → protein process it becomes easy to see ourselves and all other living things, as descended from the earliest-known life forms. What is not so easy to see is how those forms came into being. A cool examination of the process, as it operates today throughout all life, forces the observation that it is a remarkably clever, but at the same time most curious, way of creating living organisms. If an unusually imaginative and informed biochemist were to start with a bunch of raw chemicals and try to invent a way for them to organize themselves into protoplasm, would it ever cross that scientist's mind that this way was *the* way?

Not likely. How life's blueprint is assembled and then turned into tissue is one of the most startling and bizarre processes that one could possibly dream of. How on earth did it get going? If the full story of evolution is ever to be told, that question must be answered —or at least a good attempt must be made to answer it.

In the history of evolutionary studies, the tried-and-true method for looking into the past has been, for more than two centuries, a search for objects that say something about that past. Indeed, suspicion that there was a past was fueled originally by the discovery of old strata and strange bones. Fossils were the building blocks with which evolution's first timid and flawed structures were built. So, if we are to look for clues to the beginnings of life, should we not seek them in the earliest known life forms?

Easier said than done. We can learn a good deal about dinosaurs by reassembling their bones. We can even learn something about ancient worms and jellyfish that have left their imprints in mud that has turned to rock. But when we get to the lowest level, the level of bacteria, we strike out. Fossil bacteria do exist, but their shapes are so simple and so monotonous that we can learn nothing from them. When we try to look inside them we draw a blank. The interior parts of bacteria do not fossilize any better than the hearts, the lungs, and the blood vessels of higher animals. So far, no scientist has managed to extract any information from the interior of a fossilized bacterium. All that remains is its microscopic boxlike cell wall, and that tells us nothing.

That is very distressing, because all our knowledge of evolutionary processes, all our observations of how simple forms give rise to more complex ones, all the evidence of the rocks, tells the same story: older is simpler, and the oldest of all is the simplest of all. So, if we are to go looking for our earliest ancestors, it follows that we will probably find them in bacteria. But if we cannot get a handle on what ancestral bacteria were like, how will we ever learn what lies back of them?

There are two ways of attempting an answer. One is to work from the bottom up, so to speak, by trying to recreate the conditions that are believed to have existed on the surface of the earth before there was any life there, and then to devise a logical scenario for the steps that might have produced life.

The second way is from the top down: to concentrate on the molecules of living bacteria and try to work out their relationships —and most importantly their ancestry—by comparing those molecules and making a laborious catalog of their differences. Since those differences represent evolutionary change, vague lineages and ancestral ties begin to swim into view as the catalog of comparisons and differences grows. This is new work of excruciating difficulty. It will be dealt with in the next chapter. Right now, we would like to examine the problem from the bottom up.

How did life start?

That question actually has three parts. First we must make the assumption that it did start, that there was a time during the early development of the earth when there was no life here at all. Then we must examine the conditions that are thought to have existed at that time to see if the materials needed—the chemicals that make up all living things—were lying about, handy for use. Finally we must produce a model that gives a reasonable explanation of how simple, lifeless compounds could have been organized into far more complex "living" ones. We put the word "living" in quotes because, if we could go back to see that process actually taking place, there would almost certainly be a long period of time during which we could not decide whether those emerging collections of molecules were alive or not.

The first two parts of the question are much easier to answer than the third. Taking the very first, there is strong evidence that there was no life here at the beginning of the earth's history. That history, according to most cosmologists, started about 4.5 billion years ago. What things were like on the surface of the earth then can only be guessed at, because nothing recognizable from that turbulent time remains. But during the next half-billion years, as the presumably hot surface cooled and large shields of rock began to consolidate themselves, the story of observable geological events began. Modern dating methods, based on the decay rate of radioactive elements in those ancient rock shields, tell us that they are about four billion years old, the oldest known things on the planet. None of them show any evidence of having contained the slightest sign of life at that early time, even when subjected to the most painstaking microscopic examinations.

But, turn to Australia, and to a 3.5-billion-year-old rock shield there, and a dramatically different picture emerges. The Australian rock is teeming with the fossil remains of primitive bacteria. On that evidence we can assume that life on earth emerged some time between 4.0 and 3.5 billion years ago.

The second part of the question also has a fairly straightforward answer. All the elements that make up the earth today were there from the beginning. They were swept up from space as the solar system was formed, the best guess being that much of that material was the debris from unstable large stars that had exploded elsewhere. It is only in those giant furnaces that the elements that are here now — everything from light things like carbon to heavy things like iron — can be created.

Were those things readily available? Almost certainly yes. In what proportions or in what combinations they lay about cannot now be exactly determined. But the logical assumption is that what is here now was here then, variously distributed: a surface of bare rock and gravel, an atmosphere of mixed gases, and the seas — wherever they were — containing a broth of dissolved minerals.

That, in summary, was the picture drawn by Harold C. Urey, a chemist who had won a Nobel Prize in 1934 for his discovery of heavy hydrogen and had gone on to theorize about the origin of the sun and planets. He had also worked out the probable mix of gases in the primitive earth atmosphere and had begun to speculate how, in such an environment, life might have appeared. For anything to have happened, he realized, there would have to have been some energy input. He reasoned that sunlight or lightning might have provided that. These were the intellectual ingredients that Urey — or anyone at the time — would have needed to take a crack at the big question.

In 1950 Urey gave a course on the origin of the solar system at the University of Chicago. At one of his lectures he suggested that an interesting experiment would be to create in the laboratory an environment like the early earth's, add some energy, and see what happened. One of his graduate students, Stanley L. Miller, picked up the idea, and gained instant scientific renown. We can see him now, in photographs taken at the time, a slender young man with glasses, wearing a white lab coat and standing next to the apparatus he built: two glass flasks connected by glass pipes. A routine picture, surely, of a young scientist at work, but for those who knew what Miller was doing, one that carried a terrific wallop. For what he was doing was trying to take the materials of a lifeless world and from them create the building blocks of life.

Miller's first step was to reproduce Urey's proposed early atmosphere. Into his laboratory apparatus he put a mixture of water, hydrogen, ammonia, and methane. Under one of the glass flasks he put a Bunsen burner to boil the water and push its vapor, along with other gases, through to the second flask. There they got their "lightning" in the form of a healthy 60,000-volt electrical spark. So energized, the gases circulated round and round through one flask and then the other, each time being jolted by the sparking electrode. On each trip they passed by a cooling device and a small trap meant to catch any larger molecules whose construction the electricity might have stimulated. At the end of a week Miller cooled off his apparatus and analyzed the contents of the trap.

As he had hoped, the trap contained more than the gases he had started with. The electricity had helped bind certain molecules together in new and larger configurations. Among them were four amino acids.

Twenty different amino acids, it will be recalled, are the bricks that make organic skyscrapers. They are the subunits of the infinitely larger and more complicated molecules that are proteins. Proteins, Miller did not have to be told, are what constitutes living tissue. When in further experiments he found more amino acids; when others duplicated his work and found even more; when they found that one didn't need electricity to trigger the reactions in the apparatus, that one could do it with ultraviolet light as a substitute for sunlight; when they began finding short strings of nucleotides, the building blocks of DNA and RNA, there was a rush to proclaim that life had been—or at least could be—produced in a test tube.

Not quite. It is a long trip from bricks to skyscrapers, from amino acids to proteins. To build a proper protein there must be a proper agency to bind thousands of amino acids into a precise chain and then hold them together. In Miller's broth there was nothing to do that. Nothing—unless those small scraps of RNA could do it.

That scarcely seemed likely. RNA, as noted, is the messenger that carries and transcribes the instructions coded on much longer strands of DNA which exists only in a more complex double helix. Without any DNA at all in Miller's soup, with only a few nucleotides that might join chemically to form a very short strand of RNA, the problem of instruction becomes acute; the strand isn't long enough to say anything.

Returning to the model of the bicycle works, it would be like having to function with only a tiny fragment of one blueprint, something so small that all a worker could read from it might be "pai—." With a somewhat larger blueprint it might be possible to complete that work to "paint," but still the factory worker would be left wondering what to paint, what color to use, even where to find a brush to do the job. That is why the DNA strands in living organisms are so long, millions of nucleotides long. They have a great many things to say.

Put bluntly, the enthusiasm of the press at the time Miller's experiment was published was overblown. He had discovered something important; all science agreed to that. But it was not life in a test tube. All he had succeeded in proving was that small bits of RNA and a few amino acids could be created in an otherwise sterile primordial sea. He hadn't proved that they could endure. He hadn't proved that they could join in other ways to make more complicated and more meaningful molecules. He certainly hadn't proved that information could be stored and kept—that a meaningful molecule, if it did manage to get assembled, might be assembled correctly again. On the contrary, one of the most disturbing aspects of the problem was that in the primordial sea there undoubtedly were bewildering numbers of reactions taking place, under the influence of sunlight and in obedience to the laws of chemistry. Over and over again, molecules joining and breaking apart, a mindless churning with nothing gained.

Unless—unless the chemical bonds that held some units together were stronger than those holding other units together. That would be selection of a sort. Since we know today that certain chemical bonds *are* stronger than others, it is reasonable to assume that, over time, certain molecules would outnumber others by a wide margin.

It was such an assumption that caught the mind of a German chemist, Manfred Eigen. It occurred to him that one fragment of RNA, haphazard as its birth might have been, might endure better than another. From that first small assumption he and three other scientists—William Gardiner, Peter Schuster, and Ruthild Winkler-Oswatisch—worked out a scenario for the development of life. They explained it in a landmark article in the *Scientific American* in 1981.

They conceded at the outset that a clear tracking-back with fossils, as has been done over and over again to trace the descent of many higher forms of life, was impossible. "The fossil record, as far as is known, decayed or was wiped clean by later generations of life," they wrote. In place of fossils they proposed a set of rules that could explain, given a jumble of chemically *possible* combinations in a lifeless sea, the persistence of certain ones that were not only possible but *right* for life.

They began their cogitations with RNA, knowing (from Miller's experiments and from others that they had done themselves) that RNA could be created in a lifeless environment. They made the assumption that the earliest form of genetic material was RNA and not DNA. While DNA is the mediator of all forms of higher life, it has a more complex structure than RNA and depends on RNA for its functions. Clearly RNA had to precede it.

The trouble with "out-of-the-blue" RNA was its shortness; it was never more than a dozen or so nucleotides long, and averaged even fewer. Eigen (from here on his name will also represent his associates) could assume this shortness as a given because no laboratory-made RNA was ever any bigger, and there were good chemical reasons why it could not be. This forced him to face the aforementioned dilemma: such short fragments of RNA are too small to contain useful information for making complex enzymes and other proteins. Somehow those primordial blueprints were going to have to get larger.

Eigen was aware of ways to get somewhat larger blueprints. The presence of metals in the soup would help. It was known, for example, that lead was useful. Add it to the mix and RNA chains of up to 40 or 50 units are possible. Zinc is even better; it helps hold together chains of up to 150 units. Better, but still not good enough to code for the kind of enzymes that are associated with true life processes. One such enzyme that Eigen knew about, an extremely simple one from a virus that made RNA, contained more than 4,000 nucleotides.

The size dilemma remained acute: RNA too short to instruct for making the large complicated enzymes that are required to make larger RNA. A classic chicken-and-egg situation. Since each needed the other, which came first, the chicken (RNA) or the egg (enzyme)?

A tough question, and one that cannot be answered until one knows how to get a metaphorical "chicken." How, indeed, in that crazy jumble of raw chemicals, does one get something even remotely like one? Eigen had an answer of sorts to that question. He knew that short strands of RNA could and did appear *de novo* — that is, all on their own — in a chemical soup. They didn't quite qualify as "chickens"; they were too small and simplified. They needed time and nurturing and large additions to themselves before they could be regarded as reliable egg-laying reproducers. And again, one would need a proper "egg" to hatch out a chicken like that, but one didn't have such an egg either.

What Eigen did have was those short strands of RNA and also some amino acids. The former could be called protochickens and the latter protoeggs. His conclusion was that neither came first. They came together. Both were present in the primordial soup and they helped create each other in a kind of bootstrap operation.

Taking the problem a step at a time: how does one go about making a start on the "right" protochicken? "Right," of course, would be the RNA sequence best able to hold itself together, duplicate itself, grow.

From the point of view of a protochicken struggling to form itself and hang together amid all the other chemical jostling going on around it, "right" could simply be sticky, the tendency of some chemical links to be stronger than others. Since some of the ingredients that make up RNA (the four bases G, C, A, and U) do stick together better than others, a chance combination of those could—even long before the start of life—have an evolutionary advantage. In the random process of molecules' getting stuck and unstuck, natural selection would already be at work, operating at the level of chemical laws. Strips of RNA made primarily of G–C combinations, which are stronger than A–U, would tend to win. In other words, Eigen guessed, competition among molecules existed from the start. It is odd to speak of things like "selection," "competition," and "success" taking place among inert chemicals. It is even odder to think of the immensely intricate biological processes that control our bodies today as the end product of a tendency for two nucleotides to stick together better than two others. But it was from that simple beginning that Eigen found a logical trail leading from non-life to life.

As we have noted, Eigen knew that chains of up to 150 nucleotides could occur naturally with the right ingredients present. From other experiments he learned an important fact: natural selection was at work already at this level. In a particular soup a particular sequence of nucleotides would tend to emerge. Not always the *identical* sequence but an almost identical one. What showed up in those experiments was a group of similar sequences like a population of people: individuals who were slightly different from one another, but all of them definitely people. In the same way the RNA sequences fell together in a definite group. Eigen named this grouping of distinctive nucleotide sequences a quasi species, and set about thinking how it might improve itself.

The problem was that even among sticky sequences of RNA, there was a tendency to break up, particularly if the sequences were more than 150 nucleotides long. However, at that upper level of size there existed enough information in the sequence to code for very simple protoenzymes. That is, it *might* be possible to code for a useful enzyme if the sequence in the RNA didn't change. The trouble was that it often did change. It lacked stability. And the longer the sequence got, the greater the chance for accidental change in the order of nucleotides in it. Therefore, RNA not only had to be able to grow larger, but it had to be stable in order to preserve itself.

At that point Eigen turned to the experiment that had revealed the existence of quasi species in RNA. Suppose, he said, that one member of the quasi species—call it "A"—coded for an enzyme that stabilized and encouraged another member—call that one "B." Suppose, further, that B coded for an enzyme that helped A. They would stabilize each other in a mutually reinforcing feedback loop.

Feedback loops are common. Anybody listening to a badly rigged microphone in an auditorium has experienced one. Distortions introduced into the circuit feed on themselves to the point at which they become audible and the loudspeakers begin to emit a scream. Another example is seen at a cocktail party where a man finds himself attracted to a woman who is attracted to him. If they allow their behavior to betray how they feel, something will begin to build. He makes eye contact. She senses his interest, and the spark in her glows brighter. It shows in her face as she smiles at him. That signal further encourages him—and so on, up to a point where they find themselves seriously involved and ultimately, perhaps, in a stable marriage.

A similar stable marriage could occur between the mutually supportive RNA sequences A and B in the model just described. Not only would the sequences depend on each other, but, like all good marriages, they would have to learn to live together. Competition for making enzymes from a finite supply of chemicals in the soup would have to be regulated so that A and B both got what they needed. For if one got too much the other might not be able to find enough, and the "marriage" would collapse. Only together could both grow. As they grew they could store more information and eventually be able to code for the larger and more complex proteins that are necessary for the building and regulating of true living organisms.

Eigen has named his feedback loop of RNA sequences a "hypercycle," and has stated that the existence of hypercycles is the only way to explain the subsequent growth and stability of small RNA sequences. Without hypercycles there would have been nothing but an aimless scramble of dwarfed quasi-species members milling around, unable to get any bigger because of being unable to say anything that would lead to their getting bigger.

The hypercycle model is backed by many laboratory experiments and by some sophisticated mathematics, both beyond the scope of this book. It also provides for the eventual appearance of DNA, again a process too complex to try to recapitulate here. It finally focuses on the formation of cells. With a cell, something that definitely can be recognized as a living organism emerges.

As to how cells (those little boxes that wall off and protect what is inside of them) might have been formed, Eigen does not say. He is painfully aware of the complexity of cell walls and admits that their emergence must surely have taken a long time and presented prebiotic chemistry with some dreadfully difficult problems. What is known about cell walls is that they are constructed mostly of fats, and that fatty compounds are chemically possible in prebiotic soups. Some experiments have even demonstrated that when such fatty substances are heated they tend to curl up into little droplets. Any bit of RNA, enzyme, or other soup ingredient that happened to be enclosed in such a droplet would have a protective coating and would be isolated from other hypercycles, free to follow its own evolutionary destiny.

How true is all this?

Not true at all if our criterion is whether or not it has been proved. However, it is a plausible way of explaining some profoundly important steps in the grudging movement of chemicals away from their inanimate nature and toward increasingly intricate rearrangements of themselves, to a point where some of the new substances (RNA) can instruct for the making of others (enzymes), and life as we know it can begin.

The strength of Eigen's argument is in its logic. It is buttressed by experiments. It does not violate known natural laws; indeed it depends on them, particularly on the laws of chemistry. Its weakness is that, aside from the initial step, none of the next ones have been shown to have happened. Nor perhaps will they ever be shown; we do not expect that fossil traces of them will be found. A hard fact about living organisms is that their complexity makes them efficient. They are great vacuum cleaners of the environment. They eat up everything in sight. Once such relatively complex things as bacteria appeared on the scene, the ancestral, slow, blundering quasi species could not compete. Their descendants would consume them (or preempt their ingredients) as fast as they were formed. The result: no quasi species anywhere now. A slate wiped clean. An ancient process no longer able to get started because it has been replaced by a better one.

Another objection that some scientists raise to the Eigen model is that, despite the rigor of its mathematics, some of those mathematical steps may have statistical flaws. So, what is there to replace it? Three things:

First, there is belief in divine Creation: God did it. If that is so, there is no point in trying to investigate further. What God did is a matter for faith and not for scientific inquiry. The two fields are separate. If our scientific inquiry should lead eventually to God, to questions so large that they cannot be examined coherently, that will be the time to stop science. But we have not yet reached that point. There are many things for us yet to learn, so we may as well continue.

Indeed, if we truly believe in God, we should recognize that He gave us the brains to conduct scientific research. Not to do so — not to use the marvelous gift of intelligence with which He has endowed us — would seem to be disrespectful of God. My own belief, which Don shares, is that people who narrowly deny the findings of science do not understand God and do not truly love Him. It is possible to love God *and* to practice science. That is what so-called scientific Creationists seem unable to do. They substitute dogma for science. They pervert science by insisting that the earth is only about ten thousand years old and by denying the existence of fossils. They wind up with an ugly and pinched view of the world, of life, and ultimately of God Himself.

A second alternative to Eigen's explanation for the origin of life on earth has recently been proposed by A. G. Cairns-Smith, a chemist from the University of Glasgow in Scotland. He thinks that the whole system of carbon-based organic molecules — the RNA–directed process of protein and enzyme production — is far too complex, too highly tuned to account for the start of life. By the time you get to RNA, he maintains, you are already too far along the path. You are into high tech when you should be thinking about low tech. He uses the analogy of the evolution of weapons to make his point, observing that it is impossible to explain the development of a machine gun by looking at a wooden spear. He asks: how does one get to metals and explosives from the concept of using a pointed stick as a weapon? He replies: one doesn't. It is too great a jump. Similarly, in looking for the beginnings of life one must throw away the idea that it is associated with RNA and carbon-based compounds. One should look to something simpler, something perhaps entirely different.

What would Cairns-Smith substitute? His candidate: clay. He believes that a very good argument can be made, given the wide distribution of this material, for a model based on clay and its common ingredient, silica, with its propensity for forming crystals. Let carbon and its complexities wait, he says; they can come later. Let's first look at simple clay.

Cairns-Smith starts his argument by listing three attributes that would be required of the first organisms:

1) They could evolve.
2) They were low tech.
3) They were made of simple, available ingredients.

Clay meets all those requirements. It "evolves" in the sense that in its regular and steady formation of crystals mistakes sometimes occur. The crystal-forming process misses a beat, takes an unexpected turn. The result is a different shape, a different fitting together. Those mistakes persist as the crystal continues to grow and then to break into pieces. In time mistakes occur in the mistakes. Certain shapes are favored by the rock environments into which the crystals edge and squeeze. That is a kind of primitive evolution, says Cairns-Smith, evolution regulated by something he conceptualizes as "crystal genes."

As for its being low tech, crystal formation in clay is about as low tech as you can get. And — requirement 3 — the chemicals that go into that low-tech operation are certainly available. Silica is the most abundant substance on the earth's surface, being made of oxygen, the most common element, and silicon, the next most common.

So, with a kind of primitive crystal gene at work, what else is needed for the development of life? Here Cairns-Smith quotes Hermann Muller, whom we met earlier as the discoverer of radiation as the source of mutation in fruit flies. "Nothing else is needed," said Muller back in 1926. In other words, we already have in our evolving crystals a kind of prelife object. Call it a structure, call it an organism, it doesn't really matter. It is capable of growing, capable of replicating itself, and, over time, capable of change — low tech chugging along in its quiet fashion. The way to leap into high tech, according to Cairns-Smith, is not to evolve into it but to acquire it from outside. You can't make a machine gun out of bits of wood no matter how hard you try. Entirely new technologies must come along. What came along in the primordial sea or the primordial clay pit may have been some bits of RNA and some primitive enzymes, self-made with the help of sunlight, that settled down in those low-tech clay structures. Being more precise than crystal genes, they eventually replaced them.

In a bicycle factory undergoing modernization, with computer-controlled robots replacing the workers who formerly read the blueprints, there can be no evolution of worker into robot, no halfway stage that is part worker, part machine. The robot is the product of a new technology, introduced suddenly. It doesn't emerge out of a worker, it shoulders him aside. Similarly, RNA, with its greater potential for accurate instruction, would have replaced crystal in a clay "organism" and ended up running that organism under a new set of more sophisticated rules.

In summary, chemical evolution would have given way to organic evolution.

Cairns-Smith's proposal is an ingenious and imaginative one. Its strength lies in its attack on the weakness of the Eigen model: RNA does have to be fairly complex to make useful enzymes. Its weakness is that it is entirely speculative. Whereas Eigen does have some RNA, Cairns-Smith has no crystal genes. At the moment they are imaginary, he admits. On the other hand, he says, they may only be rare and hard to find. Alternatively, they may be common but unrecognized. "Take your pick," he suggests, and then goes on to say that "the most critical experimental challenge now is surely to discover crystal genes—not just one kind but many kinds." In that search we wish him well. Meanwhile we will stick with Eigen and RNA.

A third way of explaining how life got started on the earth is to say that it came from somewhere else. This idea is called panspermia. It was first aired in 1908 by the Swede Svante Arrhenius, who had previously won a Nobel Prize for his work in chemistry. His thought was that spores from outer space had populated the earth. The idea is not as silly as it may sound. Respectable scientists still espouse it, although they argue among themselves as to what got here and how. Some, like the astronomers Fred Hoyle and Chandra Wickramasinghe, say that life is constantly being blown here in the form of small spores or microbes that have had their birth on other planets or in other galaxies. They point to the fact that bacteria are highly resistant to radiation and argue that that must result from their having evolved in places where there was much more radiation than is present today on earth.

Others point to meteorites and to the remarkable fact that amino acids have been found in them. One may quibble that the amino acids are not like those found in earthly proteins, but one cannot duck the resulting question: how did *any* amino acid get into a meteorite unless there was the possibility of life where the meteorite came from?

Finally, we return to Francis Crick, the redoubtable coconstructor of the double helix. If Crick speaks shouldn't we listen? Yes—but with a hand on our wallet. Crick's thesis is that life on earth may well have started with microbes or spores but that those microbes were brought here by aliens from outer space. They were dumped here—whether accidentally or deliberately he doesn't say—rubbish left over from some intergalactic picnic.

There are two things wrong with all panspermia ideas. First, there is no evidence to back them up. The arguments that are brought forward to buttress them can be undermined by counterarguments. For example, bacteria are resistant to radiation, not because they needed to be during their journeys through space but because they needed to be right here on earth. In the early days of earth's history, when bacteria came into being, the level of solar radiation was high because there was not yet a protective layer of oxygen in the atmosphere. Oxygen, which now shields higher organisms from the destructive radiation of the sun, is a late addition to our air, contributed as a waste product by bacteria themselves.

As to amino acids in meteorites, they may have been picked up in the atmosphere when they entered it. If not, even if they were formed elsewhere, that does not prove that life arrived on their wings. Amino acids can be made in laboratories here, so why do we need them from somewhere else? It is not amino acids that make life; it is their organization by RNA and DNA.

In the end panspermia doesn't really explain anything. It just moves the mystery a step farther away, to another planet, another star. Life itself remains as enigmatic as before. For now, what we are left with as our best guess seems to be some kind of synthesis of life right here, from ingredients we know were available here and that are still being used by all living things. The Eigen model may not be exactly right—it could even turn out to be fundamentally wrong—but it is the best thing available at the moment.

Whatever model we choose, there remains a breathtaking paradox: on the one hand the endless complexity of the process, on the other the simplicity of the principle. To make everything that can be called alive, to monitor the development of every fern and feather on earth, to direct their growth, to enable them to function, to replace worn-out parts, to turn things on, to turn them off—for all those activities to be orchestrated by just four kinds of small molecules is awe-inspiring. It is that magisterial power of DNA, the power to direct, that commands our attention. Its molecules are absurdly simple. They are not alive. In many ways they resemble crystals. But they can do things that no crystal ever dreamed of. It is the limitless number of rearrangements of them as the DNA strands get longer that provides for the complexity. They wave their instructional wands, RNA jumps to work, ribosomes are activated, proteins grow, enzymes snip and patch. Lo, there emerges a bacterium, a flower, a fish, a Frenchman.

17

Carl R. Woese: A Look from the Top Down.

When we approach the enigma of the start of life from the bottom up, which was what we did with the Eigen model, we are rather quickly frustrated because we cannot rise beyond the original experiment. We can demonstrate that the building blocks of life can appear out of raw chemicals. But, with our present scientific capabilities, that is about as far as we can go. We can hypothesize, but we cannot make. Nursing into being a gene, a cell nucleus, a cell wall—all those things are still beyond us.

Therefore it may be instructive to tackle the problem from the other direction, from the top down, by starting with existing simple organisms to see if we can find something simpler: or at least if we can take apart the very simple for clues to what a simpler ancestor may have been like. Working in both directions, from the bottom up and from the top down, it may be possible someday to bridge the gap that still exists in our knowledge of what went on between nonlife and life.

"The simplest living thing," says the biologist Mahlon Hoagland, "is a cell." And the simplest kind of cell is a bacterium. Therefore our search from the top down should start with bacteria in the hope of finding something even simpler, something out of which bacteria themselves came. The search will be exceptionally difficult—and in the end will yield only partial answers—because we will be looking for something of which no visible trace has existed for more than three billion years and perhaps for as many as four billion. It may seem ridiculous even to consider such a search, but the work of Carl R. Woese, a bacteriologist from the University of Illinois at Urbana, makes it anything but. During the past twenty years his attempts to unravel the evolutionary history of bacteria and establish the relationships between different bacterial groups have shed light on three matters that are crucial to any understanding of how life may have begun:

First, his work has revealed the existence of a group of bacteria that are as different from other bacteria as the latter are from plants and animals.

Second, that discovery compels the reorganization of all life forms, past and present, into a family tree unlike any of the traditional ones commonly drawn by biologists.

Third, out of that reorganization comes a strong suggestion that there was a single ancestor to all modern forms. There are also some powerful hints as to what that ancestor may have been like.

All three insights have come from grueling analysis of minute bits of RNA in living bacteria.

How knowledge about such huge topics can be extracted from bacteria is a complex matter. It involves several intertwined threads of inquiry which we will do our best to sort out in this chapter. Perhaps the best way to proceed will be to take up in order the three topics noted. As we go, we will be exploring a new science now in the process of being invented by Woese and others who are chasing DNA and RNA sequences for clues to their history: molecular evolution.

Woese was not thinking about a new science when he started out. His ambition was more modest. He was interested in how bacteria had evolved their own way of making the DNA → RNA → protein transaction (they do it somewhat differently from higher forms) and found it necessary to learn something about bacterial evolution in general first. He quickly discovered that virtually nothing was known about it. Willy-nilly, he became a bacterial evolutionist.

But how does one pursue that trade? The tried-and-true way of tracing ancestral relationships is to look at fossils. A paleontologist, attempting to reconstruct the evolutionary history of the horse, can look not only at modern horses but also at the fossils of a whole assemblage of extinct horses. By comparing types and times of existence, he can begin arranging all his fossils in sense-making evolutionary sequences and emerge finally with a pattern that has a retriever-sized four-toed animal at one end and a far larger one-toed animal at the other, with a plethora of intermediate types branching into numerous lineages in between.

Unfortunately, as we said in the previous chapter, fossil bacteria are almost worthless in such a search. They come in three basic shapes, round, rod shaped, or twisted. All are found in fossil deposits by the billions. But those shapes alone tell us nothing. How, then, does one sort out a bewildering variety of living forms when there are no fossils to show how they have evolved or what their ancestral links may have been? There is only one place to look, and that is inside living bacteria. We go there on the assumption that their genetic material contains clues to their ancestry.

To the nonscientist it may seem bizarre to suggest that the genes of living bacteria can tell us such things, but science has already slipped through that door and is getting a glimpse of the furniture inside. There are evolutionary secrets there, and it is now beginning to be possible to read them in the nucleotide sequences in DNA and RNA. In modern bacteria are strands of genetic material that have remained fairly constant for billions of years. Therefore, when we look at the RNA of a modern bacterium we are also looking at its remote ancestor.

Bacterial RNA is not totally unchanged, however. If that were the case, all bacteria would be alike and there would have been no evolution. There is change and there is stasis. Both are observable in the RNA of living bacteria. The unchanged parts reflect common ancestry. The changed parts reflect evolution.

We have said this so often that it is surely tiresome to hear it again. However it is important: DNA and RNA direct the making of proteins. Some proteins are so basic to all of life that it cannot be conducted without them. Therefore the genetic sequences that dictate the manufacture of those basic proteins cannot change much or the organism will die. It follows that those highly conserved sequences must have been present from the beginning, or near the beginning, of the assembling of the earliest life forms. Conversely, sequences that instruct for less critical functions, or perhaps for no function at all, can change rapidly and drastically. The trick in delving into the ancestry of bacteria is to find sequences that do both, for if one is to construct ancestral trees, one must know not only what is old (that is, existed in a common ancestor) but also what is new (changes that have taken place since and account for the variety in existing forms).

So, a student of bacterial evolution must start with the living organism and work backward into a kind of void. As he goes he will follow a logical principle: the traits that are the most common in all forms will probably be the oldest, handed down on all branches from a common ancestor; the most widely diverse traits will be the ones that have evolved the most recently. Following that principle, a molecular evolutionist should begin to be able to puzzle out relationships among different groups of bacteria and eventually organize them into a family tree.

That is what Carl Woese has been doing for the last couple of decades. On the way he has come up with a totally unexpected discovery. He and some colleagues have succeeded in identifying a group of bacteria that are as different from other bacteria as they are from higher forms. "It was like going into the backyard and finding something that was neither a plant nor an animal," he said. It was that discovery that led to his redrawing of the tree of life and made possible later speculations about the nature of a universal ancestor. More on that extraordinary discovery in a moment. If we are to understand it we must first know what Woese was doing in his laboratory and how he did it.

Obviously he needed some bacterial DNA or RNA and the laboratory means to slice it into fairly short sections so that those sections could be compared for similarities and differences. But when he started his search, science's ability to detect differences in DNA sequences was in its infancy. He was forced to turn his back on DNA's hideously long and complex sequences and look for something simpler. He found it in the ribosomes of bacteria, those small organelles scattered through the cell and acting as protein-assembly sites. Each ribosome has its own RNA, known as rRNA. There are two major sequences of it. One, known as 16SrRNA, is about 1,500 nucleotides long. The other, 23SrRNA, is about 3,000 nucleotides long. Woese selected the smaller and simpler of the two. It suited his purpose because it was small enough to submit to the only analytical tool available to him at the time and yet large enough to contain a spectrum of nucleotide patterns that changed at widely different rates. In that spectrum he hoped to find traces of ancient ancestry (in the parts that changed the least) and evidence of more recent evolution (in the parts that were changing the most rapidly). He would make a catalog of those different parts by snipping them into small bits (a crude way, but the only one he had), using enzymes as his scissors, and then begin to trace relationships between various bacteria through the similarities and differences in their bits.

So Woese snipped—and snipped again until his RNA samples were cut into a great many very small pieces. He had to do that, because, with the analytical tools then available to him, he could not have recognized differences in longer pieces. And differences were what he had to have; otherwise his work would have been meaningless.

To get a better picture of the difficulty he had in recognition, let us imagine a long tippy rowboat containing two kinds of oarsmen, fat ones (represented by black dots) and thin ones. The fat ones sit on one side of the boat, the thin ones on the other. It is obvious that the

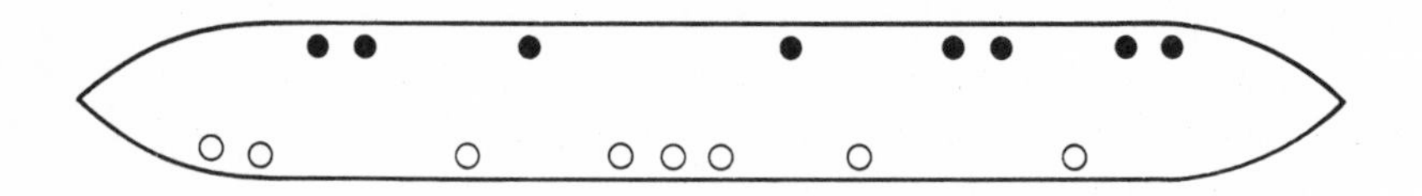

boat will tip one way or the other, depending on how many of each kind of oarsman are in the boat. Theoretically it should be possible to tell by the way the boat is tipped who is in it. Actually, however, this becomes harder and harder to do as the number of oarsmen increases, since they counteract each other and only an average "tippiness" can be detected. But, take a small section in the middle of the boat, with only three oarsmen, and it begins to be possible to identify them, since there are only four ways that fat and thin rowers can be distributed:*

* It is true that one or two rowers can be seated in ways other than those shown (i.e., move them farther forward or backward), but that does not affect the amount of tipping.

In the first example the boat will tip strongly to the "fat" side, less so in the second, perhaps sit on an even keel in the third, and tip the other way in the fourth.

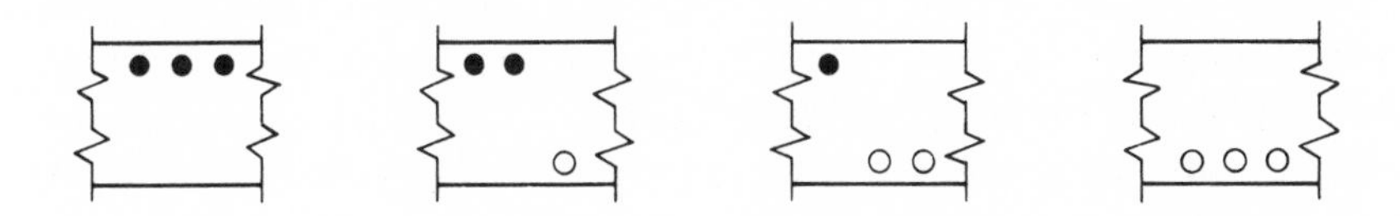

This rowboat can be compared to a section of RNA in which the fat men represent the CG and GC nucleotide pairs, and the thin ones the AT and TA pairs. Instead of putting them in a boat, Carl Woese put them on what amounted to a wet blotter with a weak electric current running through it. This gave one edge of the blotter a positive charge and the other edge a negative charge. The process is known as gel electrophoresis. Since the charges holding the CG and AT nucleotides together are slightly different, they will tend to migrate across the blotter and take up slightly different positions on it.

By those microscopic movements Woese was able to distinguish one bit of RNA from another, but only if it was very short. With longer strips he could not do that; the charges would tend to cancel each other out and he would get a blurred answer. That is why he had to go to the arduous step of obtaining extremely short bits of RNA. Unfortunately those tiny bits said nothing about themselves beyond the fact that they were different from each other.

"The result was like characterizing a play by making a catalog of the different words in it," said Woese in the fall of 1986, when he visited Martha's Vineyard and I had a chance to talk to him. "When you cut a play up that way you have no play; you don't know what it's about. But you do know that it's different from another play. You can tell a play by Shakespeare from one by George Bernard Shaw. Each has some unusual words that may appear only once. Those are important because they identify the play, even though they don't tell you what's going on in the plot. To learn that you would have to be able to identify longer sequences—like sentences, or even paragraphs. We couldn't do that at first. We did later."

With his catalog of bits from the 16SrRNA molecule, all chopped up into pieces small enough so that he could tell one from another, he began comparing samples of various bacteria. Since he needed a yardstick of some kind against which they could be measured, he also threw in rRNA samples from some larger, more complex, but still single-celled organisms that were not bacteria. As a first step he would measure the bacteria against them, expecting that the bacteria would fall on one pile and the more complex organisms on another. Almost immediately he ran into a most disconcerting problem. The rRNAs of the various bacteria — as expected — were quite different from those of the nonbacteria. But some of the bacteria rRNAs turned out to be radically different from each other as well. Instead of falling into one neat pile of bacteria, they appeared to fall into two piles, forcing on Woese the unsettling idea that bacteria themselves, the most primitive organisms known, come in two fundamentally different forms, probably three, when he thought further about it, because the nonbacteria presumably had a bacterialike ancestor of their own which would have been different again from the other two.

What is so startling about that? It is startling because it flies in the face of an idea about the proper way to classify all living forms that has been held by biologists for more than half a century. To understand that, we must backtrack for a moment to the beginnings of classification, back to Linnaeus, who started it all, for a look at how he organized things.

At the very base of Linnaeus's system, the first and most fundamental line that taxonomists then drew was the difference between plants and animals. Every living thing was one or the other, and that majestic first division prevailed unchallenged for well over a century. Forget for a moment a few smudgy little things like slime molds that weren't exactly either, that seemed to skip nimbly from one to the other by acting like plants for part of their life cycle and then like animals during another part. With the exception of those borderline things, everything seemed to be clearly an animal or a plant. The difference was obvious. Plants could make their own food from the gases and chemicals around them by using a green substance in their cells, chlorophyll, to tap the energy of the sun. Animals could not do that. To live, they either ate plants or they ate each other. In a broad sense all animals are parasites; they have to have plants to survive. Plants don't need animals. They are the bottom line; they can do it on their own. With that fundamental distinction drawn, the two kingdoms of plants and animals were established and remained in place for a long time.

But what about bacteria? Linnaeus had never heard of them, but the taxonomists who followed him had. Science was soon inundated by swarms of bacteria. A decision had to be made as to how to classify them. That turned out to be awkward. To most biologists it did not make sense to call them plants because many of the ones being discovered then did not have chlorophyll.* On the other hand they could not be called animals because of their unanimal way of getting their nourishment directly from chemicals.

Clearly there would have to be a reorganization of Linnaeus's kingdoms. In 1866 an idea for a third kingdom was proposed by the German biologist, Ernst Haeckel. He suggested that all single-celled microbes be lumped together in a kingdom he called monera—neither plant nor animal.

Haeckel's proposal worked for a while. But it, in turn, became an uncomfortable squeeze as more was learned about the differences among monera. There were many kinds, and some were not bacteria. Just because they were all very small did not justify their being put in one basket. Bacteria, in particular, began to seem increasingly different from some other things in the basket. Ultimately it was realized that Haeckel's three-way split would not do. Bacteria *were* different, radically different from other plant and animal cells. When all three were compared under the microscope, the one-celled plants and animals that were being lumped with the monera turned out to resemble each other closely in many fundamental ways. It was the bacteria that were the oddballs.

To begin with, a typical one-celled animal or plant is about a thousand times bigger by volume than a bacterium. Looking inside, one can quickly see the reason why. Plant and animal cells contain a great deal of plumbing and hardware that are missing in bacteria. Most prominent in that jumble of extra equipment is the cell nucleus. Plants and animals have it, bacteria don't. That difference

* It would later be learned that many bacteria do have chlorophyll. But, with one exception, it is somewhat different from the chlorophyll of true plants. That exception occurs in a group known as cyanobacteria. Their chlorophyll is the same as that of plants, and for a long time it led to their being regarded mistakenly as plants and being classified under the name of blue-green algae. All this contributed to confusion in the placement of bacteria and delayed recognition of their true and unique nature.

alone is enough to stand the old Linnaean dualism on its head. Plants and animals—however different they may turn out to be as end products—are very much alike in their cellular organization. By that yardstick they belong together, not apart. In due course they were so classified. For more than half a century science has recognized that the first and most basic distinction that can be made among living organisms is whether or not they have a cell nucleus. To reflect that distinction, two groups have been named: prokaryotes (organisms without a true cell nucleus) and eukaryotes (organisms with a cell nucleus contained within its own membrane).

In this widely accepted scenario all prokaryotes have been considered to be one kind of thing. That is why Woese's discovery that they are not is so unsettling. It means that the prokaryote-eukaryote division of all life, so laboriously and so recently arrived at, must be scrubbed. Such things are monumentally shaking to the orderly arrangements of science. But, once encountered, they must be dealt with. In the face of a good deal of opposition, Woese has announced that the different RNA patterns of bacteria force the conclusion, however much anguish it may cause, that there are three branches at the bottom of the tree of life, not two.

The bacteria that enabled Woese to arrive at that truly startling conclusion were rather startling themselves. Long known as oddities in the bacterial world, they had not been lumped together comfortably before because their habits were so different, wildly different, each from each. What they heretofore had lacked was a coherent, binding commonality. Woese found that in their RNA. It revealed a relationship. Without that, their links to one another had gone unrecognized.

One group, for example, is found in the mud at the bottoms of stagnant bogs, in sewage-disposal plants, and in hot vents at the bottom of the sea. Oxygen kills those bacteria; they flourish only where oxygen does not penetrate. In their life processes they give off a gas, methane. It is that stuff, "marsh gas," that one sometimes sees bubbling up from the bottom of a swamp. Put a match to it and it will burn, just as the gas in a gas stove burns. The name of the methane producers: methanogens.

Another group is equally bizarre. Its members live in water that is saturated with salt. They are found only in places like the Dead Sea and the Great Salt Lake, environments that quickly kill almost anything else. Those salt lovers are known as extreme halophiles.

Members of a third group are remarkable in their ability to withstand a high degree of heat. In fact they thrive on it. Some members — *Thermoproteus* is an example — are found in hot sulfur springs at temperatures that approach the boiling point. Their name: extreme thermophiles.

Odd life-styles all. So odd that previous biologists had not succeeded in linking these bacteria in any way. Now, however, they are linked; they have certain RNA characteristics in common that other bacteria lack. Different RNA, of course, implies different chemistries, different everything. Alerted by those RNA differences, investigators quickly began finding all kinds of unexpected differences in the bacteria themselves. They found novel cell walls, novel lipids (fatty compounds); they found unfamiliar enzymes; they continue to find new things. It is now clear that the life processes of these bacteria are fundamentally unlike those of the other "common" kind. They are a distinct group of living forms, distinct enough to give them the right to a collective name. Woese has named them archaebacteria. The others, the "common" kind, he has named eubacteria. For this pioneering work Woese was recently awarded one of the celebrated MacArthur "genius" fellowships.

Archaebacteria, well named. "Archae" means primitive or ancient, and that is what they seem to be. As one becomes familiar with them their peculiarities begin to make a kind of sense. They seem to be saying: "This is what bacteria were like when the world was young, when it had not gotten to be like the world of today." Probably the two most significant differences between the present world and that of four billion years ago are that it was considerably hotter then and that it lacked free oxygen in its atmosphere. Any life forms that were emerging at that time would have had to be able to transact their metabolism at very high temperatures and without oxygen. It is those two abilities that many of the archaebacteria have in common.

If the world was hot when it was young, and if its atmosphere lacked oxygen (few scientists doubt those assumptions), and if we have identified a group that fits that environment, should we not consider those heat-loving, oxygen-hating archaebacteria as representatives of the original, most primitive kind of bacterial form, ancestral to all others? Should we not regard them as the dominant, perhaps the only, kind back then? Now that the earth has cooled, should we not regard them as anachronisms, as holdovers from the past, hiding out in obscure places that are still hot enough for them?

All that is logical, says Woese, but it is wrong. While it is true that archaebacteria appear to represent an extremely primitive type of organism, that does not mean that they are ancestral to the others. On the contrary, his RNA analysis keeps insisting that the two types of bacteria are fundamentally different, so different that eubacteria cannot be descended from archaebacteria.

That being so, it is necessary to explain where the other group, the eubacteria, came from. How is it that they tolerate oxygen and do not like heat? Those traits make them seem like younger organisms; they have a whiff of the new world about them — and yet their rRNA says they are old. If they are not descended from archaebacteria, what is their origin?

That question, initially troublesome, goes away when we begin looking at eubacteria from a different perspective. Traditionally we have thought of them as an oxygen-tolerant group, because the ones we normally see are the ones that live where we do, in an oxygenated atmosphere on the earth's surface. But, go down into the ground a few inches and we begin to encounter anaerobic eubacteria right away. That changes our concept of eubacteria; most of them are anaerobes too. And it produces a larger insight. Woese now believes that all bacteria were once anaerobes. They had to be; there was no free oxygen in the air. They had to make their living without it, in other ways. What makes the eubacteria different is that they apparently were more flexible, more innovative, better able to adapt, when it became necessary, to a new kind of atmosphere and a cooler environment. The archaebacteria, more conservative and stodgy, comfortable with the old, did not change much. The two are clearly different today. Their RNA says that they have been different for a long time.

Woese's research allows us to look at heat, too, in a new way. Before archaebacteria were identified as a group, scientists needed an explanation for the capricious way in which the ability to withstand heat was scattered through apparently unrelated groups in the bacterial world. The explanation most often given was that various groups had adapted to it independently. If a certain kind of bacteria occurs today in hot sulfur springs, it must have gradually found its way there, becoming more and more tolerant of heat over time as it worked its way into a niche otherwise unoccupied and therefore wide open to any organism capable of exploiting it.

According to Woese, that is exactly backward to the story revealed by his research. It lumps the heat lovers together for the first time and thus makes possible a more sensible scenario: they were all originally spawned in hot water. They liked it then and they like it now. Indeed they had to like it then because there was no other environment; it was either heat or nothing. Present differences among the archaebacteria are explained by evolution that took place later. Their fiery inheritance they have retained. Nor should we be surprised by that. Just because we, and all the higher life forms we are familiar with, perish immediately in boiling water, does not mean that life cannot be conducted there. It can; living archaebacteria prove it. Their ancestors, all of them, lived that way.

Thanks to RNA matching, this resorting of archaebacteria and eubacteria has built an increasingly clear picture of their present nature and probable past. It fits Woese's molecular findings and it fits what he sees in the behavior of various archaebacterial strains today. What it does not fit at first glance is eubacteria. They do not like heat. And yet, according to their RNA they are old, probably as old as the archaebacteria.

There is no problem here. Eubacteria were born in heat also. If they do not tolerate it today, it is because they have lost the ability. Woese now believes that all bacteria were originally thermophiles. His work produces a sharper and sharper picture of the archaebacteria as the conservatives that have stayed pretty close to their origins, and the eubacteria as the progressives. As the earth has cooled, so have they. But even among those venturesome eubacteria there is one laggard which, for reasons of its own, has clung to old ways. It is *Thermotoga,* a heat lover. Interestingly, it is the oldest known among the eubacteria, the most primitive, the least evolved, and presumably the closest to the primordial eubacterial type. It seems to be saying: "Take a good look at me. We eubacteria were all like this once. The rest may have gone up the road, but I'm the one who held on here where it's nice and hot."

Isn't that what evolution is all about? The environment changes, and certain organisms change with it. What Woese has managed to come up with in his RNA studies is a reasonable explanation of how and why the various kinds of bacteria are the way they are today. It is a remarkable achievement. He had to cut through a great deal of evidence that for a long time seemed to support the opposite of what now appears to be the true state of things.

"It wasn't easy," said Woese. "When I started my work I didn't know there were two major groupings of bacteria. Like everybody else, I assumed that they—the prokaryotes—were a single large group ancestral to the eukaryotes."

He gave a wry smile. "That's not so. I had to find that out for myself. Now I have to work against entrenched attitudes. Biologists are stuck in the old prokaryote-eukaryote dogma, just as I was at first. That prevents them from appreciating the importance of treating archaebacteria as distinguished from eubacteria. People keep telling me: 'They're different, but they're still prokaryotes.' That's all very well, but it's the wrong way to think about them. As long as people think that way, they won't understand the enormous significance of those differences. They won't understand evolution."

For example, they won't understand the evolutionary history of photosynthesis, another vital function that separates archaebacteria from eubacteria. Photosynthesis is a pioneering way for certain cells that contain chlorophyll to use sunlight as an energy source. This is a complex and sophisticated process that is scattered widely among bacteria. Before Woese, the first impulse of the classifier was to lump all the photosynthesizers in one group and all the nonphotosynthesizers in another. That was about as useful as putting all red birds in one group and all nonred birds in another. So a second idea surfaced; photosynthesis must have been invented over and over again on different bacterial lines.

But can such an intricate process, a truly unique and revolutionary one, come and go so easily? Common sense says no, and Woese's RNA analysis supports that. His studies have revealed that chlorophyll-based photosynthesis occurs *only among eubacteria!* It apparently was developed very early in their evolutionary history. Its spotty appearance since among the eubacteria is explained by its having been lost on various lineages over the course of time. In the beginning, Woese reasons, all eubacteria were probably photosynthetic. Today only some of their descendents are.

By contrast, none of the archaebacteria are photosynthetic in the generally accepted sense. Chlorophyll is unknown among them. The few that do achieve a form of photosynthesis use a different mechanism to capture and transform the sun's energy. The chemistry of the two processes is totally different. It marks a deep and significant separation between eubacteria and archaebacteria.

To summarize, and to emphasize the first point made on page 297, Woese went looking for a way to detect differences in the RNA of various bacteria, hoping to work out evolutionary relationships among them. In doing so he stumbled over the hitherto unsuspected fact that bacteria came in two fundamentally different forms. When he began sorting them out according to that division he discovered things about them that made even more clear their basic differences. The archaebacteria thrived in heat, could not stand oxygen, and were not conventional photosynthesizers. Eubacteria, on the other hand, were divided between heat lovers and heat haters, divided again between oxygen lovers and oxygen haters, and once again divided between those that could photosynthesize and those that could not. Clearly they were the more enterprising group of the two, straddling two worlds, born in the ancient one but adapting to the modern one. The archaebacteria, by contrast, were anchored in the old world. To the extent that the world remains old, that is where they are found today.

Two types—yes, indeed, and very different. But do not let those differences conceal two stunningly important traits that they have in common, also as revealed by Woese's research. Both were originally heat lovers (thermophiles); both were originally anaerobes (oxygen haters). That is the kind of primitive organism we could expect the old earth to have produced. And that, it becomes increasingly clear, is exactly what it did produce.

These revelations about bacteria are given us *after* they are sorted out according to the basic division dictated by Woese's RNA research. They underscore that division, they strengthen it. RNA prescribes a pattern of relationships; lo, the behavior of the beasts confirms its logic. That is the power of Woese's work, its fascination. It bears out what we said earlier in this chapter: by looking inside living bacteria we can find clues to their true relationships and their evolutionary history. Now it becomes possible to sketch out a family tree based on these new perceptions, and from there to speculate about what the common ancestor to all known life may have been like. Thirty years ago this would have been inconceivable.

Even today Woese's knowledge is painfully skimpy. It consists only of what he has been able to sweat out of the 16SrRNA and 23SrRNA molecules in the bacterial ribosome. But the ribosomal RNA in a bacterium represents only two genes. The bacterium itself has more than two thousand genes. He has now set his sights on that larger target, the entire bacterial genome, and has begun the long task of organizing a project to sequence it all. This will be a mammoth undertaking requiring the work of a team of biologists, computer scientists, engineers, and mathematicians. It could take as much as ten years to complete and will cost many millions of dollars. Its goal: by studying the evolution of all the genes in the bacterial cell, to learn which are the primitive families from which all modern genes have descended; and from that deduce what the ancestors of bacteria were like — in the older, simpler days before they ever aspired to be bacteria. What a mind-boggling idea: to probe the genetic blueprints of living bacteria to learn something about the nature of an organism, an ancestor that was almost certainly like nothing alive today and that vanished from the earth four billion years ago.

Will that ever be done? Theoretically, yes. A technique for identifying long sequences of RNA has now been perfected. Woese no longer has to chop up the 16SrRNA molecule into little pieces in order to recognize them — to destroy the play in order to read the words. With the new technique he has succeeded in sequencing the entire molecule; Act One of the ribosomal play is now readable. He is now at work on Act Two, sequencing the longer 23SrRNA molecule. When that is done two genes will have been sequenced. For the entire bacterium there will be a mere two thousand to go.

Now, a closer look at the family tree of bacteria.

"I gave a lecture at the University of Illinois," said Don. "I met Carl Woese then and visited his lab. To me, a bone man, it's hard to believe what he can pry out of bacteria. I'm used to something you can get your hands on, like a fossil. You can measure a fossil. You can compare it to other fossils."

"Woese measures," I said. "He just measures different things. He also compares."

"I know that. And I know he's been able to sort out bacteria in a new way. We're all sorters at heart. But when I sort fossils I can begin to make a family tree. I can relate a human to an ape by perhaps finding a fossil that can stand in as a common ancestor to both. Maybe I can find some older fossils that will enable me to relate apes to other mammals, even to fishes."

"Woese makes trees too. He goes a step farther. He makes a family tree for all of life."

"Then he has to be guessing."

"Aren't you?"

"Yes, in a way. But our guesses are based on real things, things that once existed, that still do as fossils. I have them here on my shelf. Woese doesn't have those."

"He would dispute that," I said. "His real things are RNA sequences. Just because you can't see them doesn't mean that they don't exist. We've devoted nearly half of this book to explaining what DNA and RNA sequences do. They're real, and some of them are old. They haven't changed significantly in hundreds of millions of years, maybe billions of years. They are making some of the same proteins and enzymes that they were making a billion years ago. When Woese draws on that evidence, sorts out those relationships, he can make a family tree too."

"I'd like to see it."

I showed it to him. It is reproduced on this page. Its pattern is the fruit of studying similarities and differences in the rRNAs of many kinds of bacteria, and also in those of eukaryotes. The insights that

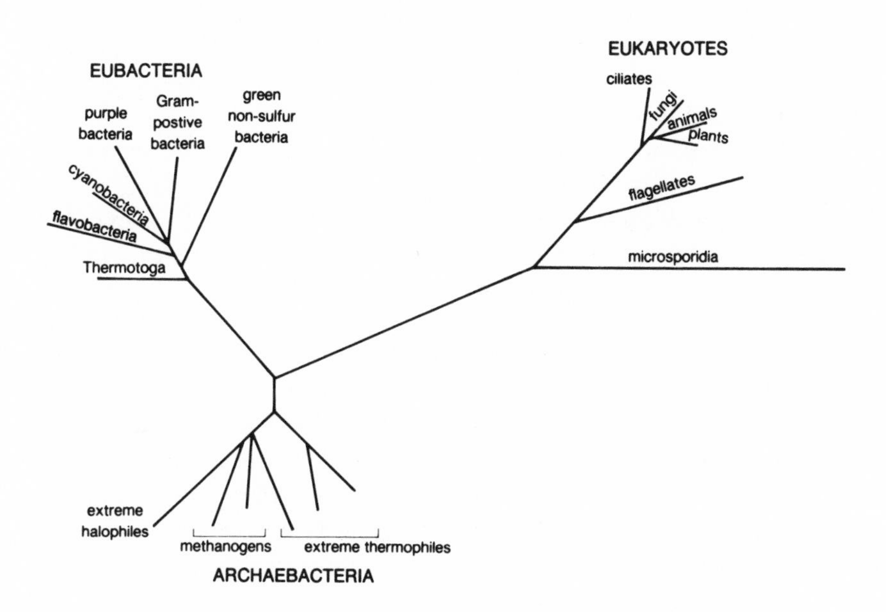

have come from that study—all the matching and comparing—have forced on Woese the shape of the tree he has drawn. It does not make sense on a molecular level when drawn any other way. A mapmaker—when he learns that New York, Chicago, and Atlanta are each a thousand miles apart—must respect that triangular relationship. If he were to put New York a thousand miles west of Chicago instead of to the east, it would be too far from Atlanta. When the distances between other cities are similarly cross-referenced they fall into an increasingly rigid pattern; it becomes impossible to move any of them without disrupting the distances to some others. So it is with the differences and similarities of RNA bits. Measure and compare enough of them, and they, too, begin to fall into patterns. First, the three kingdoms emerge: the archaebacteria, the eubacteria, and the eukaryotes—that third group which certainly had a bacterialike origin of its own before giving rise to all higher nonbacterial forms. Further sorting breaks each kingdom into subgroups and families, and a new and universal tree of life begins to take shape.

To a nonbacteriologist the family groupings on two of the main branches may seem meaningless, just clumps of bacteria with unfamiliar names. Be assured that they are all different and deserve those names, those placements. To appreciate the full implications of this tree, though, the reader should look at the eukaryote line, our own, to see how the organisms familiar to us—the large and visible denizens of our world—are related, how far removed they are from the others, and how recently sprung.

Woese's tree, stark and simple though it may appear, is worth careful study. Many insights may be found in it. The most obvious, surely, is how shallow the earlier divisions made by older taxonomists are now revealed to be. What Linnaeus thought was the fundamental fork that separated all of life (the difference between plants and animals) turns out to be a surprisingly recent branching. Before it stretches an immense span of time during which life forms of which we as yet have no inkling came and went.

A second insight stretches the history of life in another way. The reader should note the position of cyanobacteria on the eubacterial branch of the tree. Cyanobacteria are found today in places like the hot springs in Yellowstone Park. They grow there in clumplike masses virtually identical with clumps made 3.5 billion years ago by fossil bacteria. Those ancient clumps are known as stromatolites; biologists assume a direct relationship between cyanobacteria and stromatolites.

Don squinted at the tree. "But the cyanobacteria are not the oldest on that line."

"That's the point. All Woese's cutting and matching has shown him that there are older eubacteria, at least three kinds. Now that there's a link between a living bacterium and a fossil at 3.5 billion, the ones lower on the tree must be older yet. They're pushing the start of life deeper in time."

"How deep?"

"He doesn't know yet. In a sense he's just getting started. His data base is still very small, only a couple of genes."

Don looked at the diagram again. "That oldest one on the eubacterial line, *Thermotoga,* does he know how old it is?"

"No. But he knows something else about it that's very revealing. In addition to being the oldest, it's the champion heat-lover among all the eubacteria. Does that suggest anything to you?"

"Yes. It suggests that the older you are the hotter you probably like it."

"Right. As we said, it also suggests that the eubacteria as a group are old and that in their earliest days they all liked it hot. It's part of a growing body of evidence that convinces Woese that all bacteria were originally thermophiles."

"That's really neat," said Don. "What else does the tree show?"

"It shows the amount of evolution on the various branches."

"How does it do that?"

"By the length of the lines on the branches. The longer the line, the greater the amount of evolution. The archaebacteria show that. Their lines are very short, indicating that they have evolved very little from an earlier ancestral state. By contrast, the eukaryote branch has very long lines. It has undergone a tremendous amount of evolution. It leads to all higher life forms, to you and me. That's the branch with the cell nucleus."

"So, where does the cell nucleus come from?"

I had asked Woese that question, and he had replied that no one knew. It remains one of the grander mysteries of life. All he could say was that the eukaryotes—as bearers of a nucleus—were probably older than scientists used to think. "Much older," he said, touching a pencil to the point on the tree where the microsporidia branched off on the eukaryote line. Microsporidia, he explained, are microscopic curiosities that have long intrigued biologists. They are extremely primitive entities, living as parasites on a wide variety of larger organisms. Although they lack many features of the more advanced eukaryotes, they do have a cell nucleus, the earliest-known appearance of that structure in any organism. If plants and animals separated from one another roughly a billion years ago, said Woese, one could see how much earlier the microsporidia lineage emerged, way down the tree, as far back as two billion years ago, probably more. Back of that he loses the cell nucleus.

"I wonder if he will ever find it," said Don.

"If he does, it will have to come out of a lot more research."

"Perhaps that will also fill in that long empty space on the eukaryote branch. There's a couple of billion years there with nothing on it at all. Where do the eukaryotes themselves come from? How did they get so complicated?"

I had asked Woese that question also, and gotten a surprising answer. The best evidence is that they did it, not by slow evolutionary steps, but by swallowing some of their smaller bacterial contemporaries and then using them and their genes to build with.

"So that's how you and I got started," said Don. "We were bacteria that ate other bacteria."

"Probably." On that critical point in the emergence of higher life, Woese had told me to think about the ancestral eukaryote as a largish flexible cell with other, smaller, cells inside it. Suppose that an early eukaryote that could not perform photosynthesis—one that derived its energy in another, less efficient way—happened to absorb a small bacterium that could photosynthesize. The latter would be a self-contained ready-made energy machine. That chlorophyll unit that got into the larger cell would be for it the equivalent of electricity suddenly brought to a nonelectric world. Efficiency would have gone up immediately. The ingested cell would have found a safe home, and the host would have gained the ability to derive energy from sunlight. The small cell would be a chloroplast, and its host, by definition, a plant.

"You mean to say that one bacterium swallowed another and they became a plant?"

"That's putting it a little one-sidedly. The small one may have invaded the big one."

"How does Woese know these things?"

"He looks at the genetic material inside those smaller structures in the cell. Each tells its own story. He looks at chloroplasts. He looks particularly at mitochondria, the tiny organelles in the cell that convert the oxygen we breathe into useful chemical energy. Chloroplasts and mitochondria have their own DNA and RNA. Biologists have learned that it's not related to the DNA and RNA of the host cell. They brought it with them. In it is written their evolutionary history. Woese will tell you that both chloroplasts and mitochondria were once free-living organisms. There is no question that those little things, still bacteria-sized today, were once bacteria themselves."

"Do you know what all this boils down to?" said Don. "Those enduring fragments of DNA and RNA are the equivalent of minute fossils."

"In the way he looks at them, they are."

Don bent to look at Woese's tree again. "We've come quite a way, haven't we? I mean — look at that tremendous stretch out to higher life on the eukaryote line. Animals, plants, even fungi; they all got started in what seems like the last couple of minutes compared to everything else."

"That's right. It's enough to make Linnaeus's jaw drop, isn't it? That's how your mind has to open up when you find yourself playing the new molecular ball game. When you think how different a cocker spaniel is from a mushroom and then look and see how closely related they actually are, mere twigs at the end of one long branch, only then do you begin to take in the immense gulfs that separate the three basic forms."

"What else?" said Don. "This is fantastic."

"A couple of things. First we can see the history of the planet in some of these evolutionary events."

"Come on."

"I mean it. You can see changes in the earth coinciding with changes in life forms." Woese had pointed that out to me by noting where the flagellates split off on the eukaryote line of his tree. That point, he said, marked the appearance of an entirely new kind of life. The flagellates were a great deal more complex than anything that had evolved before. Their appearance obviously had some connection with the beginnings of the oxygenation of the atmosphere, which took place at about the same time — about two billion years ago, according to geologists. Woese then moved his pencil a little farther along on the eukaryote line, almost to its upper end, to where there was a violent explosion of all today's higher forms. That marked the time, about 1.5 billion years ago, when the sea finally became oxygenated. One cannot overemphasize the importance of those two events in thinking about evolution, he said. They had a profound effect on it. Before them, only bacteria existed as far as is yet known. After them, everything else.

As for the appearance of oxygen, it must be remembered that the primordial earth contained no free oxygen at all. That is not to say that there was not plenty of oxygen around. It was abundant in another gas, carbon dioxide, and it was an ingredient of many minerals. But none of that oxygen was free. It was being used, locked tight in various chemical embraces. There was no extra supply of it floating around in the air as there is now. But that situation began to change with the emergence of certain eubacteria containing chlorophyll. They gave off oxygen in the conduct of their life processes, just as plants do today. The appearance of those chlorophyll-bearers was one of the great turning points in the history of life, for it would lead from a world of simple one-celled organisms to a world containing all of the higher forms that now exist.

Being efficient energy converters, the chlorophyll-bearers were extremely successful. Over the millennia they gave off enormous amounts of oxygen. At first, as fast as it was produced, it was absorbed. Oxygen combines readily with many other elements, like iron; we can see the ceaseless industry of those long-gone bacteria reflected in the immense bands of iron oxide that exist in the earth's crust today. It was only when those storage reserves were filled up that there began to be a spillover of oxygen. When there was no longer anything lying around for it to combine with, there began to be a gradual buildup of oxygen floating free in the air.

At that point the archaebacteria, to whom oxygen was a deadly poison, were faced with a dilemma. Either they had to learn to handle oxygen or they had to hide. Since they lacked chlorophyll and thus had no use for oxygen and no way of coping with it, most chose to hide. That is why they are found where they are today, in muffled corners of the earth where oxygen does not go, conducting their chemistry in the old way, many of them using sulfur as their currency instead of oxygen.

"Well," said Don, "that's a logical scenario. Chlorophyll is invented by eubacteria. They begin to pollute the atmosphere with oxygen. Some of them learn to use it. The archaebacteria don't; they stay put. I guess the ancestral eukaryotes learned to use oxygen too. I mean, here we are, breathing the stuff. We can't survive more than a minute without it. I wonder what kept bacteria as single-celled organisms and allowed us to get so big and so complicated."

"A cell nucleus," I said. "Also more DNA. I guess that would have given us an opportunity to achieve greater complexity. And we got even more DNA and even greater efficiency when we began swallowing other bacteria. You might say that, given oxygen to build on, all it took to develop higher forms was greed, an appetite for other bacteria."

"Did Woese tell you that?"

"No. He's far too cautious. But the implication is lying right there in the family tree he has drawn and in what it shows about the three different forms of life. He sorted them out. The rest more or less follows."

"It's a great tree," said Don. "But there's one thing about it that I don't like. There's no root, no common ancestor at the bottom."

That brings us to the third major point in Woese's work. First he identified archaebacteria. That permitted the organizing of all early life into three major branches and to the drawing of a tree that showed that relationship. What remains is the root, the ancestor that connects the three.

"No proper root there," said Don. "The three branches just shoot off any old which way. Why isn't there a single ancestor down at the bottom, the way there is in conventional trees?"

"Because Woese doesn't know where to put it. He doesn't yet know enough about it. Was the ancestor more like the archaebacteria, more like the eubacteria, or more like the eukaryotes? He doesn't know."

"Obviously it was like the archaebacteria. They're the most primitive ones; the tree shows that they've evolved the least. They're the closest to the common fork, down where it's hot, where there's no oxygen."

"That's true. But just because the others have evolved faster and moved farther doesn't mean that they weren't originally just as primitive as the archaebacteria. I have to keep reminding you that Woese's research says that they were. Therefore the best he can do is make a kind of general connection for the three branches. That's why molecular evolutionists have to draw those funny fanned-out trees."

"But he does have them all connected at one point. That implies a single ancestor."

"Yes."

"Was there one?"

"Again Woese doesn't know. But he thinks there was."

The argument for that requires going back again to his rRNA research and to what has been learned lately about the chemistry of the three branches. All three are different but are also alike in a number of fundamental ways. Each has had an extended period of development on its own line. Each has accomplished its own evolutionary explosion, radiating into many types but retaining features peculiar to that group. The picture Woese gets is of three variations on a theme, three biochemical organizations that, although no longer identical, are similar in some of their deepest characteristics. That similarity makes it overwhelmingly likely that they came from the same source. It is the biologists' principal argument for there having been a single ancestor to all three forms of life.

What was the universal ancestor like? It was probably an organism that depended for its existence on some kind of transaction between nucleic acid and proteins. Every organism on earth does that. Therefore the ancestor probably had the same two-sided relationship.

For all of life, one side, the genotype (the information center) directs the growth and activities of the other side, the protein-based phenotype (the living organism). The two sides exist in delicate and intricate harmony. For any organism to survive through time, it must be able to remake its genome over and over again with great accuracy. In the analogy of the bicycle factory, the integrity of the blueprints in the central office must be preserved. That must be done despite constant perturbations: mutations that occur from time to time, frequent mistakes that happen when the genetic material assorts into daughter cells or recombines. Therefore there must be a sophisticated editing or screening process to ensure that integrity. It is accomplished partly by the manufacture of precisely engineered enzymes and other proteins whose job it is to ensure the accurate copying of the gene, and partly by the elimination of big or bad changes that might lead to the crippling of the organism. Small "neutral" changes can be retained. As we have seen, they add to the variability of the gene pool and may come in handy if future change in the environment calls for some evolutionary agility in meeting that challenge.

Most of the time, environmental challenges are small or slow, and a population of organisms will slide easily in one direction or another as the mix in its gene pool changes slightly to accommodate to the changing environment. That is evolutionary fine-tuning. It goes on constantly. No population of organisms is ever totally tuned to its environment. The very fact that each consists of a lot of not-quite-identical individuals guarantees that some will be slightly less fit than others. Thus, the process continues in an unstressed way. Stephen Jay Gould alluded to this process recently when he wrote: "Imperfections are the primary proofs that evolution has occurred, since optimal designs erase all signposts of history."

Among higher animals the genetic structures that regulate their shape and their lives are huge. They must be, in order to provide the instructions for making the seemingly endless supply of proteins designed specifically to carry out the multitude of functions that those higher organisms require. Down among the bacteria the process is the same but occurs on a more limited scale. They are more simply made, have fewer functions, and as a result need less DNA. But, skimpy as it may be, bacterial DNA is as sacrosanct as those larger chains. It makes its own proteins as scrupulously as they. It is that scrupulosity in the gene–protein relationship, even at the bacterial level, that seems far too complex, too precise to stand in as a model for an ancestor. Something simpler must have preceded it. Woese starts his speculations about the universal ancestor with that assumption.

It was a heat lover, he feels sure. If its bacterial descendants were born in hot water, surely it was born in water just as hot or hotter. It was an anaerobe; there would be no free oxygen to trouble it for another couple of billion years. Its genetic component was less developed than that of bacteria, therefore simpler and smaller. It would also have been less accurate, for smallness (simplicity) and inaccuracy go together. Get small (simple) enough, and you can no longer manage to produce the varied menu of large proteins you need to keep you in the exact shape you should be in. The result: an ancestor population of high genetic variability, many of whose individuals would seem so peculiar as to raise questions about their ability to endure as a species. Too furious a spouting of variants could doom any species.

But, in an environment that is changing violently—as may well have been the case four or more billion years ago—would not variability be a good thing? Any population would have been exposed to extreme environmental pressure. If conditions are that tough, shouldn't there be as many "hopeful monsters," as one biologist put it, out there as possible?

"Yes," says Woese. "In an environment that is stretching an organism to its limit, there will be great difficulty in surviving. Forget about the kind of fine-tuning that you would encounter in a stable environment. With stability a population has time to make the small delicate adjustments that its members normally engage in. When things get really bad they can't afford that luxury. It becomes a matter of brute survival. I liken it to a human population in wartime. In peace, all sorts of niceties in dress, manners, and other conventions come into play, but in war, I repeat, it's brute survival. Everything else is stripped away and you're down to basics like getting enough food."

The universal ancestor, because of its assumed small library of genes, was probably a hundred or a thousand times more error-prone in its genetic replication than a modern organism is. That would have resulted in a lavish scattering of novelty across the landscape and would have increased the probability that at least some of it might stick. But, again, this also raises the question of how, in such an unstable population, there would have been enough unity, enough coherence—indeed enough genetic material—to make even minimally functioning individuals. Woese has a possible solution to that problem. He suggests that the blueprints in the ancestor were not strung together in long strands (chromosomes), as they are in us, but existed in many copies of short separate genes. There would have been some good ones and some bad ones in such a collection. The way to integrity would not be to repair the bad ones but to get rid of them.

Getting rid of a gene is not easy when genes are connected together in long chains, because when the bad gene is thrown out the whole chain goes with it. But if genes are separate, that could be done more easily. One could visualize a recognition process within the cell whereby a badly flawed copy of a gene would be distinguished from the good one and discarded, enhancing the viability of the individual.

Clearly those early genes were not like genes today. In addition to thinking that they were physically separate from one another, Woese feels that they played direct functional roles in the cell, as RNA now does. Therefore he agrees with Manfred Eigen and others that RNA, and not DNA, came first in the earliest organisms.

The case for a primitive, inaccurate, rapidly evolving ancestor fits well with what biologists observe of the three primary branches of Woese's tree. Each branch has its unique biology. Cell architecture is unique, metabolism is unique, and so is the way genetic information is organized. The kind of variation that later occurred within each of the three kingdoms is minor by comparison. What apparently was going on prior to the three primary splits was a rapid churning of life forms that could and did produce unusual evolutionary changes and also produced them at unusual speed. There would have been a step-up in the quality or kind of evolution (its mode) and in its rate (its tempo). It would have introduced a great deal of novelty at the time the three major kingdoms were in the process of formation. This would explain how, given a welter of options, each kingdom could have ended up with a slightly different life process, its personal, tailor-made set of biological solutions, each coming from a smorgasbord of opportunities presented by a small and inaccurately functioning ancestor. Indeed, if the ancestor had been capable of greater accuracy, there might not have been three descendant kingdoms; there would not have been so much to choose from.

Looking at the three kingdoms as carefully as he has, Woese can see no other way of explaining their origin. They are basically too different for any one to have evolved directly from any of the others. And yet they go about their business in a sufficiently similar fashion, as to suggest beyond doubt that all three learned it from one teacher.

So, a single ancestor—a heat-lover and anaerobe. Perhaps one that originated in hot sulfur springs, using the heat there to metabolize the sulfur. What a strange way for life to start. And, if true, how far that steaming, bubbling world is from the picture of a placid little warm pond that Darwin gave us more than a century ago. How far also from the cool chemical soup that Stanley Miller brewed. Miller was concentrating on the ingredients of an early atmosphere rather than its temperature. He heated his mixture only to get it moving through his apparatus in vapor form so that he could condense new compounds out of it. His experiment was based on the proposition that an abundance of the right chemicals and a little push from sunlight would do the job. It now appears that near-boiling water was a necessary ingredient. All the best evidence points that way.

We end this chapter without having found a bridge over the gap between incipient life working its way upward through a knitting together of increasingly complex chemical organizations, and real life working its way downward toward chemical organizations too simple to be regarded as alive. In both cases we have run up against the problem of smallness of the genetic arsenal. Manfred Eigen, with his hypercycles, suggested a way of getting around that original smallness and instability. Carl Woese acknowledges that the universal ancestor, whatever it may have been like, must also have had to deal with this problem: very small and unstable genes. It is interesting that each man, coming from a different direction, has arrived at the edge of the chasm with the same dilemma in his hands. Meanwhile work at the molecular level continues at an unprecedented pace. As information of a kind not even dreamed of a few decades ago continues to pour in, we can see the gap narrowing. Will it ever be closed? Who can say?

"Don't bet that it won't," said Don. "During my short lifetime, what we have learned from fossils has completely rewritten our ideas about human evolution. There's no telling what the molecules will do for all of evolution."

DNA and the Fossil Record

The great tragedy of science—the slaying of a beautiful hypothesis by an ugly fact.

—T. H. HUXLEY

I know my molecules had ancestors. The paleontologist can only hope that his fossils had descendants.

—VINCENT SARICH

18

What Old Bones Have to Say About Human Evolution. What Molecules Have to Say.

The evolutionary thrust of the molecular story is so recent and so dazzling—and so complicated—that one tends to get bogged down in it. Up to our necks in nucleotides, we may forget that most of the work in the field of evolution has been grounded in fossils. We have said almost nothing about fossils so far. Now it is time to do so.

We have also left untold the story of human evolution. We have done that deliberately. The best current knowledge of how we evolved depends on a blend of fossil and molecular evidence. It produces a better understanding of our ancestry than either could have done by itself. There is a bone story to be told; then it must be edited by molecules. So, now that we have struggled through to a little knowledge of molecules, let us look at bones. Let them be human bones.

"Don't forget," said Don, "fossils for a long time were the only hard evidence we had that evolution had occurred. Without them, how would you follow the paths that older organisms took to reach the forms of their living descendants?"

"You wouldn't."

"You can't invent an ancestor by looking at yourself. You just can't guess which way to look, what kind of animal to draw. You have to go out and find one. Molecules may be great for tracking relationships, but they will never produce a Miocene ape to tell you what your umpty-ump-great-grandfather looked like."

"Never."

"Who could ever have dreamed that dinosaurs came and went. I mean, they just *vanished.* It would take a pretty drunken imagination to invent one. You won't find it in a test tube."

"That's for sure."

"I want to emphasize this. The story of human evolution has been told by finding bones. It's been a long hard trip because the bones have been so scarce. We've made some horrible mistakes along the way. We'll make many more. And I'll concede that the molecules clarify some things that bones never will. But I repeat, you cannot tell the story of evolution without a lot of old bones. You have no way of knowing what was there."

The trouble with old human bones was that at first there weren't any. When the *Origin of Species* was published in 1859 the human fossil cupboard was empty. This caused problems for Darwin, who had mountains of fossil evidence from plants and animals to shore up his theory, but not one trace of a fossil human. That lack, plus his reluctance to toss humans on the hot fire that he knew his book would ignite, forced him to limit his speculations about mankind to one tepid sentence, and we repeat, "Light will be thrown on the origin of man and his history." For someone like Darwin, who required evidence, the evidence was not there.

We said "not one trace." Actually there was one. Part of a skull and some limb bones of a curiously primitive human had been found in a cave in the Neander Valley in Germany three years earlier. But they were of no help to Darwin, for there was no agreement at all as to what the fossil—later named Neanderthal Man—really was. Most of the scientists who took the trouble to look at it concluded that it was a damaged or pathological specimen of a modern man. There were some other conclusions about it:

- It was the skull of an elderly Dutchman.
- It was a Cossack soldier who had gotten lost while chasing Napoleon's army, wandered into a cave, and died there.
- It was a hulking Celt who resembled a modern Irishman and had a low mental organization.

- It was a deformed modern individual. He had suffered rickets as a child and arthritis as an old man and had taken some heavy blows to the head in between.

"So what was Neanderthal Man?" I asked.

"That's a good question," said Don. "At first he was nothing, one fragment of some kind of peculiar hominid.* With just one thing on the table, what do you compare it to? Yourself, obviously, because there isn't anything else. And what do you look at? The differences. When you find differences, you tend to emphasize them. And when other Neanderthalers began to turn up with those same differences, scientists emphasized them more and more. Gradually they came to the conclusion that there actually was a primitive ancestor to modern men fifty thousand years ago, maybe a hundred, maybe two hundred, they couldn't tell."

"The dating was that vague?"

"There was no dating. The kind of dating we do today, using radioactive elements, was still a hundred years in the future. The only way to calculate the age of a Neanderthal fossil was by crude geology—where in a layer of sediments it was found, how deep in a cave floor—and by the kinds of animal bones it was found with. You don't get actual dates that way. All you get is 'older' than something else, or 'younger.'

"Anyway, a hundred thousand years ago the climate in Europe was different. There were a lot of large mammals wandering around: woolly mammoths, giant cave bears, woolly rhinoceroses, and saber-toothed cats. Neanderthalers hunted and killed those animals. They used spears and stone tools. They were men, definitely, but they were *primitive men;* the emphasis was always on that. It explains the pictures you've been looking at in books for the last seventy or eighty years. Neanderthalers were seen as low-grade brutes. They had beetling brows, they slouched and shambled along. They were the model for all the comic-strip characters who beat women over the head with clubs and drag them off to their caves by the hair."

---

* We define hominid again: a member of a group of erect-walking primates that includes modern humans, their extinct ancestors, and some extinct, but also erect-walking, cousins—the australopithecines. We will meet the australopithecines later in this chapter.

"Neanderthal wasn't all that primitive?"

"Not at all. At first he was thought to be. He got only grudging acceptance as an ancestor, and some people wouldn't concede even that. They shoved Neanderthal to one side and concentrated on another early human type that was beginning to show up in increasing profusion in southwestern France. That was Cro Magnon Man, first discovered in a cave in 1868. He was younger than Neanderthal, about thirty thousand years old, and had a much more elaborate culture. He made better stone tools and was responsible for the magnificent wall paintings that have been found in caves in the Dordogne Valley in France."

"Cro Magnon Man was like us?"

"He *was* us. Today Cro Magnon Man is labeled a *Homo sapiens,* just as you are. He wasn't so-labeled right off the bat, because there were very small skeletal differences between him and you. But so are there between many races of men around the world right now. Local populations produce local variations, just as they did with tortoises in the Galapagos. The trouble with humans is that they move around a lot more than those tortoises did. They don't stay isolated in small populations and build up their differences. They travel, they fight, they capture women, and they interbreed. That tends to smooth out the bumps. But never entirely. There was surely less moving around fifty thousand years ago than there is now. And when you take that into account you begin to wonder if even that old shaggy Neanderthaler wasn't just a slightly peculiar local type hanging out in southwestern Europe. That began to seem more likely as fossils were discovered to the east. There the beetling brow and heavy bones were found to be much less conspicuous. Finally—and this is the late news flash on Neanderthal—he, too, is *Homo sapiens.* He gets the subspecies name of *H. sapiens Neanderthalensis.* We get the name *H. sapiens sapiens.* What it took to bring all that into focus was the discovery of something on the human line that was really primitive."

The something "really primitive" that Don was referring to was discovered in a riverbed in Java, an island in the East Indies, by a young Dutch doctor named Eugène Dubois. Having read Darwin and believing in evolution, Dubois examined the Neanderthal skull. Convinced that it was a human ancestor, he was seized with the dream of finding something older, a missing link between apes and man. Java was then a Dutch colony. Knowing that there were apes there, he managed to get himself posted to Java.

In 1893 Dubois reported the discovery of part of a skull and some teeth. Those, he said, were from the missing link, something so primitive and apelike that it could stand nowhere except on the borderland between apes and men. He gave his find an appropriate name, *Pithecanthropus erectus,* using two Greek words—*pithecos* for ape, and *anthropos* for man—to nail down his contention. The species name, *erectus,* was justified by a leg bone found near the skull. It was that of an upright-walking man. Then, as if that were not enough, Dubois announced that *Pithecanthropus* was five hundred thousand years old.

Its great age and primitive look shot *Pithecanthropus* into the headlines, and it quickly earned the popular name of "Java Ape Man." Once again science was so struck by the differences between it and modern man—and also between it and the Neanderthal and Cro Magnon fossils—that at first they refused to find room for it on the human line at all. That pleased Dubois. Then they dismissed his half-million-year age claim. That displeased him. Finally Sir Arthur Keith, a leading English paleoanthropologist, gave the specimen a thorough review. He announced that it qualified as a primitive human being of some sort but was not a missing link. That so disgusted Dubois that he hid *Pithecanthropus* under the floor of his dining room and for thirty years refused to let anyone look at it.

Meanwhile other fossils of a similar type were being found. Inevitably each got its own name. There was Heidelberg Man, dug up in Germany, Peking Man (quite a few of them) in China, others in other places. All had things in common. Their bones were uniformly heavier and more massive than those of modern man. That was particularly true of their skulls, which also had pronounced eyebrow ridges. Clearly these were extremely powerful people, if indeed they were people. They made the brutish Neanderthalers seem positively effete by comparison.

That they were people was eventually established. Several of their dwelling sites were discovered in Asia. Study of those made it clear that their occupants had been hunters, that they made stone tools and knew the use of fire. It was also established that they had walked erect—something that critics of Dubois had questioned, pointing out that the leg bone Dubois depended on for his claim had been found some yards away from the skull and probably belonged to a different creature.

All this took time, of course, and a great deal of arguing. But when the beetle-browed specimens were studied en masse, it began to become clear that the human line did go back half a million years. There were men back then—not apes or ape-men, but *men.* These primitive people were scattered widely throughout the Old World. Although they had local physical differences, those were no greater than what might be expected, given such a wide distribution. In due course they were lumped together into one species that is now universally conceded to have been ancestral to modern man. It bears our genus name: *Homo.* Its species name is *erectus,* denoting what was then seen as the first appearance on the world stage of a mammal that walked upright, a trait considered to be unique among mammals, a mark of humanness.

The idea that erect walking began with *Homo erectus* did not last long. In 1925 a voice spoke unexpectedly from South Africa to announce the discovery of a much older, much more primitive fossil in a limestone quarry at a place called Taung. The specimen was the well-preserved front of a child's skull and for many years bore the name of "Taung Baby"—something of a misnomer, for, after its discoverer, Raymond Dart, had spent four years delicately picking bits of rock away from the fossil bone and separating the jaws, there was evidence that the "baby" was about six years old. It is now known as the Taung Child.

But a six-year-old what? Dart was an anatomy professor, and he thought he knew. He announced to a dumbfounded world that an erect-walking, apelike creature had been strolling the African veldt a million years ago. He named it *Australopithecus africanus,* the southern *(australo)* ape *(pithecus)* from Africa.

If people had been slow to buy Dubois's *Pithecanthropus,* they were not buying *Australopithecus* at all.

"That's the odd side of this business," said Don. "We're all out there looking for older hominids, but whenever anybody finds one, he gets the cold clammy hand."

"Not so much now," I said.

"That's because there have been so many finds in the last few decades that the people who were yelling the loudest began to get shell-shocked. They learned to go a little slower on their guillotining of new evidence."

"They also know more," I said.

"Yes. And that's important. Anybody in this business who knows his anatomy can recognize a hominid pretty well now. Those basic early bafflements are over. The picture is filling out. It's harder to shock people now. Back when Neanderthal Man was first discovered, the idea of a human ancestor fifty or a hundred thousand years old was enough to knock your hat off. Today we take ten or twenty times that without blinking. But when Dart spoke up he was doubling the life of erect-walkers in one gulp. People couldn't stretch their minds enough to accept that. Also, who said that this thing of Dart's walked erect? Only Dart, and nobody believed him. He was an obscure man from South Africa running up against all the big guns in Europe. And the biggest, Sir Arthur Keith in England, said that Dart had found a young chimpanzee.

"Dart knew better. Even allowing for the fact that young chimps and young humans look a lot more alike than the adults do, he could see by the teeth of the Taung Child that it was no chimp. Furthermore, chimps are apes. Apes live in tropical forests. There are no tropical forests in South Africa and haven't been for millions of years. That's why no ape fossils have ever been found in South Africa, only monkeys. Dart knew his monkeys well; he was familiar with fossil baboon skulls, and he could see immediately that the Taung Child didn't resemble a monkey at all. Since it could be neither ape nor monkey, and since he could not bring himself to call it a man, a *Homo*—it seemed far too primitive for that—he was forced to find a new slot for it. He decided to create a new genus: *Australopithecus.* Readers had better get used to that name because australopithecines are going to turn out to be very important to the story of human evolution."

Predictably, Dart got nowhere in his efforts to establish the credentials of his extraordinary find. As we have noted, one of his difficulties sprang from his claim that the Taung Child walked erect. He had deduced that from the way its spinal column left its skull, from a hole at the bottom rather than at the back, suggesting that the Taung Child's head had been perched on top of an upright body. But those deductions fitted none of the then-current ideas about primate evolution. When Dart took the fossil to London it was ignored. That infuriated another South African, Robert Broom, an eccentric itinerant doctor who had devoted a lifetime to the study of African fossils; he had become a world authority on the continent's ancient mammals and reptiles. He was greatly excited at the discovery of a hominid in South Africa, and disgusted that Dart, after several years of promoting the Taung Child, had given up the fight and returned to teaching anatomy. Broom became, as it were, Dart's "bulldog," as Huxley had been for Darwin. He entered into acrimonious correspondence with the men in London; then, hearing about other possible hominid sites in caves in South Africa, he began investigating himself. In 1936, in a cave at Sterkfontein, he found a hominid fossil. He found others in other caves.

Were they like the Taung Child? Yes—no—yes. They came in fragments, crushed parts of skulls, pieces of limb bone, splintered jaw parts, many teeth. Sorting them out took a long time. Broom gave *his* names to them.

After World War II had come and gone, paleoanthropologists in Europe and America finally consented to take a serious look at what by that time had grown to be a respectable collection of bits and pieces. The conclusion was unavoidable. Australopithecines were legitimate. They were not apes. They walked erect. They could well be human ancestors, so old and so primitive that they could not be called humans. "Not yet human, but on the way," was how one anthropologist put it. In that sense they were the missing links of earlier scientific hope—with fossil apes in their past and true men in their future.

Broom may have done most of the hard work in establishing the australopithecines, and he was certainly the better scientist. But to Dart goes the everlasting credit for discovering the Taung Child and recognizing its true significance. It is entirely proper to say of him that he triggered the modern era of paleoanthropology, which for the next sixty years would be centered on learning more about the australopithecines. Dart lived through that entire explosive period,dying in November 1988 at the age of ninety-five.

Dart had said that the Taung Child was a million years old. Broom could neither confirm nor deny that. All the South African fossils came from caves. They were embedded in a mixture of sand and pebbles cemented by limestone into a rock known as breccia. It is impossible to date breccia. It is like a pudding whose ingredients have come from different places at different times and are now all mixed up. Even crude attempts to date hominid fossils in breccia by associating them with adjacent mammal fossils are questionable. An antelope may have fallen into a cave at an entirely different time from that of the arrival of an australopithecine. Thus the best that could be said of the australopithecines was that they were old; nearly all the mammals found in caves with them were extinct species.

Later it became possible to date them more precisely. The mammal fossils with them were found to occur in other parts of Africa where dating is possible. Dart's guess of a million years turns out to be better than anyone thought at the time. Today it seems short. Australopithecines appear to have become extinct a million years ago and were living for at least two and one-half million years before that in some parts of Africa.

One great puzzle about the fossils of Dart and Broom (each went on to find more) was that they appeared to come in two kinds. There was a smallish, slightly built kind: the Taung Child and others like it. There was also another bulkier type with slightly different features. These included extremely large and thickly enameled molar teeth, and a bony ridge on top of the skull similar to the crest on the skull of a gorilla.

Males and females? Somebody suggested that; others took it up, recalling that there is much size difference between the sexes in many primate species, a phenomenon known as sexual dimorphism. But that phenomenon would explain the two types of australopithecines if only males lived in one cave and only females in another a mile away, for this was how the fossils were found. That idea died. Eventually, with the finding of still more specimens, and with the knowledge that the caves were of different ages, two species were identified: *Australopithecus africanus,* Dart's "gracile" australopithecine, the small type; and *Australopithecus robustus,* the bulkier type.

But were they human ancestors?

For a time there was no answer to that question. One way of getting at it was to find out if australopithecines made and used tools, since that ability has always been considered a human accomplishment. Dart thought they did. After examining a great many broken animal bones found in caves with the australopithecine fossils, he came to the conclusion that they had been broken deliberately and used as clubs, spears, knives, and so forth. He coined a clumsy name for that use: the osteodontokeratic culture (from the words for bone, tooth, and horn), and went on from there in lurid language to describe the australopithecines as murderous little fellows who differed from apes in their propensity for killing. They were, he said, "carnivorous creatures that seized their living quarries by violence, battered them to death, tore apart their broken bodies, dismembered them limb from limb, slaking their ravenous thirst with the hot blood of victims and greedily devouring livid writhing flesh."

He found a disciple in a traveling American writer, Robert Ardrey, who unfortunately picked up the killer-ape idea and incorporated it in a best-seller, *African Genesis.* His book promoted the idea that modern man's propensity for warfare and killing stems from an aggressiveness developed when an erect ape first picked up a club and discovered that it could be used as a weapon.

The osteodontokeratic culture idea did not wear well. But the search for tools — either weapons or everyday implements — went on. Or, to put the inquiry the other way around, where stone tools were found, might not one find hominids there?

One place where stone tools were known to occur was in the Olduvai Gorge, a dry riverbed in Tanzania, more than a thousand miles north of the southern caves. For decades a husband-and-wife team, Louis and Mary Leakey, had been working there. Over the years they discovered a great many animal fossils, some new to science. They also found bushels of tools — but no hominids until late one afternoon, as Louis lay in his tent with a fever, Mary spotted a skull peeping out of the eroding wall of the gorge.

It was an australopithecine, the first ever to be recognized from outside South Africa. It was a robust one, but so much more robust than the others that Leakey decided it deserved to be put in a genus of its own. He named it *Zinjanthropus,* and "Zinj" became famous as the most rugged crested hominid ever seen, with back teeth like paving stones.

Was this crested creature the tool user? Apparently so; there were tools all about it.

Three things quickly happened.

First, a new dating technique that depended on the decay rate of radioactive potassium in volcanic rock was applied to the deposits at Olduvai. It gave Zinj an age of 1.8 million years, confirming at last the claim that Dart and Broom had been making for the great age of their fossils. What a triumph for them!

Second, a more leisurely examination of Zinj made it clear that its extreme robustness was more a matter of degree than of kind. This *was* a robust australopithecine; it did not belong in a separate genus. However, it was not exactly like the South African *robustus*. The two did not overlap geographically, and they had small but consistent physical differences, a fact made clear when more of the super-robust specimens were found by Leakey's son Richard at Lake Turkana in northern Kenya. Ultimately they were determined to be close—but separate—species. Zinj now bears the name *Australopithecus boisei,* honoring Charles Boise, who had been an early financial supporter of Louis Leakey.

Third, *Boisei* turned out not to be the tool maker. Soon after it was discovered the Leakeys made another sensational strike at Olduvai Gorge. They made several, in fact: skull and jaw parts of four individuals that were not the robust type. They were more manlike, had larger brains than any australopithecine. Those, said Leakey, were men, the oldest men ever known, 1.6–1.8 million years old. Erect men and erect ape-men apparently walked the Gorge together; they may even have looked each other in the eye. Anthropologists had to swallow hard again, for here were two lines of hominids, not a single one as they had always thought. There had to be two, because the two sets of fossils were contemporaries found in deposits of the same age; the one could not be ancestral to the other. Each moreover seemed to be evolving in a different direction.

Leakey named his new find *Homo habilis,* or handy man, to underscore that it was this creature, and not *boisei,* who was the tool maker. That simplified the picture markedly, for the idea of having such an inhuman creature as *boisei* on the human line had never gone down well. But even with *H. habilis* as a replacement ancestor, things were still far from satisfactory. Critics said that the bits and pieces of the *habilis* skulls were too fragmentary for its brain size to be calculated accurately. Fit them one way, and you got a fairly large brain. Fit them another, and you got a brain in the australopithecine range. As a result, *habilis* had to endure a shaky, quasi-legitimate status in the family of man for a number of years.

Whatever *habilis* might turn out to be, the century after Darwin had produced a phenomenal growth in our understanding of our human roots.

"Remember what Darwin said way back in the nineteenth century?" asked Don. "He predicted that the ancestors of man would be found in Africa. That was an inspired prediction because he had almost nothing to go on. All he knew was that men bore a close relationship to chimpanzees and gorillas and that those big apes came from Africa."

"He might as well have guessed Java," I said. "That's where Dubois went to find the Java Ape Man, and that's where another big ape, the orangutan, lives."

"Yes, he might have. But he didn't. And he was right; chimps and gorillas are more like humans than orangs are. Darwin sensed that—although there are people who still argue that orangs are closer. Now it looks increasingly likely that they're not. Their DNA molecules say so; we'll get to that in a moment.

"Meanwhile we have the *australopithecines!* There they are; you can't duck them. In Africa, nowhere else. And we have *Homo habilis* there too, the oldest man in the world. What a time to be a paleoanthropologist. I couldn't wait to get to Africa myself and find something, because I could see there were big unsolved problems still out there. We had all those great fossils, but there was no general consensus on them. Dart and his allies clung to *Australopithecus* as our ancestor. The Leakeys threw *Australopithecus* off the line entirely and pinned their flag on a separate, as yet undiscovered, older ancestor for *Homo.* With things as they were, it was impossible to say who was right. Clearly we needed to find older fossils to answer the sixty-four-dollar question: were australopithecines ancestral to human beings or weren't they?"

What had been accomplished in the century after Darwin is best shown in the two diagrams that follow. The first has some serious mistakes, but it is still far more elaborate and informative than anything that Darwin's contemporaries might have drawn. It is derived from a tree made by Sir Arthur Keith in 1931 and represents the thinking by the best paleoanthropologist of that decade. It sorts out correctly the relationships between New-World monkeys, Old-World monkeys, and apes. It also adds humans to the tree, and attempts to find branches for the then-known fossil human types: Java Man, Peking Man, and Neanderthal Man. It even finds two spots for the newly discovered *Australopithecus:* Dart's preference on the human line; and Keith's own placement of it on the ape line, which reflects his view that the Taung Child was a chimpanzee. The

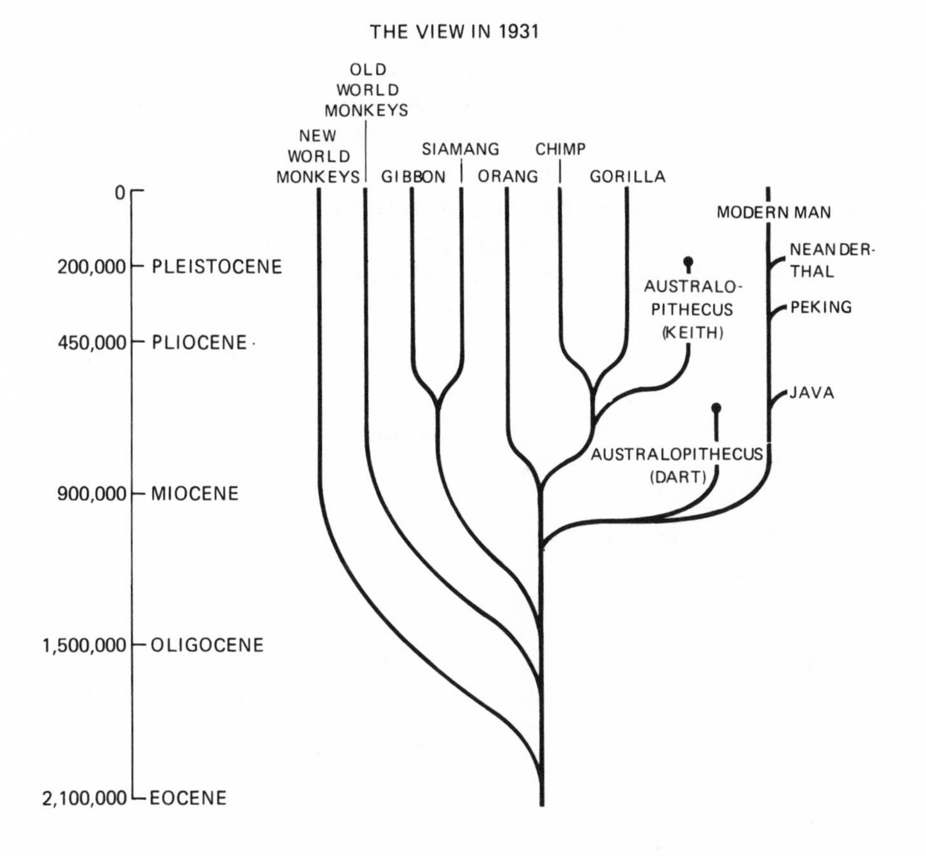

great flaw in the Keith tree is its absurd constriction of time. We know today, thanks to radiometric dating, that the Miocene age started twenty-four million years ago, not one million. We also know that *Homo* already existed about two million years ago, which was when Keith thought that the entire primate line still had a common ancestor. In that respect he was off by a mere thirty or forty million years.

The second diagram represents the most commonly held view of thirty years later. By that time a great deal of new fossil evidence had accumulated, and the tree of the 1960s reflects that. It recognizes the small gracile *africanus,* of which Dart's Taung Child was the type specimen, as the probable ancestor of humans. Its credentials were good. It was the oldest erect-walking fossil hominid known and the most manlike. Through better knowledge of the South African caves it was now acknowledged that *africanus* was older than the two robust types and might have evolved into them. The conviction grew that the latter's apparent primitiveness was the result of a tendency to depend increasingly on a diet of coarse vegetable matter. That would explain their cranial crests, their huge back teeth,

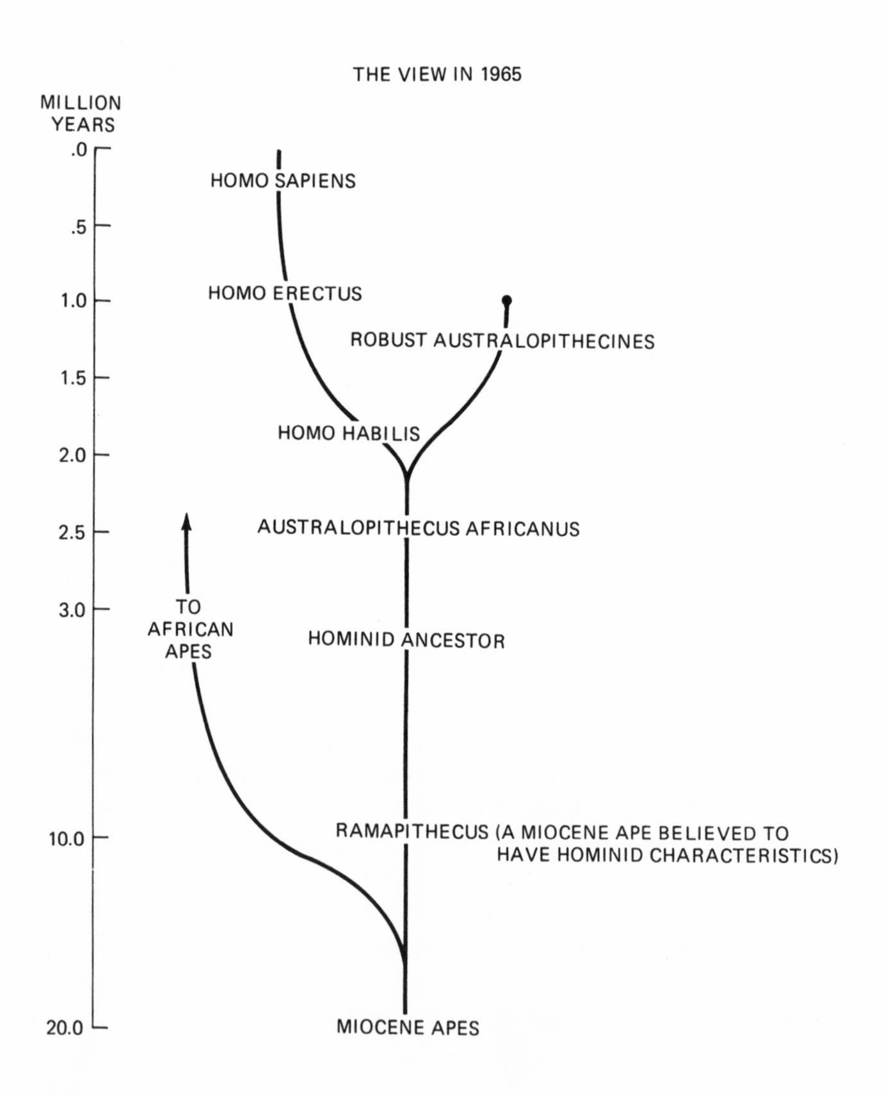

and oversize jaws. In their case, what looked primitive might well have been a growing evolutionary specialization away from *africanus.* On this evidence the diagram of the 1960s is drawn to show *africanus* as the ancestor of the other australopithecines and of *Homo habilis* as well.

Whatever the case, all australopithecines seemed to have gone extinct by one million B.C. For those seduced by the concept of the killer-ape as the prototype for the emerging killer-man, the reason for that extinction was obvious; the emerging *Homo* simply exterminated the hapless and dimmer-witted robust. For others the answer was not that simple. Today nobody knows why the australopithecines went extinct, although there is a growing feeling that they became too narrowly specialized and failed to adapt to changes in the environment.

Looking toward the bottom of the second diagram, we must drop back about fifteen million years before we encounter the next older candidate for hominid ancestry. Here we run into *Ramapithecus,* one of the younger of a large group of primitive apes that swarmed in the forests of the Miocene age, a warm period that began a little over twenty million years ago. During the 1950s and 1960s that group was subjected to increasingly close scrutiny by two paleontologists, Elwyn Simons and David Pilbeam, both then working at Yale.

The Miocene apes are a tantalizing group, tantalizing for being so dimly seen. They were scattered widely over Europe and Africa, with important finds having been made in northern India and Pakistan. Twenty million years ago they were an abundant and extremely successful group of mammals. But as time went on—with the climate changing and the tropical forest shrinking—they became less so. By ten million B.C. they were becoming increasingly scarce. By about nine or eight million B.C. they were on the verge of disappearing.

Fossils of those late Miocene apes are correspondingly rare. But, rare or not, logic compelled a search among them for a hominid ancestor on the ground that they were the closest thing, both in time and in their physical characteristics, to the australopithecines and humans that would evolve later. In that search, largely conducted by Simons and Pilbeam, the choice fell on *Ramapithecus.* Simons reconstructed a complete *Ramapithecus* jaw by making a mirror image of a half-jaw in the Yale fossil collection. So reconstructed, it showed the teeth running in a curve around the mouth. This is a human trait, as any reader can learn by standing open-mouthed in front of the bathroom mirror. Modern apes—and the other Miocene apes as well—have rectangular tooth rows; the teeth on opposite sides of the jaw are parallel to each other. On the evidence of that distinctive curve in its tooth row, *Ramapithecus* became the ancestral candidate and is so shown at the bottom of the diagram on page 338.

"That, very roughly, was the state of the art when I hit Africa in 1970," said Don. "We had a logical but unproven arrangement of hominids after about 2.5 million and a very questionable ancestor at about 15 million. Nothing in between. Boy, was that ever going to change."

Some of the changes would be wrought by two young men, by Don himself and by Richard Leakey, the son of Louis and Mary Leakey. Both were in their early thirties with their careers still ahead of them. Richard had grown up in the shadow of his two world-famous parents. He had become so fed up with a family life drenched in paleoanthropology that when he finished high school he stopped his education to become a white hunter, a leader of shooting and photographic safaris. He became very good at that, so good, in fact, that when his father proved too old and lame to become the Kenya member of an international field expedition being formed in 1967 to work deposits along the Omo River, just north of Kenya in Ethiopia, Richard was recruited to replace him.

Joining the Omo expedition, Richard quickly found himself playing a backup support-and-supply role. He was a mere high school graduate among a group of card-carrying Ph.D.'s who were doing all the interesting work. Dissatisfied, he left the Omo group and took off on his own to investigate some eroded badlands in an uninhabited and virtually unexplored region of northern Kenya on the east shore of Lake Turkana. Almost immediately he began finding hominid fossils. He became an instant paleoanthropologist and has been a prominent one ever since.

Don got to Africa himself in 1970. He was still a graduate student and had not yet finished his doctoral dissertation. He spent three field seasons working at Omo with his professor, Clark Howell of the University of California. Howell also took him to South Africa, where the two men devoted a month to examining all the hominid fossils that had been found by Dart and Broom. It was an unforgettable experience for Don to handle those legendary finds. Doing so, he began to get a good sense of what australopithecines were like. They were not like chimpanzees, he saw quickly (his doctoral paper was about chimpanzee teeth and he knew them well); how could Sir Arthur Keith ever have thought they were?

Then he got a chance to go to Nairobi, to meet the Leakeys and see some of the equally legendary finds they had made. There were the *habilis* specimens from Olduvai Gorge. Equally exciting were some new super-robust individuals that Richard had brought down from Lake Turkana; they had not even been published yet. Don was inspired by that rush of thrilling material. It was hitting him almost faster than he could take it in.

On his way home in 1971 he stopped off in Paris and met a young French geologist named Maurice Taieb, learning from him that there was a place in Ethiopia where fossils were abundant. When he was asked by Taieb to join him in forming a small survey party to go there, he stopped working on his thesis.

"Looking back," said Don, "that was one of the most foolhardy things I have ever done. Suppose it turned out that there were no hominids in Ethiopia. What would I do then? Without a doctorate I would have no academic future. I had little experience in running expeditions. I had no money and had to scrounge small grants and donations from a few friends. I knew nothing about Ethiopia or its customs. I didn't even know my partner, Maurice Taieb, very well."

The idea of finding something overbalanced all that. The dream, the golden hope, lay in one thing he had learned from Maurice: the deposits in Ethiopia were very old. Don had examined some pig fossils Maurice had brought back with him and recognized them to be the same as pigs he had seen at Omo. The Omo pigs were three million years old, which meant that the Ethiopian deposits were at least a million years older than those at Olduvai. Any hominid he found in Ethiopia would push the human ancestral line deeper into time than it had ever been pushed before. Whatever he found would have to be something new, something special.

What Don did find, at Hadar in the northeastern Ethiopian badlands, was something that probably no young paleoanthropologist will ever duplicate. His discoveries there, coupled with those of Richard Leakey on both sides of Lake Turkana, have revolutionized our way of looking at our ancestry.

Don's first trip to Ethiopia with Taieb was not a real expedition; it was a fishing trip. The two men traveled about with a guide, covering as much territory as they could in a couple of weeks. But what Don learned was important. The deposits were huge; they extended for hundreds of square miles. They were old, in the two-to-four-million-year range. And they were full of fossils. He found no hominids on that first superficial survey, but the region was so strewn with other animal fossils that he decided to organize a full-fledged field expedition for the following year.

In 1973 Don was back in Ethiopia, this time as coleader of a joint American-French team of specialists. He would go back again in 1974, 1975, and 1976, after which time the sheer volume of his finds forced him out of the field and into the laboratory for the long task of analyzing his fossils and writing the paper that would result in the introduction of the species *afarensis* to the scientific world. A necessary choice, but a hard one, because the surface has been barely scratched in the Ethiopian badlands; there is much more out there to be found, Don is certain.

"We found a knee joint our first year," said Don. "It was of an upright-walking creature and therefore had to be a hominid, although we could not tell what kind. All that we knew was that it was very small. You can't make much of an animal out of one knee joint. What is the rest of it like, you ask yourself. How about its head, its hands? We answered those questions with a crash the next year. We found a whole series of wonderfully preserved jaws, and, of course, we found Lucy. If I never do another thing as long as I live, I think my place in paleoanthropology will be secured by the finding of that one fossil."

We wrote on the opening page of this book that Lucy was the most famous hominid fossil ever found. The reason is that she packs so many "firsts" or "mosts" in her small frame. Here are four:

- At the time Don found her she was the oldest known hominid —by a good half-million years.
- She was an upright walker, no doubt about it. Her pelvis and leg bones proved that. Again, the oldest known.
- Her head was the most primitive and one of the smallest-brained of any hominid known.
- Her skeleton, when assembled, was the most nearly complete of any "old" hominid fossil.

That last point, her completeness, was one of the most important things about Lucy, and one of the most astonishing. Hominid skeletons of that great age and quality are simply unknown. Everything else comes to us in the form of fragments: skull parts, jaws, an arm bone here, a leg bone there. From such bits and pieces paleontologists can put together a whole individual of sorts, but it must be a patchwork job. The human skeleton is a fairly flexible piece of architecture. A composite made from a score of individuals who may have been separated geographically by many miles and separated in time by thousands of years, and who may even be of opposite sexes, can only end as a somewhat blurred and unconvincing whole.

With Lucy, the whole—or more precisely, the half—is right there on the table. Actually she ends up being more than half. Since we are bilateral creatures, a missing arm or leg piece from one side can be found in its mirror image on the other. Taking Lucy that way, we have her nearly whole, except for some missing skull parts that Don would dearly love to have. All in all, an almost unbelievable find.

Lucy's spectacular age and even more spectacular completeness have tended to obscure the importance of a strike made the following year that in some ways is even more revealing than she is. That is a cache of bones of an entire group of individuals who apparently died all at once in a catastrophe of some kind—perhaps a flash flood. Their bones were leaking out of an eroded hillside in profusion when Don's expedition found them. All in all, several hundred teeth and pieces of bone were recovered. From duplication of parts it was possible to conclude that the group had consisted of at least four children and nine adult males and females. This was a large enough sample to provide a sense of the variability that existed among those long-vanished creatures. It also gave Lucy something to be compared to; if hers had been the only fossil, it would have been impossible to decide whether she was a representative specimen of a species, all of whose members were small, or a female of a species whose males were larger. The hillside collection confirms the second possibility. Don has christened it "The First Family."

Unable to return to Ethiopia after his first four field seasons, and with a mountain of fossils to study, Don recruited a friend, Tim White, to help him analyze them. Tim had worked in the field with both Richard Leakey and his mother, Mary, and he knew australopithecine fossils well. He was just what Don needed for help with the Ethiopian collection. He was then completing his Ph.D. Don had found a job at the Cleveland Museum, and the two met in Cleveland every chance they got. After prolonged study they decided that the collection represented a new and primitive species of australopithecine, that it was probably ancestral to all other australopithecines, and probably ancestral to *Homo* as well. They also learned that Don's fossils were virtually identical to some more meager specimens that had been discovered at about the same time in Laetoli, Tanzania, where Mary Leakey had a dig and where Tim had worked on an evaluation of those specimens.

Don and Tim, with a French colleague, Yves Coppens, named the new species *Australopithecus afarensis,* honoring the Afar desert in Ethiopia where it had been found. *Afarensis* was introduced to the world in a landmark paper that they published in *Science* in 1979. It had taken them two years to complete their analysis. Most reactions to the *Science* paper were favorable. But Mary Leakey rejected it out of hand. She did not approve of their lumping her Laetoli fossils with Don's Hadar ones. She also insisted that no australopithecine should be placed on the *Homo* line. That was the thesis earlier enunciated by her husband, Louis Leakey. Mary and her son Richard clung to it even after Louis's death in 1972.

Today Lucy and her kind are generally accepted as a true species. *Afarensis* is real, most paleontologists now concede, although there are ongoing arguments as to how later types connect to it.

Confirmation of the reality of *afarensis* came, surprisingly, from Mary Leakey herself. At her site in Laetoli were found some ancient footprints in a shallow ash layer from a nearby volcano. These were carefully excavated and preserved by Tim White during the season that he worked at Laetoli. They are among the rarest and most astonishing vestiges of human bipedal ancestry ever found: more than fifty impressions made by two individuals, one smaller than the other, walking along together. Criss-crossing the same ash layer are the prints of birds, elephants, and numerous other mammals. The hominid footprints are the same as would have been made by Lucy if she had had flesh on her bones. They are also the same as those a modern human would make—but they are 3.5 million years old.

That those footprints should have been preserved at all is a miracle. That they are precisely datable is another. That they confirm the existence of an upright-walking hominid at that great age is the crowning gem in that paleoanthropological diadem. Lucy's bones said she walked upright. The Laetoli footprints confirm it. And thus, feet clinch the unity of the Laetoli and the Ethiopian fossils.

With the Ethiopian collection in hand, and with better dating

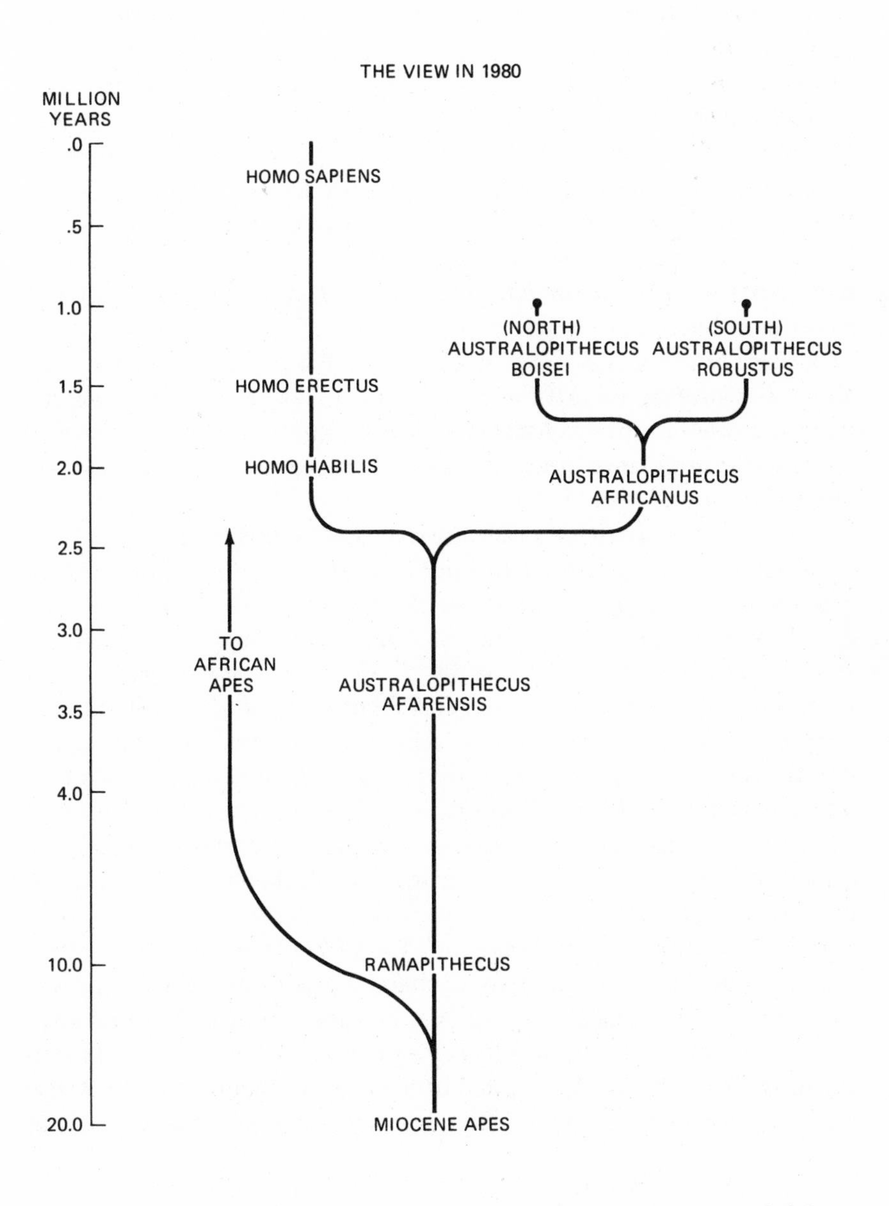

worked out during the 1970s, it became possible to revise the family tree shown on page 338. Here is what we might call the version of 1980, the one that seemed to make the most sense to most paleontologists at the turn of the decade.

The 1980 tree includes Lucy and her kind for the first time, placing her as the presumed ancestor of two descending branches. One branch leads to an increasingly large-brained line of *Homo*s. The other, continuing to be smaller-brained and retaining more of its australopithecine traits, leads to the three already known australopithecine types. The question of whether or not *africanus* is an ancestor to *Homo* can be better answered now. So long as there was nothing older, it was the best candidate. Now there is something older, and *africanus* can be looked at a little more dispassionately. Its retained australopithecine traits become more intrusive and are exaggerated in its presumed descendants, *boisei* and *robustus*. *Africanus* is probably not a human ancestor; its new position on the right-hand branch of the tree shows that.

"Please understand," said Don, "this isn't necessarily the last word. It's simply the most parsimonious and sensible arrangement that we could come up with at the time, given the fossils that we had. Find something that doesn't fit—adjustments will have to be made."

"How do you really feel about the Leakey position, given your expected bias as the finder of Lucy?" I asked. "What about *Homo habilis* and even older *Homo* ancestors?"

"The case for *Homo habilis* is strong," said Don. "After all, Richard did find the first good *habilis* skull. He did it in the face of a lot of withering criticism of his father for having invented the species on the basis of some poorly preserved fragments from Olduvai Gorge. People kept insisting that the so-called *habilis* was really just a few *africanus* fragments and didn't deserve being elevated to the status of a species. That argument was still rumbling when Richard produced his fantastic *habilis* skull from Lake Turkana. That shut them up pretty fast.

"On the other hand, the case for a long *Homo* line with no australopithecines on it is not strong. Richard seems to think that if he did it once he can do it again—find an even older *Homo*. There's room, there's time, for him to do it. There's nearly a million years between *afarensis* and *habilis,* with virtually no fossils ever found to fill that hole. It's one of the things I'm aching to do. I happen to believe that when a so-called linking ancestor is found, it will turn out to be something that connects *afarensis* to *habilis* through a mixture of traits, some kind of an early australopithecine verging in the direction of man. Richard thinks otherwise; he'd kick away all the australopithecines.

"Maybe he'll turn out to be right. But I don't think so. I've been living with australopithecine fossils for twenty years and I know them pretty well—what's going on evolutionarily, what the relationships are. But in the end, the only way to prove which of us is right is to go out and find the evidence."

True to that principle, Richard Leakey did go looking for evidence. His situation has been somewhat different from Don's. Commitments back home, particularly those involving the start-up of a new Institute of Human Origins, have kept Don out of the field. Richard is a Kenyan citizen and Director of the Natural Museums of Kenya. He can prospect wherever and whenever he wants in that country. He was grounded by ill health for a few years, having had to endure a kidney transplant, but is now back in the field, working deposits on the west shore of Lake Turkana. All his previous finds have been made on the east shore of the lake at a place made famous by him: Koobi Fora. Both sides of the lake are about as somber and desolate spots as can be found anywhere in Africa. They are arid wastes of gravel and lava rock, old lake beds dried up a million or more years ago, low stony hills eroded here and there into gullies by long-vanished rivers. This is a desert region, dry as a bone except for a very narrow green strip that edges Lake Turkana, and it is ferociously hot.

"That's also a good description of Hadar, the spot where I worked in Ethiopia," said Don. "The two places are very much alike. They are both part of the huge continental rift system that runs up across Africa and into the Red Sea. There are volcanoes all along it, and considerable earth movement. The rift goes down right through Olduvai and Laetoli and continues south. Hominids apparently walked up and down the rift for millennia, living along the river mouths and the shores of lakes. That's another argument for lumping the Laetoli and the Hadar fossils. They had a million or more years to spread up and down the rift. What's illogical is to say that they didn't."

By another irony, Richard Leakey has actually clarified Lucy's status with new finds at West Turkana.

"I disagree with Richard on much of what he says about australopithecines and early *Homo,*" said Don, "but I have to hand it to him as a fossil finder. His pioneering work at Lake Turkana gave our entire business a terrific boost. Consider these finds," and he ticked off three:

One: Richard has found enough *boisei* specimens to establish that form as a valid species. It is now a very well-known species, probably the best known of any early hominid, thanks to him.

Two: He put *Homo habilis* firmly on the map. Remember, it was his father who first proposed *habilis* as the earliest human on the basis of somewhat fragmentary and doubtful specimens from Olduvai Gorge. Richard removed that doubt in 1972 when he found a superb skull of the same type at East Turkana, a magnificent specimen that proves beyond question that his father was right, and that the human line does go back to two million.

Three: Now he's at West Turkana. Another pioneering effort, and another smash success with another wonderful fossil skull. His label for it is KNM-WT 17000 (the letters stand for Kenya National Museum, West Turkana), but it is popularly known as the Black Skull because of its dark color. It was found in 1985 by Alan Walker, a professor at Johns Hopkins and a long-time field colleague of Richard's. Walker's wife, Pat Shipman, also a paleoanthropologist, was a member of the field team and helped recover and reassemble it. Later she wrote an article for *Discover* magazine, stating that the new skull "in a single stroke, overturns all previous notions of the course of early hominid evolution."

"When we first talked about writing this book," said Don, "you asked me what was new in the early hominid line. The big news is the Black Skull. It's a very intriguing discovery. But I don't think it overturns things to the extent Pat Shipman says. On the contrary, it may actually clear up matters by allowing us to improve the tree on page 345."

"What is the Black Skull?" I asked.

"It's a super-robust, very primitive australopithecine." He went on to explain that it resembled *A. boisei* in a few respects and that Walker and Leakey had placed it in that species. But the trouble with that placement is that the skull turns out to be nearly half a million years older than any other known *boisei* specimen.

"Half a million years," said Don. "That's well out of the *boisei* time range. When you get a surprise like that you begin to look at the skull more carefully, and you begin to discover a series of primitive characteristics in which it resembles Lucy and the Hadar fossils more closely. Richard Leakey and Alan Walker found those when they studied the skull. They published a detailed description of it."

The heart of the Leakey-Walker paper was a listing of the Black Skull's characteristics broken down into the things it shared with all australopithecines, the things it shared only with one or another type, and the things that were unique to it.

It is important in studies of this kind to determine which features are "primitive" and which are "derived." "Primitive" refers to those that it shares with an ancestor and carries down relatively unchanged. "Derived" means evolved, altered. If two fossils share several derived features that are not seen in any other closely related type, it is an indication that the two are extremely close and that one may descend directly from the other.

"You weigh, you compare, you jiggle, you juggle," said Don. "You trace the primitive traits down until you lose them. You try to follow and lump together the derived ones. You note what's shared and what isn't. Sooner or later you begin to get a sense of what goes where. That's when it's possible to start sketching out a family tree. It's what Tim and I did with the Hadar fossils. We checked them against all the known australopithecines and the early *Homos*. The Hadar ones were so old and had so many primitive features that they had to be ancestral. Some of those features went toward the younger australopithecines and gradually changed; they became more and more exaggerated as time went on. Bigger back teeth, a dish-faced profile, a very large jaw. Others were retained in the human line. A few changed scarcely at all. You and I have in our teeth and jaws some characteristics that we get almost directly from Lucy. That's why, in the end, we felt compelled to draw two lines of descent, one to australopithecines and one to humans. Nothing else would satisfy. Now we have the Black Skull, and it's time to take another look at that pattern."

Under the Leakey-Walker analysis, and as noted, the Black Skull quickly revealed itself to be, in its numerous derived features, a super-robust australopithecine. In fact it had two characteristics that it shared uniquely with *boisei,* suggesting a close affinity to that species despite being so much older. However, it also had twelve primitive features that it shared uniquely with the even older *afarensis*. Such a large number of differences from the familiar younger *boisei* prompts the thought: *boisei*—? Perhaps, but not quite *boisei,* not yet *boisei.* In addition to being too old, it is too different. Leakey and Walker did not go into that. They limited themselves to describing their fossil. However, Pat Shipman, in her *Discover* article, went right at the problem of where the Black Skull fitted.

"It's a funny article," said Don. "Some parts of it are bang-on, absolutely right. She spots the Black Skull as something going toward *boisei.* She argues, I think correctly, that it can't be ancestral to or descended from any of the South African types. But she ignores its close affinity to *afarensis.* Instead of noting all the similarities that her husband listed, she gets into the old Richard Leakey argument that *afarensis* isn't a valid species. She says that there are big ones and little ones up there in Ethiopia, two kinds, maybe three —because, like Richard, she'd like to find an old *Homo* back there too."

"Certainly not three," I said.

"Of course not. There aren't any *Homo* fossils in the Hadar collection. Not two either. When you've got a population whose members all fall into an acceptable range of variation, who are alike— and I mean *alike*—in every respect except for their size, then you've got to assume that the big ones are males and the smaller ones are females. Sexual dimorphism is a characteristic of higher primates; you should expect it. You have big male baboons and little females. You have big *boisei* males and little females, big human males and little females. Why not big *afarensis* males and little females?"

"Why not?"

"Because if you accept that, you wind up with only one species back of three million. You wind up with *afarensis* as the probable ancestor of everything that follows. As I keep saying, Richard won't buy that."

"So, where does the Black Skull fit?" I asked.

"It fits right between *afarensis* and *boisei.* It shares unique features with each. It's a perfect transitional form. It's what you would expect to find if you were trying to get from one to the other. In fact, it isn't even a new fossil. Camille Arambourg and Yves Coppens found a mandible of the same age in southern Ethiopia in 1967. They even suggested a name for it, *Australopithecus aethiopicus.* But most people ignored it because there was only that one mandible, not much to float a new species on. Also, the French have a tendency to oversplit."

"I never heard of Arambourg's mandible," I said.

"Not many people have. It's been largely forgotten, because there wasn't enough of it to prove anything. But with the Black Skull, it instantly assumes some importance. Now there's something to compare it to. And there's so much information locked in the Black Skull that it would be permissible to name a species after it. I'd name it *aethiopicus,* just as Arambourg did. It has the projecting face of *afarensis,* but it also has some derived features of *boisei,* like a larger jaw, larger teeth, and a massive crest. It's a mix of the two."

"Parts of this skull were evolving faster than other parts?"

"Yes. It's a case of what is called mosaic evolution. Once the animal starts concentrating on a diet of coarse vegetation—maybe as a response to a change in the climate, maybe for some reason we don't know anything about—once that starts, then there's strong selection pressure to develop a big jaw, big back teeth, and a bony ridge on top of the skull, a sagittal crest. You'll need that crest because the head is small. You'll need a lot of bone surface to anchor the big muscles you're going to develop to do the kind of heavy chewing you're getting into."

"So you make the skull surface bigger with a sagittal crest," I said.

"You do. *Boisei* has it particularly. The Black Skull has a big one, but Pat Shipman doesn't read it right. She doesn't see the resemblance between the back of Lucy's head and the back of the Black Skull. She thinks Lucy's put together wrong. She thinks the *afarensis* skull that Tim and another anthropologist, Bill Kimbel, reconstructed is actually the cranium of one species and the face of another. No one else thinks that, but Pat seems to. And if you start thinking like that you're going to wind up not knowing what the Black Skull is, even though the evidence is staring you in the face."

If Don's assessment of the Black Skull is correct, a useful clarification of the tree on page 345 becomes possible. Two things now make the 1980 scenario incorrect. One is the Black Skull itself. It cannot be descended from *africanus;* it is just as old, and it is quite different. The other point has to do with *africanus* itself. Being the first australopithecine ever discovered, it played for a long time the common-ancestor role. That was logical. The logic became stronger when South African cave evidence revealed it to be older than the robusts. Also it was considered to occur all over Africa, since fossil teeth assumed to be those of *africanus* kept cropping up in places as far away as Omo in Southern Ethiopia.

But the Omo fossils were mostly single teeth. Although it is virtually impossible to identify a hominid species on the basis of a single tooth, that was routinely done at Omo with the scattered teeth found there. With no other species to choose from, an Omo tooth almost automatically got an *africanus* label.

"The people who were finding those teeth had to do that," said Don. "Where else would they put them? They weren't robust, so they had to be gracile. *Africanus* was the only basket that would hold them."

"What about Lucy?"

"Exactly. Those teeth would fit Lucy. But when they were found, there was no Lucy. She hadn't been discovered yet. Now that we have Lucy in the north, it's no longer necessary to look for *africanus* up there to explain the teeth. A consensus is beginning to grow that there never was an *africanus* in the north. If that is so — and I now believe it to be — then we can draw a really interesting, and in my view a highly plausible, tree.

"I'd call that the view in 1990. We're a little ahead of ourselves on the date, but by 1990 I think that the identity of the Black Skull and the geographical distribution of *africanus* will be agreed on. At last we have a clean and logical place for all the fossils. *Afarensis,* as the oldest yet known, remains the common ancestor. From it we can draw three lines of descent — three, not two; we can thank the Black Skull for that — each with an intermediate link. There is an australopithecine line in the north, with the Black Skull as the link between *afarensis* and *boisei.* There is a separate australopithecine line in the south, with *africanus* supplying the link between *afarensis* and *robustus.* Finally, there is a *Homo* line running from *afarensis* to *habilis.* What a neat and wonderful arrangement. There's only one thing wrong with it."

"What's that?" I asked.

"We don't have the *Homo* link. That's the one thing that we can't fill in. That's what I'd give my right eye for, to get back to Ethiopia and find the missing piece of the puzzle. We know that there was a very early *Homo* up there. We've found stone tools at 2.5 million. Those are among the oldest stone tools in the world by about half a million years. Who made them? It had to be a human being. No australopithecine ever made a stone tool; there is no evidence for it anywhere. Therefore, it has to be up there in the north, between two and three million years ago, that humanity started. And my guess is that it started out of something like Lucy."

"Something halfway between Lucy and *Homo habilis?*"

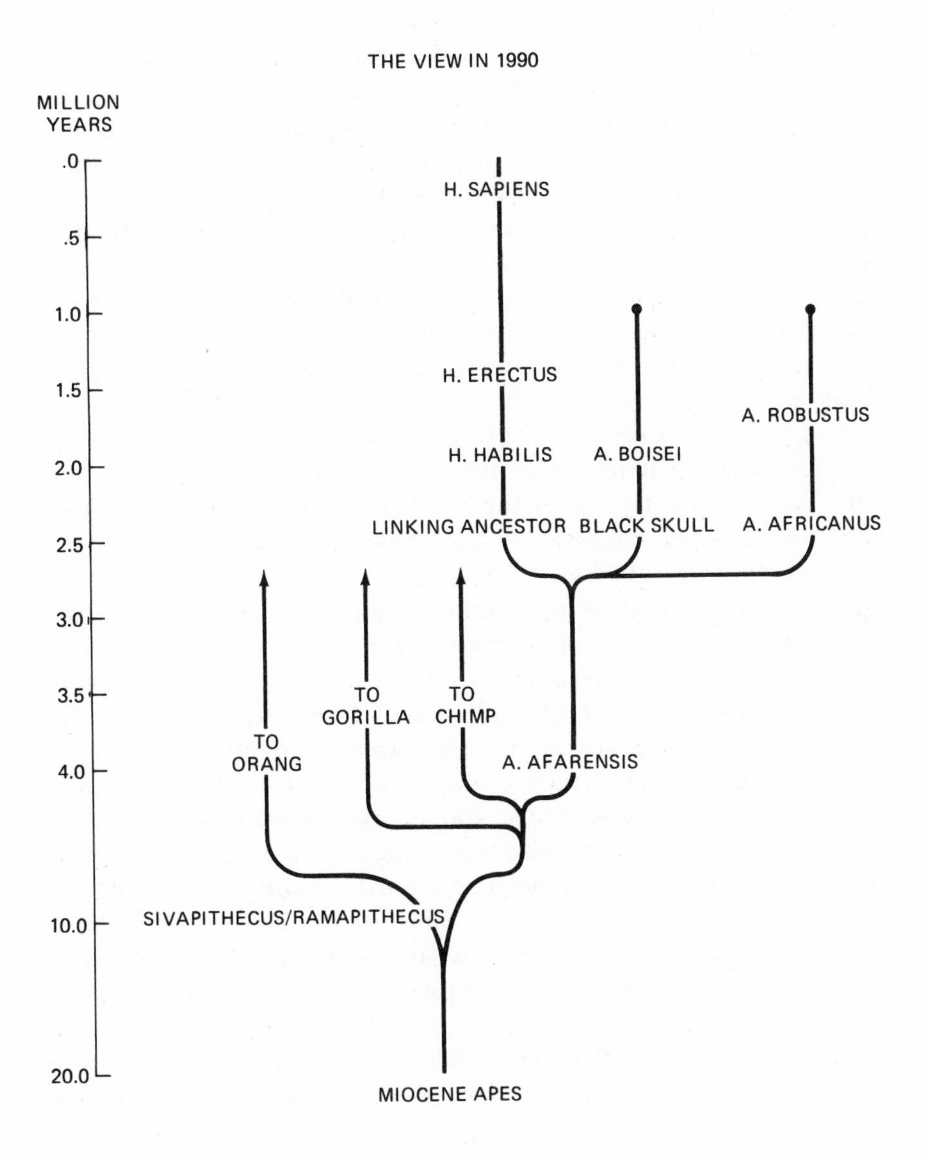

"Presumably," said Don. "But don't bet on exactly what. This is a humbling business. It's full of surprises."

Pausing to review the material covered in this chapter, we find it hard to realize that none of it—absolutely none of it—was known by Darwin. The entire story of hominid evolution has been written since publication of the *Origin*. Most of it has been packed into the last three or four tumultuous decades. It is now overwhelmingly clear that, as Darwin predicted, Africa is where humanity originated. Today, for the first time, we can begin to see with some accuracy the evolutionary path that we have traveled.

We have told that story, as we said we would, with fossils. Now it is time to make an independent nonfossil check to see if the story is accurate, or whether it must be edited by the evidence of molecules.

It turns out that a drastic editing is in order. To the dismay of those who get their evidence solely from fossils, the entire lower half of the tree on page 345 (the view in 1980) must be changed. To understand the severity of the change, that tree should be compared to the one on page 353 (the view in 1990). The differences at the bottom will jump right off the page. The old idea of a human-ape split going back for twenty million years is wrong. Apes now loom perilously close on the line of man's ancestry. The molecules decree that. They usher in one of the most intensive and hotly debated controversies in hominid evolution since that precipitated by the discovery of the Taung Child in 1924.

The idea that humans and apes were separated in their evolution for fifteen or twenty million years has for several decades been one of the linchpins of paleoanthropology. Elwyn Simons and David Pilbeam have devoted their professional careers to the development of that concept. As all the Miocene apes were sorted out under Pilbeam's careful hand, they seemed to fall more and more logically into two groups: those whose traits verged toward the traits that modern apes show, and those that verged toward humanness. One powerful bit of fossil evidence that identified humanness in one ape was the previously mentioned *Ramapithecus* half-jaw that Elwyn Simons had turned into a complete jaw by casting a mirror image of the half he had and fitting the two together. The result was a curved "human" tooth row.

Ramapithecus, on that evidence and in default of a better candidate, emerged as the most manlike of all the Miocene apes and was cast in the role of human ancestor. From the size of its jaw a rough size estimate for the entire animal could be derived; it stood about three feet tall. Little else was known about it. There were no post-cranial bones to say whether it walked upright or not, or what kind of hand or foot it had. Whether it was arboreal or terrestrial was not known. It bore a suspiciously close resemblance to another slightly larger ape, *Sivapithecus,* which occurred in the same places and at the same time. But *Sivapithecus* had the boxlike tooth row of an ape and was considered not to be on the human line. All the other Miocene apes were either older and possibly ancestors to *Ramapithecus* and *Sivapithecus,* or else were considered to be more apelike.

In 1967 a young biochemist, Vincent Sarich of the University of California, Berkeley, announced that that scenario was all wrong. He had been comparing samples of the blood protein—serum albumin—of gorillas, chimpanzees, and humans, and his tests told him that the three had a common ancestor as recently as five million years ago. To paleoanthropologists that was a preposterous figure. Still more preposterous was Sarich's contention that the two apes and the human stood in triangular relationship with each other: man was as closely related to the chimp as the chimp was to the gorilla. Could any paleoanthropologist, familiar with Pilbeam's work and bound to the old dates that were clamped to that work, believe that? Could he suddenly look over his shoulder and see chimp and gorilla looming so closely in a brotherly relationship with him? Not possibly; everything known about the bones and about geology said fifteen or twenty million years, not five. Nobody paid attention to Sarich.

Almost nobody. A few did, notably Sherwood Washburn, also of California-Berkeley. Washburn is one of the most distinguished primatologists in the country. He pioneered the practice of sending his doctoral students into the field to study apes and monkeys in their natural state and has spawned an entire generation of gifted young field scientists. He is also profoundly interested in the evolution of primates. When Sarich spoke, Washburn blinked but did not turn away. He knew that another Berkeley scientist, a molecular biologist named Allan Wilson, was doing the same work, and that he and Sarich had reached similar conclusions and had joined forces. He knew that they were doing something new and exciting: molecular matching. It was the molecules that had given Sarich that ridiculous answer. Sarich was prepared to stand by it, and was saying in effect: "Prove that the molecules are wrong."

This bizarre turn in the hominid story was not a total surprise to Washburn. He was aware of other similar molecular pioneering, notably that of Morris Goodman of Wayne State University in Detroit, who in 1962 had already determined that the African gorilla, the chimpanzee, and the human were more closely related to each other than they were to the Asian orangutan. As a result, Washburn was able to see past Sarich's troubling date for the ape-human split to the thrilling and novel methodology behind it. He became convinced that this was the wave of the future, and took up Sarich's case vigorously, confident that either the fossil date of twenty million years or the molecular date of five million would have to bend.

"Washburn was right," said Don. "But the bending was going to have to be something approaching a crackup. There were fifteen million years there that somebody was going to have to swallow."

Sarich's work was based on the same proposition that Carl Woese's work with bacteria was based on: changes in their molecules build up over time in all organisms because of mutations in their genetic structure. That being so, it should be possible to determine how closely related species are by measuring how different their molecules have become. Two species that have split recently—such as horse and donkey—will have little time to pile up differences in the molecular structure of certain proteins such as serum albumin. Others—horse and frog, for example, with a couple of hundred million years separating them—should show many more differences. The idea was not new. It had been suggested as early as 1904 by the immunologist G. H. F. Nuttall. But it was not then taken up seriously by others, partly because laboratory methods for accurate measurement of difference were crude, and partly because nothing was yet known about the relationship between genes and proteins. The double helix of Watson and Crick lay in the future. Genes were unknown. As a result, their role in making proteins was also unknown. Blood proteins in two animals might be different, but what that might say about their relationship was impossible to determine.

It was only after those locked boxes were opened that people like Goodman, Wilson, and Sarich were stimulated to put the new knowledge to work. By using better laboratory techniques, they would look to the molecules to see whether and how evolving relationships could be traced. Wilson and Sarich had a strong faith that that could be done. Furthermore they believed that molecules would ultimately prove more reliable than fossils. They regarded the latter as suspect, dangerously incomplete in some cases, and always potentially misleading because of being differently sized and differently shaped from individual to individual. A protein molecule, they knew, was a far more stable thing than a bone. Catch a molecule in some kind of measuring trap, and the same answer would come out, clear and accurate, time after time. So they caught molecules, they measured differences. So did other scientists. The trouble was that those reliable molecules kept coming up with terrible answers. Wilson and Sarich were faced with a terrible dilemma.

How they dealt with that dilemma we will get to in a minute. Meanwhile, what, exactly, was it that Sarich and Wilson were doing? They were taking advantage of a phenomenon that Nuttall and other immunologists, as far back as 1904, had recognized in their work on infectious diseases. This is: when a foreign substance such as a disease germ penetrates a human—or a gorilla—the immune system immediately begins to manufacture something designed to neutralize or kill it. That process, at the heart of modern disease control, is now well understood by science. It is known that the immune system of any higher organism is adept at defending itself by producing antibodies that are tailor-made to repel specific invaders. These are not broad-scale defenses, but rapier-sharp weapons, each aimed at combating a particular invader. That specificity is what made Sarich's work possible, for it gave him a way of measuring differences in blood proteins with great accuracy.

His way was to inject some human blood (he used his own) into a laboratory rabbit. The rabbit, reacting to the foreign protein, immediately began manufacturing an antibody to combat it. This was not just any old antibody; it was a Sarich-aimed, human-resisting one. Sarich collected a supply of it. Then, in a test to see how effective it was, he exposed it to another sample of his blood, and learned that the rabbit had done its job well. The reaction of the antibody to human blood was 100 percent. Next Sarich mixed some of the rabbit antibody with a sample of chimpanzee blood. There was only a 96 percent reaction, demonstrating that, over time, the chimp blood protein had become slightly different from the human one; the failure of the antibody to react totally to the chimp blood protein had revealed that fact with great precision. He repeated the experiment with gorilla blood, and got the same 96 percent reaction. Then he reversed the procedure. Injecting chimp blood into a rabbit to produce a chimp-designed antibody, he found that the human reaction to that was 96 percent.

Does this mean anything? Perhaps all animals have a 96 percent congruence in their blood proteins. Logic tells us that they do not. If Sarich's basic idea was correct—that evolution is responsible for the 4 percent difference between a human and a chimp—then it follows that the difference between a human and a horse will be greater.

When Sarich got around to making further comparisons, that assumption also turned out to be true. Since then hundreds of cross matchings have been made and the evolutionary distance between many species calculated. Now the difference measurement that was begun between men, apes, and monkeys has been extended to include dogs, sheep, camels, elephants, and goes on to birds, amphibians, fishes, and insects. This is an absolutely astounding development. It means that a worker sitting in a laboratory, surrounded by a hundred unlabeled samples of serum albumin from a hundred different animals, can sort them out into sensible relationships without knowing what they are. If enough cross-checks, simple measurements of the amount of difference between the samples are made, a web of relationships that can fit logically in only one framework will result. If those relationships—those positions in the web—are marked down on a large sheet of paper, they will fall together into what emerges as a family tree. Only then—after it is done—need the investigator look at the names of the animals whose serum has been used, to see how closely that tree resembles one made by study of the animals themselves.

It is at this point that paleontologists must sit up and take notice, for the family tree drawn by serum albumin studies is a virtually exact match with one that would have been made by examining the bones, skin, size, shape, and behavior of living animals.

The tree drawn from molecular evidence differs from that drawn by paleontologists only in that it is more precise. It is able to tell us things that they cannot. Are humans more closely related to mice or rabbits? Paleontologists cannot say for sure, but the molecules can. The answer is mice. The molecules also tell us that pigs are more closely related to whales than they are to horses—and so on. What is more, those relationships hold up when other molecular measuring methods are used. Since Sarich started his work, it has become possible to do more direct analysis by identifying differences in the amino acids that make up the blood protein. Finally, and still more recently, it has become possible to go to the source itself, to DNA and RNA, and note the differences in nucleotide sequences there. That last step is the most telling of all. It puts in our hands the very blueprint that codes for amino acids and hence for all of life. It gives us at last the fine print by enabling us to pinpoint the precise submicroscopic spots at which differences exist among different species. It spawns a new science: molecular anthropology.

Molecular anthropology had a slow and difficult birth. Most paleontologists ignored it, although those who did not were forced to concede the logic of its position: over long periods of time, differences in serum albumin, in amino acid sequences, and in DNA will pile up. Therefore those differences can be used to determine relationships between species and also can be used to draw family trees. A tree drawn by two biologists, Walther Fitch and Samuel Margoliash, based on differences in an enzyme (cytochrome C) is shown on page 360, and reveals some interesting things. It shows, for example, that the commonly accepted relationship between birds and certain reptiles is incorrect. Birds, long held to be in a class of their own, now seem to fold more closely into the reptile group than previously had been thought. A pigeon and a penguin are more closely related to a turtle (that is, have a closer common ancestor) than the turtle is to a snake. That comes as a great surprise, for turtles and snakes are both reptiles; a bird has always been thought of as something else. In some future classification will birds end up as reptiles? Conceivably they will.

Another nagging question that Sarich himself solved with molecules was: are pandas bears or raccoons? The answer: the giant panda is a bear; the lesser panda is not; it stands very close to the raccoons. Prolonged study of the physical characteristics of the two animals had failed to provide a satisfactory answer. The molecules promptly gave it.

Matters such as these fascinated taxonomists. But they were quick to notice that there were no dates on the kinds of trees that Fitch and Margoliash were drawing. "How," they asked, "does any molecular

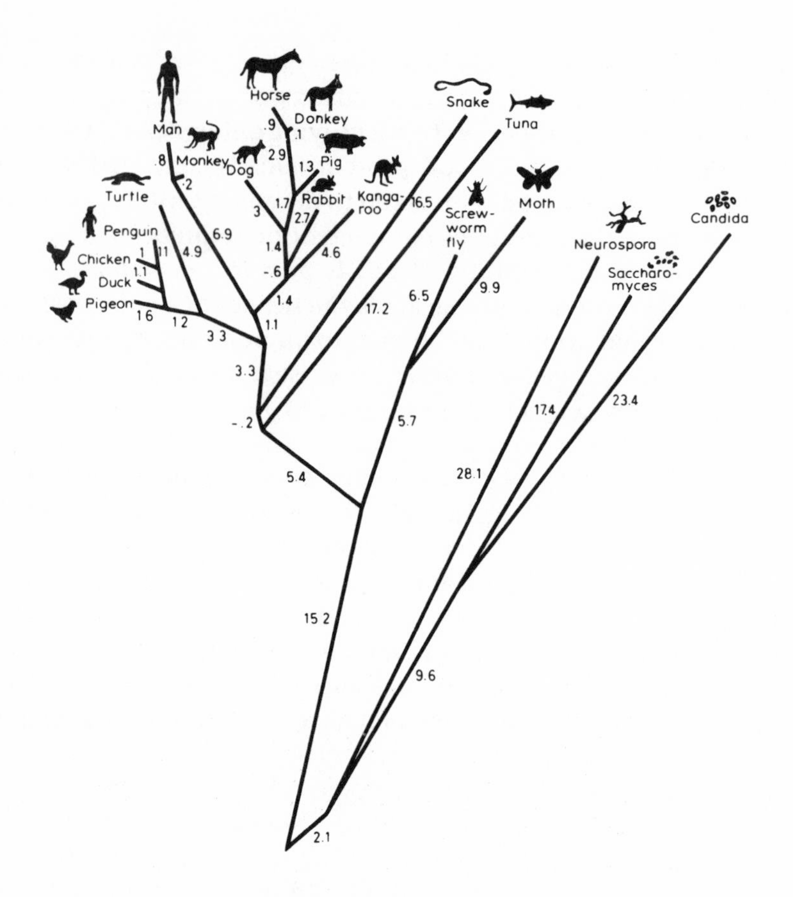

A MOLECULAR FAMILY TREE based on differences in an enzyme (in this case cytochrome C) was constructed in the late 1970s by Walter Fitch and Emanuel Margoliash. The numbers represent differences that the various species have accumulated over time in their genetic code; the bigger the number, the greater the evolutionary distance. Since evolution is presumed to run at approximately the same rate on all lines, the total change from any species to a common junction point with another species should be the same—and they are. Add the numbers running down from man to where his line joins the turtle fork. The total is 8.8. Now, measure back to that same point from the turtle. Again it is 8.8. Elsewhere on the chart, though the numbers are not always identical, they are remarkably close, attesting to the overall evenness of evolutionary change on all lines.

Illustration by George Kelvin from *Scientific American*, September 1978, page 69.

tree builder have the right to determine the time of a split between two species? Particularly, where does Sarich's ridiculous date of a five-million-year split between apes and humans come from?"

Sarich and Wilson had an answer for that. They asserted that molecular differences build up at a regular rate and therefore constitute a kind of "clock" that can be used to measure time. This idea did not originate with them. It had been suggested by Emile Zuckerhandl and Linus Pauling in 1962 (Pauling seems to have thought of just about everything in his long and illustrious scientific career). Wilson and Sarich decided—with an encouraging push from Washburn—to see if the suggestion could be put to practical use.

By way of understanding the principle of the molecular clock, picture twin brothers about to go out into the world to make their fortunes. They leave home wearing identical sets of clothes, but they travel different roads. As they go, their clothes begin to wear out and they have to buy new ones. They begin by buying new socks after one year, new shoes after two years, and so on. The possibility of their buying identical articles is remote; they are in different cities, with different tastes and different incomes. Therefore, by examining their wardrobes at any point along the road and counting the number of differences between them, it should be possible to figure out how long the twins have been separated.

Comparing the serum albumin of two species is equivalent to comparing the wardrobes of the two brothers. For the brothers, one looks at the dozen pairs of socks each had started with and notes that five pairs have been replaced by new and different ones. If it takes a year for a pair of socks to wear out, then the brothers have been separated for five years. For the serum albumin samples the same thing is done. If five differences are noted there, it means that five genetic changes have taken place since the two species were one. Allot each step a measure of time, and it follows that the date of the split can be calculated.

To set their molecular clock, Wilson and Sarich realized that they would need a yardstick against which their work could be measured, something outside their own calculations that could be agreed on by everybody. They went directly to the book of the paleontologists and said in effect: "We will hang all our dates on one that you yourselves have picked, thirty million years—which is when you say the split between monkeys and apes occurred. Certainly you can't complain about that."

Having chosen that date, Sarich then calculated the distance between living descendants of the two lines. Assume (for simplicity's sake) that thirty units of molecular difference have accumulated between a modern monkey and a modern ape during those thirty million years. Then the clock is running at one tick (or one molecular change) every million years. So, if there are five ticks (or five units of molecular difference) between African apes and humans, there are five million years separating them from a common ancestor. What Sarich had done was to say: "I have used your date of thirty million to derive my date of five million. I can't help it if the molecules give an answer you don't like."

That did not satisfy the paleontologists at all. They pointed out that the discrepancy between the fossil story and the molecular story was so huge that the only explanation could be that the molecular clock was not running smoothly. If it was not, the conclusions of Wilson and Sarich were worthless. Indeed, when Morris Goodman had gotten his first shocking ape-human results, he had bowed to the heavy weight of fossil evidence and had concluded that, for reasons unknown, the molecular clock was not steady but had been slowing down on the ape-human line. Only in that way could the five-million-year date be explained away. Find a number for that slowdown, factor it in, and the split might be moved back again, close to fifteen million years, which was where the fossils said it should be.

Wilson and Sarich did not accept that explanation. "We saw the problem clearly," said Sarich recently. "We saw that one could make a choice between 'there's been a slowdown in protein evolution among higher primates,' and 'the paleontologists don't know what they are talking about when it comes to higher primates.' Our solution, after having isolated the problem, was the only really original contribution we made to this business."

That is an understatement. True, they did build on ideas of others, but they did it in the face of almost universal disbelief and enough vilification to have discouraged most young scientists. Furthermore, they perfected laboratory techniques of great sophistication that enabled them to come up with a standard of measurement that was reliable and could be applied to any two animals to show to what extent their blood proteins differed. The science writers John Gribben and Jeremy Cherfas, in describing this laboratory work, have called it "a masterpiece of practical biochemistry." It does not bear further description here, but it ends up by producing a simple figure for the so-called ID, or immunological distance, between any two species.

With the reliable tape measure of the ID in hand, Wilson and Sarich could test the regularity of the molecular clock. Theory said there was regularity, but it had to be proved. They proved it by making a great many cross-checks between species and then comparing the numbers. They learned, for example, that a human differs by the same amount from an orang that a gorilla does, which suggests that the molecular clocks of human and gorilla have been moving at the same rate since they split from the orang. If they had not, the ID figures would have not been the same.

To broaden their test, Wilson and Sarich looked deeper into time, far enough back on the family tree to locate a common branching point not only for humans and apes but also for horses and dogs. If their thinking was right, they could expect to find that the ID numbers on the various lines descending from that ancestral point would also be the same. In other words, if the time from a common ancestor to a modern dog and a modern horse and a modern human is the same, and if the molecular clock is running steadily for each, the ID numbers will also be the same.* And they are. When Wilson and Sarich learned that, they could state confidently that the clock is not slowing down on the primate line. It is running at the same speed as it is on all the others. Although humans may look entirely different from chimpanzees and gorillas, those differences are superficial. Where it counts — in their genes — all three are ninety-nine percent identical.

Although paleontologists continued to ignore the molecular evidence for nearly twenty years, that became increasingly difficult to do as more, and increasingly more precise, ways were developed for measuring molecular differences. These did not always agree in their smallest details. One method would show a gorilla ever so slightly closer to a man than to a chimp, and another would show the man and the chimp as closer. But they all agreed so closely overall that the major contention of the molecular anthropologists finally

* An example may make this clearer: Assume an immunological distance (ID) between human and dog of 100. Assume also that the clock is running at a rate of one tick per million years. It follows that human and dog have a common ancestor at 100 million years. Now, if horse and dog also have an ID of 100, and if horse and human also have the same, it further follows that all three have the same common ancestor at 100 million years and that the clock is running at the same rate on all three lines.

became inescapable: relationships can be worked out; the clock runs at a steady rate; the ape-human split (like it or not) took place between four and six million years ago, probably about five million — the figure that Sarich had arrived at in 1967. But he had to endure a long time in limbo. His combative espousal of his cause did not help. Described by the science writer Roger Lewin as "huge in stature, voice and opinion," Sarich managed to irritate his opponents rather than persuade them. Something he wrote in 1971 was particularly galling: "One no longer has the option of considering a fossil specimen older than about eight million years as a hominid, *no matter what it looks like.*" In short, *Ramapithecus,* at fifteen million years, could not be a hominid even though it might look like one. Lewin commented: "A statement more calculated to raise the blood pressure of paleoanthropologists could hardly be imagined." But Sarich's bulldog attitude did not change. He was deeply outraged at the dismissal of his work, and is today. "It continues to annoy me in the extreme," he said recently, "to see that contribution trivialized by ignoring it."

The first crack in the trivialization came not — as Sarich would have liked — from his own work, but from fossils. During the 1970s several experts, taking a second look at the famous reconstructed *Ramapithecus* jaw, began to wonder if Elwyn Simons had put it together right. Put together differently, the tooth row came out like those of other Miocene apes, and not curved as in humans. Simons's reaction was to assert that the new reconstructors didn't know what they were doing and had made a number of fundamental mistakes. Then, in 1976, a larger part of a *Ramapithecus* jaw than anything previously known was found in Pakistan. Matching it with other fragments from the same site produced the first really good look at *Ramapithecus* that science had ever had. After a careful examination of it, Pilbeam was forced to conclude that it could not be pegged as more probably the ancestor of either a modern ape or a modern man. His fixation on *Ramapithecus* as the human ancestor, and particularly as the *old* human ancestor, began to crumble. It crumbled further when two enormously exciting fossil finds were reported in 1980 and 1982. The first was from Turkey. The second was from Pakistan and, of all things, was turned up by Pilbeam's own field team. Both these fossils were partial faces of that other Miocene ape mentioned a few pages back, *Sivapithecus.* It will be remembered that *Sivapithecus* was almost exactly like *Ramapithecus* except for being slightly larger and having a proper ape jaw instead of the human jaw posited for *Ramapithecus.* Now that second difference was gone.

"They're alike?" I asked.

"You bet they are," said Don. "Throw out that mistaken reconstruction of the *Ramapithecus* jaw, and except for size the two are alike. They're so much alike that they could be considered males and females of the same species. Sarich and Washburn both suggested that several years ago, but nobody listened because they weren't paleontologists. They didn't have expertise; all they had was good sense. Now several other people are beginning to think that way. Poor *Ramapithecus.* His days as our ancestor are over."

"Why?" I said. "There's got to be something back around ten or fifteen million for an ancestor. Why can't it be Siva-Rama, whichever name you use?"

"Because those two *Sivapithecus* faces finally give us enough to see what that animal looked like. Well, it looks like an orang."

"An *orang!*"

"Yes, an orang," continued Don. "And that finished off any lurking ideas there may be about an old *Homo* ancestor back there."

"I don't see why. Why can't it be ancestral to all of them, the orang as well as the chimp and the human?"

"Think about it. You can't get from an orang to a chimp. It's too late; they're already split; they're on different lines, headed in different directions. That's all been nailed down by the molecules so many different ways now that people don't argue the point any more. If *Sivapithecus* has already gone in the direction of the orang —and it clearly has—it cannot be ancestral to the others."

One of the newer molecular methods that not only confirms Sarich's work but also refines it is a technique called DNA hybridization. By this method a DNA double helix from a human is unwound to get a single strand. This is allowed to wind up again with a matching single strand from the DNA of a gorilla. That is where the name comes from; the new helix is a human-gorilla hybrid. Those two strands wrap themselves together very well, the reason being that they are nearly identical. But they are not entirely so; there will be a few places where their nucleotides do not match. Those mismatches mark the difference between a gorilla and a human, and a way has been found to measure them.

This is done by heating, which is how the single strands were obtained in the first place. At a certain temperature the helix bonds break and the strands separate. So, when a hybrid gorilla-human helix is heated, it will break apart at a slightly lower temperature than a nonhybrid helix because there are not quite so many bonds to break. The payoff of this technique is that the biologist ends up with a set of temperature breakpoints that tell him how closely related two species are.

"That's all been worked out for the apes by a couple of other guys in this field, Charles Sibley and Jon Alquist," said Don. "When they applied the molecular clock to their temperature breakpoint figures they got a very good tree for the apes. They were able to show that gibbons split about twenty million years ago, orangs about fifteen million years ago, gorillas about eight or ten million years ago, and humans and chimps about six or seven million years ago. Those figures are just about the latest that the molecular people have come up with. They stretch Sarich's a bit—but not much. What's more significant is that they show the gorilla as not being quite as close to the other two as Sarich thought. As we just noted, different molecular yardsticks have been giving slightly different answers to that relationship. All say the three are very close. So does this one, but the best answer at the moment seems to be that the gorilla is a little farther away. Three drawings will show how that thinking has evolved."

It was now Don's turn to put a pencil to a family tree. "See that little gap between the gorilla split and the chimp-human split? That's where the trouble is."

"What trouble?"

"The walking trouble. The gorilla and the chimp are both knuckle walkers. That's a unique way of getting around. You don't just suddenly start doing it. You need a lot of adaptations in your arms and muscles, and particularly in your hands. Now, if gorillas and chimps are both knuckle walkers and appear to have the same—and very recent—ancestor, it seems obvious that they inherited the trait from that ancestor. That's much more likely than that they developed something as odd as knuckle walking independently."

"So, what's the problem? The ancestor was a knuckle walker."

"The problem is," said Don, "that that ancestor was also your ancestor. You're not a knuckle walker. And it's even more difficult to

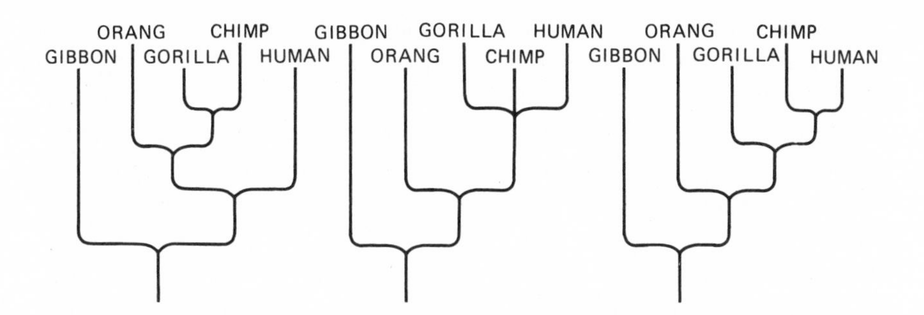

A family tree based on fossil evidence and on the assumption that humans split from the great apes 15 to 20 million years ago looks like this.

Sarich's version, derived from blood-serum analysis, suggests a three-way recent split among gorillas, chimps, and humans.

DNA hybridization suggests that the chimp–human relationship is the closest and that the gorilla split off a little earlier.

go from knuckle walking to erect walking than it is for two animals to invent it independently. Knuckle walking is a specialization, something that takes you away from the more generalized quadrupedalism that other primates have. Somewhere along the line your ancestor got up on its hind legs and became a biped. You can do that fairly easily from a four-legged ancestor. But to take a detour through knuckle walking is pretty darn difficult. In fact, I'd say it's next to impossible."

"So how did we get on our hind legs?"

"Nobody knows. There are a dozen theories, most having to do with changes in climate, different eating habits, pair-bonding, brain enlargement that leads to carrying things around, and to the beginnings of tool use. But those are all speculative. The molecules don't help at all. They're absolutely silent about how things happen. They say how we are related but they don't say what shape we are or how we got those shapes. It takes bones to do that."

We end our chapter on hominids squeezed from two directions.

There is the molecular evidence pushing apes much closer to us than we had recently thought. There is the fossil evidence of Lucy, pushing erect-walking ancestors back toward the apes. Both are saying: "Look somewhere along about here, somewhere around five million." With a small degree of nudging and accommodation, that date will probably stand up as the time when the first whispers of humanity began to be heard.

"Maybe a little more than five million," Don said. "After all, you have to develop those peculiar hominid legs, that pelvis, those weird big toes that humans have. Can you do all that in a million years?"

"Can you?"

"There's no answer to that at the moment. It's something the molecules will never tell you. That's why you'll always need people like me to go out and find fossils."

Specialization and Extinction

A brain weight of nine hundred grams is adequate as an optimum for human beings. Anything more is employed in the commission of misdeeds.

—ERNEST HOOTON

What will we do as the wisdom of the past bears down on our future? . . . It is possible to live wisely on the land and to live well. And in behaving respectfully toward all that the land contains, it is possible to imagine a stifling ignorance falling away from us.

—BARRY LOPEZ

19

Is There Danger in Being Too Smart?

The laborious climb upward from simple bacterial forms to the hideously complex organisms that make up higher life today has been accompanied, among those with a central nervous system, by a dramatic increase in brain size and by an ever-increasing ability to think. Indeed it may be cited almost as a rule that a better and better brain is a mark of ongoing evolution among those higher organisms. The reason is simple. Smarter is better. Given the potential for developing a better brain from a lesser one, there will be selection for the former, since it confers a clear advantage to its possessor in its struggle for existence.

What organism it was that first was able to engage in what we now call "thinking" is impossible to determine. There is a large gray area in the behavior of living forms, not to mention extinct ones, about which there is little agreement as to how much is so strictly programmed that it can be predicted with accuracy and how much can be changed by experience. Experts cannot even agree on whether a given change can qualify as the result of "learning" or "thinking," or whether that, too, at the lowest levels, can be attributed to the activation of genetic potential that is already there. Few will argue that the fruit fly can think. It has so few neurons in its simple little brain that there is little possibility of the fly's doing more than its genes have programmed it to do. A dog, by contrast, obviously can think some fairly complicated thoughts and also store a good deal of information in its memory. But in between — who will ever know?

The human brain is a truly extraordinary instrument. It allows a living creature, for the first time in the history of life, to contemplate itself (himself/herself), to gain some knowledge of our origins, of the planet on which we live, and of the universe of which our planet is a part. All earlier life has been utterly ignorant of all these things. The human brain does more; it allows us to alter our environment for our own benefit in ways no other creature has ever done — indeed it gives us awesome power to do that. It allows us to weigh our acts and to speculate about the future. It forces some questions: Is the human brain continuing to develop? If so, how far can it be expected to go, and what kind of thinking will it be capable of that it is not capable of now? Those questions are unanswerable for the moment, and may be forever. But there is an answer of sorts to a third question: Is there a danger in becoming too intelligent?

In 1853 a small flock of sparrows were brought in cages from England and released in the Greenwood cemetery in Brooklyn, New York. They were put there because it was hoped that those noisy, adaptable, and fast-breeding birds would eat canker worms, which were ravaging foliage in the Northeast at the time. The sparrows lived up to their reputation as breeders but not as worm eaters. They ate a few but quickly made it clear that their principal diet was being supplied by human beings, partly in the form of thrown-away bread and other food but mostly in the form of grain. The United States was then still a nation of small towns and small farms. The main source of power and transport, both in town and on farms, was the horse. Every community had its feed-and-grain depot.

The sparrows exploded. Within a few decades they were everywhere. They were belligerent, rough-and-tumble little birds. They quickly drove out many native species — some of which, like the bluebird, have never really recovered — and in so doing caused plagues of other insects that those birds normally ate.

Bounties were set on the sparrow in an effort to eradicate it. But as long as feed stores and horses — and horse droppings, which supplied an endless source of food in undigested grain — remained, curbing the sparrow was impossible. By the turn of the century it was the most numerous species in the country.

What made it so successful?

Aggressiveness and fertility surely helped, but the principal reason was that it was extremely adaptable. Apparently it could live anywhere. Hot, cold, damp, dry—it didn't seem to matter. All it needed was people and their abundant food. It was right at home hopping about under their feet. No native bird seemed as easy in that close association.

If you can adjust to such a wide spectrum of habitats, shouldn't you succeed? Compare the English sparrow to Kirtland's warbler, a little bird so rare that during the nineteenth century it was known only from a handful of specimens collected by ornithologists. Its habitat in the United States was discovered only after a half-century of searching. It turned out to be one small section of Michigan and consisted of only a few stands of jack pine of a certain height. Where the pines were too big they kept out sunlight and prevented growth of the low bushes in which the warblers preferred to build their nests. Only where forest fires had killed those large trees and thus permitted the sprouting of smaller ones did the warblers breed. Today their population is a few hundred birds. Estimates for the English sparrow at its peak ranged up in the hundreds of millions.

Why is one species so persnickety in its habits and another so tolerant? If the aim of all species is to perpetuate themselves, isn't it capricious, as well as extremely dangerous, to let yourself be crowded into a narrow niche? Yes, it is. But neither Kirtland's warbler nor the English sparrow had any real choice in the matter. Each was following on a line that accident and a sensible rule would dictate. The rule is: try to fit better and better in your niche.

The accident lies in the kind of niche you find yourself in. If it is narrow to begin with, or if it shrinks for reasons beyond your control, you are doomed to a shaky future. Competition may also force you into choosing a smaller and more specialized piece of what was once a larger and more generalized niche. That, presumably, is what happened to Kirtland's warbler. It is one of about fifty different warbler species found in North America, all of them active little insect eaters. They are the result of a rather recent, and certainly dramatic, evolutionary burst. As they have speciated they have carved up the small-insect–eating niche with great efficiency into a number of specialized slices. The various species now have different ranges, some overlapping, some not. They have different feeding habits. Some work low bushes; some work various levels in trees. Some are bark specialists; some are leaf specialists. Some prefer pines; some prefer deciduous growth. All, of course, are exploiting an environment that is only about twelve thousand years old. They are flooding into a continent that previously had a different climate and a lot of ice spread over it.

In the scramble for living space, Kirtland's warbler apparently found its best chances in the bushes under jack pines. There was no way of predicting that that habitat would prove to be fatally small. Indeed, the warbler of recent millennia was probably chasing a habitat that it was already adapted to and one that might have been much larger in another place or time. Either way, it is now locked into a way of life that it cannot change. When the habitat finally winks out, so will the warbler.

The sparrow picked better. For the record, it is neither English nor a sparrow. It is most closely related to a family of Asiatic weaver birds and probably had its own origins in the Middle East. It would have been there when stone-age dwellers of the fertile crescent first began growing crops and domesticating animals, some ten thousand or more years ago. It was a seed-and-grain eater and must have learned early that farmers are great accumulators of seeds. By sparrow standards farmers are sinfully careless with their food. The crumbs a man drops as he eats a loaf of bread, or the oats his horse nudges out of its nose bag are banquets for sparrows. Human settlements were thus ideal places for the birds. Over the centuries they adapted quickly. They spread west through Europe and eventually arrived in England. Hence their name: English sparrow. It has recently been changed to house sparrow. Why it is not called a house weaver is something that only the men who have been appointed to name birds can say.

One thundering success, one small failure. But before we pass final judgment on the sparrow we should look more carefully at it. As a seed eater it is a generalist and has a host of competitors. Each, like the insect-eating warblers, has to find a corner in the seed world that it can exploit a little more efficiently than the competition. The sparrow apparently chose people as its niche, and it turned out to have been an inspired choice, for people have flooded the world and turned much of it into farms. They took their domestic animals with them, scattering food as they went, and the sparrow tagged along.

But the ride has turned out to be a roller-coaster one. The horse was replaced by the car. Where every hamlet had its hay-and-feed store there is now a gas station. The only food that sparrows can scrounge there is the insects caught in the radiator grilles of automobiles. With the passing of the horse, their former meal ticket, they must now fall back on the natural supply of seeds in the environment. That brings them face to face with other seed eaters that have been at that game for millennia and know it better. Today house sparrows, that pernicious blight of only half a century ago, are actually rare in large parts of the country. They are still common in cities, but in nothing like the numbers they once were, when every street had its bounty of horse droppings. A few pairs live in the backyards below the apartment in New York City that I visit in winter. They are surviving—I see them every year—but I do not know if they are doing it on weed seeds in the gardens or on the bread crumbs and popcorn they pick up in the street. More likely the latter. The sparrow is tightly bound to humans. Where they do not go, it does not go. We have become its specialization.

Every organism is specialized to one degree or another. Evolution is constantly offering each an interesting trade-off: will it keep its options open and try to be a not-too-efficient exploiter of a wide range of possibilities, or will it try to be the best exploiter of one? In other words, will it share a large turf with competitors, or will it defend a little one as its very own? Competition is constant. It forces all organisms unconsciously to make choices every day of their long—or perhaps short—evolutionary lives.

Common sense tells us that the organism that can manage a wide stance and maintain it despite competition has the best chance for a long evolutionary history. Science proves it by pointing out that past evolutionary innovations, the explosions of new forms that take place from time to time—often following crises in earth history that have wiped out most living things and left large gaps to be filled—have been rooted in the least-specialized surviving types.

The paradox is that they do not stay that way. Always there is that relentless pressure to fine-tune, to get a competitive edge by getting just a little better at one thing, a little more comfortable in one place. That is what happens at the edges of populations spread over wide areas whose conditions vary. The edge dwellers find themselves living slightly different lives than those at the center. There is more pressure on them to change. That is why there are fifty species of warblers in North America and not one. They split. Today each is a more efficient worker in its corner of the realm than the more generalized ancestor was. Go back in time, and we might find that ancestor doing its not-too-efficient thing but surviving comfortably because there were not more efficient competitors around. Go farther back, and the warblers will merge with vireos, tanagers, and other families of small insect eaters to fuse into an even more generalized perching ancestor. Go back still farther, and increasingly unlikely types will be sucked in: herons, swallows, grebes, hummingbirds, ostriches. Finally: one archetypical ancestor from which all birds, all feathers, all colors, all songs, all the later crazy shapes and habits flow. Each was invented along the way as opportunities to gain a small advantage by deviating slightly from the norm were seized.

Endless opportunity. Clear short-term gain, but possible long-term danger, because there is no turning back. An organ lost or grotesquely enlarged cannot be recovered. A bat will never learn to pick up things with fingers that have turned into wings. The tunnel narrows; you cannot turn around and crawl out. It is like trying to get an open umbrella out of a chimney. Once you have started up, the only way out is to keep going. If you reach open air at the top, fine. If you don't, it's the end of the umbrella.

Science has a nice word for one class of creatures: ***obligate parasites,*** things that live in or on other things, get their living from them, and have been doing it for so long that they have lost their legs, their eyes, or whatever appendages or organs they haven't been using as parasites. Now they are stuck, they are ***obligated*** to live on their hosts, and their fates are tied to them.

I think of a bird that became extinct on the island of Martha's Vineyard when I was young. It was the heath hen, a kind of prairie chicken that had been widely spread over sandy barrens in the Northeast before the arrival of white settlers in the 1600s. The birds were so numerous, so tame, so foolishly easy to catch during their mating antics in the spring that indentured servants insisted that clauses be written into their labor contracts that they were not to be served "heath cocke" more than three times a week.

By 1900 the heath hen population had been reduced to one small colony of a couple of hundred individuals scattered over a stretch of sandy scrub oak in the center of Martha's Vineyard. There were none anywhere else. During the year they were seldom seen, but in spring they gathered on open strutting grounds, where the males displayed before the females. Local people hunted them there despite efforts by conservationists to protect them. By 1930 the flock was down to under a dozen birds. In 1933 watchers saw only one, a male. It had an aluminum identification band on its leg. It was never seen again. Somewhere, under the oak leaves, that band is lying. The bones of the bird will be gone by this time, devoured by bacteria and the acid soil. People talk a lot about the heath hen, but they never talk about the particular kind of louse that lived on that bird. When the last bird died, the last louse died.

What are the survival chances of a tiny crustacean that lives in the far back coil of a shell that a hermit crab is occupying? They are excellent as long as the crab stays healthy. The small tenant keeps out of the way of the larger one and thrives, like a sparrow, on the careless eating habits of its host. The crumbs that the crab misses keep it going. Only when the crab grows and has to move to a larger shell is its little partner exposed to danger. It must move when the crab does, timing its dash just right.

Deep in the new shell, it is safe again. Nothing jostles it back there. It has made its unlikely little niche its very own, but at a stiff price. It is wholly dependent on the crab and cannot survive without it.

The balancing act, then, for an organism, is to find a niche that is broad enough to provide a little leeway, to give it room to evolve in this direction or that, but never to spread itself so broadly that it cannot defend its turf against more efficient trespassers. By comparison with the little shell dweller, the house sparrow has a wide universe of options and would seem to have a more secure future. But don't bet on it. The sparrow is tied, too, although more loosely, to humans, and may last only as long as we do. The environment of the air is less stable than that of the sea. Crabs may outlive people. Both will some day be extinct.

Extinction is a way of life. Of the uncounted millions of species that have come into being on this planet, more than 99 percent are gone. That process continues. Two million years from now most of the species that exist today will have disappeared, either through failure to adapt to changing times, or by evolving into something else.

Small episodes of extinction occur steadily, like rain. Mass extinctions take place only once in a great while. They have occurred several times during the history of life on earth. Recent research seems to indicate that they may come at fairly regular intervals, and science is busy trying to figure out what causes them. Two early—and unique—causes that we know of were the aforementioned oxygenation of the air and later of the sea, both brought about by a proliferation of life forms that release oxygen. Other more recent causes of mass extinction may be related to the amount of sunlight that hits the earth. Fluctuations in that could be caused by extreme volcanism that smokes up the atmosphere. Alternatively they might be caused by outside forces such as the impact of a large meteorite or comet, or the earth's passing though dustier regions in the heavens. This matter is by no means clear. The result, however, is. There have been wide swings in climate, making large portions of the earth colder or drier than before. Extended hot, damp, tropical periods follow. In the slow calendar that marks earth time, these environmental changes and their associated extinctions are comparatively abrupt events. We speak of the sudden disappearance of dinosaurs and try to link that mysterious evolutionary disaster with a large and sudden catastrophe. What seems more likely is that the dinosaurs disappeared over thousands of years.

Fast or slow, dramatic changes in living conditions on earth have brought about equally dramatic changes in living forms, new flowering to fill the voids left by extinctions. Ignoring the long and intricate story of the evolution of life in the sea, we can see those dramatic jumps taking place among the organisms that emerged from it and began to invade a new environment on land. Plants appeared there for the first time more than four hundred million years ago. They were followed by scorpions, by millipedes, and later by insects. All these are creatures with hard exoskeletons. They derive from similarly structured marine ancestors.

Meanwhile an entirely new animal was evolving in the sea with a basically different structure. That was a fish. It had a backbone instead of a shell as a framework for its body, a central nervous system, and fins that, when its descendants first went ashore, would develop into four legs. By contrast, insects have six legs and spiders have eight. They came out of the sea so equipped and cannot change. Those fundamental structural differences persist, and have for hundreds of millions of years. The blueprints that direct them are so intertwined with other life-determining genetic patterns that such structures can no longer be altered without dire consequence to the entire organism. A six-legged man or a two-legged beetle can no longer happen. Like an open umbrella, each is too far up the chimney.

Six legs and a hard exoskeleton, as it has turned out, was a masterful design. Today insects are the world's most numerous multicelled organisms, numbering at least a million different species, and perhaps as many as three million. The only limitation that their structure puts on them is size. With hard shells, it is difficult for them to breathe by swelling up their bodies to suck in air. They must depend on a natural exchange of air molecules through a great many small holes in their bodies, a swap of oxygen drifting in and carbon dioxide drifting out. Since that exchange can be efficient for distances of only about an eighth of an inch, insect bodies must remain small. The largest moth, the largest beetle known, is about as large as a moth or beetle can be. That also puts a limit on the size of the insect brain. It, too, must remain small; there is no room for it to get larger. Consequently, an insect is a poor thinker; everything it does must be programmed by its genes. It can learn virtually nothing and must continue to do what it is genetically fated to do. Certain caterpillars, for example, are programmed to follow on the heels of other caterpillars. In a famous experiment, a number of them were put on the rim of a bowl. Although there was food only an inch away, the caterpillars trudged around the bowl, head to tail, until all of them starved to death.

Similarly with social insects like bees and ants. They do extraordinary things. Bees go through complicated dances to tell other bees where nectar-bearing flowers are. Ants grow crops, keep slaves, go on hunting expeditions, camp out as thieves among other ants, and so on. That may seem like a well-thought-out variety of activities. But, upon further examination, it turns out that each ant can do its own thing and nothing else. It lives on a narrow treadmill.

It was the descendants of fishes who, when they came ashore, bore with them the potential for something new. Without the corset constriction of an exoskeleton, a land-exploring fish could develop a better breathing apparatus and thus grow much larger than any insect or scorpion could. It could also develop a larger brain. This new creature was an amphibian, the ancestor of the frogs, toads, and salamanders that are still with us. With no more efficient competitors, amphibians became the dominant land animals and remained so for nearly a hundred million years. But they never succeeded in freeing themselves entirely from the water. They mated and laid their eggs there and still do.

When an improved model came along, its principal advance lay in emancipating itself from the water. It was a reptile and could lay its eggs on land, which protected them from a host of water predators. Free to range widely over the land, the reptiles did so, becoming increasingly large and agile. They took over from the amphibians in a most dramatic fashion. Their development culminated in a spectacular group, the dinosaurs, which in their turn were the dominant form of terrestrial life for more than a hundred million years.

A single sentence that notes only the coming and going of the dinosaurs cannot begin to do justice to those animals. They had a complex evolutionary history, branching into the most bizarre forms, with successive bursts and foliations of unexpected variety following each other in ways that no observer could possibly have predicted. They produced by far the largest herbivores the world has ever known. They did it several times, with *Diplodocus, Brontosaurus,* and *Brachiosaurus,* which weighed in at fifty tons, appearing at widely spaced intervals and from different stocks. They produced the world's largest and most formidable predator, *Tyrannosaurus rex.* They produced tanklike armor-plated models, and flying models with twenty-foot wingspreads. They also produced a great number of smaller, speedy, bipedal types. Who would have guessed that this panoply of dominance would crash and vanish for reasons science cannot yet explain satisfactorily? Their only survivors today are the descendants of some of those quick little bipeds. They became birds, and (relatively recently) have enjoyed an explosion of their own, now numbering about nine thousand different species. Some other reptiles less theatrical than the dinosaurs have also survived. They are the present-day lizards, turtles, crocodilians, and snakes. But like the amphibians before them, they have been pushed from center stage and into the wings by a more efficient kind of animal.

That new type — again who could have predicted it — would be a group of small scurrying mammals, as obscure among the towering dinosaurs as mice are among the men and machines of today. They did not "take over" from the dinosaurs by being smarter and eating their eggs, as has sometimes been suggested. They existed side by side with them, again for millions of years. Dinosaurs were probably not as stupid as is commonly thought, nor the first mammals as intelligent. Dinosaurs undoubtedly dined on a great many mammals, and mammals surely scavenged in many ways on dinosaurs. Yet when the crunch came, one type went and the other type stayed.

Mammals had two evolutionary advantages over reptiles. They had a higher rate of metabolism and thus could sustain their own body temperatures, regardless of external conditions. There was a cost to this; it sentenced them to a ceaseless quest for food to stoke the metabolic boilers that kept them going. But the payoff was high. They could live in cold places as well as hot ones, and they could sustain peak action for much longer periods of time than a reptile could. Reptilian life, by and large, is leisurely and lethargic. It is pursued mostly in warm places. No alligator ever seems in a hurry. It will lie in the sun for an entire day without moving and can go for long periods of time without eating. It is capable of short bursts of surprising speed but tires soon. And it must have warmth to quicken its activity. Snakes and turtles are the same. Those that live in cool climates can function well only when the day heats up. In winter they hibernate, unable to move for months at a time.

The second advantage mammals had over reptiles was in their superior strategies in raising young. Everything, of course, comes from an egg. What evolution seems to be telling us is that its so-called upward steps, its improvements, are measured by what is done with the egg once it is formed. Simpler organisms like oysters release millions of eggs into the water, counting on sheer numbers to guarantee that half a dozen will reach maturity. Any hint of parental care is totally beyond an oyster. Amphibians do better. They produce far fewer eggs but they take better care of them. They cluster them in protected spots, often gluing them to underwater objects, sometimes even making efforts to conceal them. Reptiles do still better. Alligators make nests for their eggs, guard them, and even try to regulate the heat coming from the sun by rearranging nest materials. Birds, as reptilian descendants, carry the nest-care activity still further. But all of those strategies include the built-in hazard of having to deposit an egg somewhere, and then expose it to danger for varying amounts of time. Shorten that time (that is, keep the egg safe inside its mother longer), increase the quality of parental care, and there surely will be an evolutionary advantage. That, in effect, is what mammals have done.

The earliest mammals presumably did not do it very well. The most primitive that now survive are the monotremes, of which the duck-billed platypus is the best-known example. Monotremes are classified as mammals because they are fur bearing and give milk to their young. But they still lay eggs. They have not reached the evolutionary point of bearing live young.

An improvement on the monotreme model was the evolution of marsupials, animals like kangaroos that "hatch" their eggs inside themselves but bear them at a very early stage in their development. The newborn babies of an opossum, the only marsupial native to North America, are so small that five or six of them will fit in a teaspoon. They must make their way as best they can from the mother's birth canal to a nipple. Those that don't make it perish.

A further improvement was to keep the infant inside the mother still longer, giving it a better chance of surviving at birth. Better care after its birth was a final and coevolving improvement. Mammals that follow these strategies—placentals—have become the world's most successful. They are the ones we see around us today, from mice to elephants, and we humans are included.

High metabolism has surely proved its value. We need only look around us to see how far the amphibians and reptiles have fallen vis-à-vis ourselves. But is there any proof that a longer gestation period carries an evolutionary advantage? Again, all we have to do is survey the continents to find that the answer is a resounding yes. Mammals appear to have had their origin in a single land mass that included what is now all the major continents. The breakup of that super-continent (called *Pangea*) took place some sixty million years ago, after the development of monotremes and marsupials but before a great explosion of placentals. As a result, Australia went floating off with a freight of monotremes and marsupials but no mammals. In isolation it could act as a competitive arena for the earlier two types. In time marsupials would become the dominant form of terrestrial life in Australia. The less efficient monotremes apparently never were numerous there, because their fossils are as scarce as they are themselves today.

By contrast, the marsupials were everywhere. Virtually every mammalian niche was occupied by a pouched animal, from giant kangaroos that played the role played by large herbivores like sheep and cows elsewhere, down to tiny insectivores. In between were marsupial rats, marsupial anteaters, marsupial wolves, and even tree-dwelling, vaguely monkeylike marsupials.

The superiority of placentals over marsupials is also demonstrated in Australia. When Europeans settled there they brought with them a wide range of placental mammals. These proceeded to ravage the marsupials in devastating fashion. Many Australian marsupials are now extinct. Others exist in drastically reduced numbers. Meanwhile their new competitors are flourishing. To cite a single example, rabbits quickly overran most of the continent after their introduction. Large portions of it were eaten almost bare by hordes of rabbits, which became such a serious ecological threat that they were slaughtered in staggering numbers. Never completely wiped out, they have always rebounded, even though crippling diseases were deliberately introduced in an effort to control their population. Today the rabbit is a fixture in Australia. Its marsupial counterparts are mostly gone.

Simple cause-and-effect explanations for things evolutionary are dangerous. What may seem to be a straight-line relationship often turns out to be a much more complicated one. Nevertheless, if sheer numbers, if total body mass and variety have any relationship to success, then it becomes pretty obvious that there has been an advance from amphibians through reptiles to mammals. The latter have done better. They are more numerous, they occupy a wider variety of habitats. Where there is direct competition, the mammal almost invariably wins. The same progression is seen through the three major types of mammals. Monotremes are having the thinnest time, marsupials do better, placentals are the most successful.

This rule is certainly related not only to gestation but to the fact that mammals have larger brains than either amphibians or reptiles. Any mammal is smarter than any toad. As a result, its arsenal of behaviors is larger. We have discussed the evolutionary danger of having too narrow a life-style. In that context the advantage of a large brain immediately manifests itself. An intelligent animal can often find alternative ways of making a living that a stupid one cannot.

The question we wish to raise here is: Can an over-large brain carry a hidden danger? Is it an overspecialization? The obvious quick answer that many would give is certainly not. Thinking has conferred on us a priceless adaptive advantage. Evolutionarily speaking, we are successful because our ability to think has enabled us to remain physically unspecialized. We are the supreme generalists. We prove it by our ability to live anywhere and make our living in a hundred different ways. We don't grow thick coats; we get them from other animals. We don't grow long necks; we invent ladders. We don't have teeth as big as apes do, nor are we as strong, pound for pound. We don't see as well as hawks. We don't run as fast as any large quadruped. But by our wits, and more recently by the devices we make, we can outperform all of them in every way.

The evolutionary tool that man's brain has enabled him to forge is culture. Today it is evolving rapidly, but its inventor is evolving extremely slowly. Physically humans have not changed perceptibly in at least twenty thousand years. Culturally our development has been profound. And culture, building on itself, moves faster and faster. It took early humans half a million years to progress from the use of fire to the cultivation of crops and the domestication of animals. From there it took them ten thousand years to discover and use metals, a couple of thousand more to learn to read and write, and another thousand to develop explosives, a few hundred to perfect the internal combustion engine, a hundred to tame electricity, a generation to harness the atom, and a decade to put a small computer in the hands of anybody who wanted to make in an afternoon calculations that would otherwise occupy a mathematician for a thousand years.

We have learned our place in the universe and have some clues to its origin and its dimensions. We have learned what we are made of, how we come into being, and why we die. We are beginning to understand the processes by which we get to be the shapes we are, and we are on the verge of being able to change those shapes. In the last decade we have become genetic engineers. Now, with sleeves rolled up, we are in the central office of the factory, redesigning the blueprints. We use bacteria as production centers for useful substances. We are designing better food plants by juggling their nucleotides. Gene splicing is becoming a commonplace. It is only a matter of time before we turn to redesigning ourselves.

Are we up to that?

Let us assume that genetic research is done only under careful supervision in large laboratories, and following agreed-on guidelines, what then? What kind of research will it be? Who will direct it? If it should be discovered that a genetic defect in a human embryo can be cured by a simple genetic transplant, should not that be done? From there it could be only a small step to designing people with richer blood and greater resistance to disease: better people. Should we take that step? What about the trickier problem of emotional stability? Since that, too, has its genetic component, it is not beyond the realm of possibility that an emotional profile could be designed for any individual. But what emotions will be considered the right ones? Who will decide that?

For more than three billion years life has been assembling itself in increasingly complex forms. That process—unpredictable, blundering, and chance-driven as it is—has always been propelled, as Darwin said, by two simple laws: there will be variability in populations; selection will operate on that variability. Perhaps a third law should be noted here: out of complexity come higher states of complexity and unforeseen results. One unforeseen result has been the ability to think. There emerges the exquisite irony that the human brain, the end product of a natural process, is now in a position to halt that process and begin directing it itself. Does that mean that evolution is over? Will Darwin, in the end, turn out to have been wrong? Or will he be right and the brain turn out to be just another dangerous specialization?

At the moment there is no way of answering those questions. The kind of evolution we are now experiencing—cultural—is going at such a clip and its momentum accelerating so rapidly that no one can tell what the long-term results will be, not only for ourselves but for the entire planet. All we can say right now is that our technology is evolving a great deal faster than we are psychologically and emotionally and therefore faster than our ability to handle that technology.

The phenomenon of a culture that is evolving faster than we are forces us to reconsider the question we asked a few pages back: Can an over-large brain that creates such a culture carry a hidden danger? The quick answer, as we said, is that it cannot. However, it is beginning to appear more and more likely that that answer may be wrong.

Culture has never been a serious problem before. The messes humans made in the past were not even recognized by them as problems of their own creation. And they were always parochial or regional. Early dwellers of the Tigris-Euphrates river system, as they extended their irrigation projects to accommodate growing populations, did not know that, by going into the foothills to enlarge their fields, they would ultimately destroy one of the richest lands in the world. By cutting down forests they denuded the hills of all topsoil. Seasonal flooding followed, and over a thousand or two years the entire ecosystem became unstable. Irrigation projects became clogged, productivity collapsed. Rulers, to ease pressures at home, went to war and ravaged their neighbors. Ultimately the Eden of the Bible became the stony waste that is nibbled by goats throughout much of the Middle East today. Greece is another barren land that once was much more fertile than it now is. Italy is another, Spain another. The entire coast of North Africa was a green land in the centuries before Christ. Although changing weather patterns have had something to do with this, it is now becoming clear that the hand of man can nudge the weather. Forests affect local temperatures and thus wind patterns and rainfall. They are essential to the land's ability to hold and store water. Semi-arid areas cannot stand tampering. Deserts grow.

Damage to the earth by human hands has been going on for thousands of years. We recognize it today. One would think that with a better understanding of ecology we would be capable of avoiding similar mistakes, but we do not seem to be able to. Take one example, the Amazon jungle. It is the largest green spot on earth, with the greatest variety of fauna and flora anywhere. This is such a priceless genetic reservoir that any thoughtful person should instantly recognize the necessity of preserving it. But it is being cut down; an area equal to the size of West Virginia is disappearing every year. At this rate, all will be gone by the middle of the next century. Forty percent of it is gone already.

What is going along with the forest is worth a couple of statistics. The entomologist E. O. Wilson counted forty-three species of ants on one tree in the Peruvian jungle; this is equal to the entire ant fauna of the British Isles. In another spot, measuring about two acres, a group from the Smithsonian Institution identified 41,000 species of insects, 12,000 of them beetles. All together, though it represents only about 7 percent of the earth's land surface, that jungle contains more than half of its plant and animal species, many of them cascading into oblivion before they can be discovered.

Biologists and conservationists repeatedly point out that the jungle soil is poor, that crops cannot be grown there for more than a year or two. Now farmers and land developers are discovering the same thing. Their solution is to move on and cut faster, even though professionals predict that the result could be a huge semi-desert whose effect on world climate is incalculable. The Brazilian government listens, wrings its hands, makes plans. But there is great poverty in Brazil and a population problem—and a terrible external debt. The rape of the jungle continues.

It is surely a bore to reiterate all the bad things that humans have been doing to the environment. But we wish to make a point and will run over the disasters fast. First, we should note that the troubles of the past have been almost entirely agricultural, with some long-term climatological side effects. With the industrial revolution an entirely different class of hazards has appeared, and the world has only just begun to wake up to them. They are the danger of nuclear holocaust, pollution of air and water by waste chemicals, the destruction of forests and lakes by acid rain, the danger of radiation due to man-induced changes in the atmosphere, and even the possible heating up of the earth's surface caused by pouring too much carbon dioxide into the sky.

Those menaces are growing. Lethal nuclear wastes are being stacked up in ever-increasing amounts, and we don't know where to put them. There are now more than 63,000 chemical compounds in production in the United States, many of them of proven toxicity. New ones are appearing faster than the authorities can examine them to learn if they, too, are toxic. Many of these chemicals are the waste products of industrial processes and must be disposed of somehow. In 1985, 600 million tons of hazardous waste products were produced in the United States. For years they have been dumped into the ground, into lakes and rivers, and into the ocean. The result has been a serious threat to the water supplies of American cities. The Environmental Protection Agency—the arm of the government that is supposed to look after these matters—estimates that nearly half of U.S. cities with populations of over 10,000, and which depend on groundwater, are now using water that contains pollutants.

We are not keeping up with the problem. The EPA maintains an inventory of hazardous waste sites that need cleaning up. In 1982 there were 15,000 of them. By 1987 the figure had jumped to 27,000. The EPA also keeps a National Priority List, a sort of rogue's gallery of the very worst sites. These represent serious health hazards and are earmarked for immediate cleanup. There are ninety-five sites on this "most wanted" list; so far, fifteen have been attended to.

We are scared, some of us. Most of us prefer to ignore the problem as long as it does not interfere with our lives directly. The trouble is that by the time it does affect us, the situation may be out of control. As long as we are able to, we will continue to regard it as something that can be solved with money and by dumping dirty things somewhere else. But with an immense federal deficit, money is hard to come by. Politicians in the United States are reluctant to raise taxes to pay for cleanup. And yet this nation is a leader in ecological concerns. In the Third World, where countries are trying desperately to industrialize and catch up, the last thing planners, politicians, and promoters are thinking about is cleanup.

The poor countries have a double problem. In addition to being poor they have the highest birth rates and the most pressing population problems. In most of them, in dire fulfillment of the warning made by Malthus two centuries ago, the standard of living has actually fallen during the last two decades. Basic resources are literally disappearing. In Ethiopia a third of the population chronically cannot get enough to eat. In Nepal, right this minute, people are burning up their country's last pieces of wood to cook their food. Soon there will be nothing left. They will have to import charcoal, but as they have no means to pay for it, who will give it to them?

Will the United States? Our national debt is already immense. Going further into debt at home to support hungry people around the world is not politically popular, nor is it sensible. Would it not be better to help emerging countries with nearly hopeless fiscal problems by advising them on how better to control their populations?

Demographers and economists on all six continents are agreed that the world's number one problem today is too many people. Everything in the end comes back to that. One would think that the "civilized" nations, the "educated" nations, the "sophisticated" nations would be smart enough to recognize such problems and do something about them. We like to think of our own nation as one of the civilized, educated, and sophisticated ones. And yet, in 1986 the United States Government warned all birth control and population control agencies in the country that they would lose their tax-exempt status if they did not prevent abortions from being part of the programs that they sponsor overseas. On whatever grounds it is defended, United States policy promotes population increase in the countries that are the most crowded and the hungriest.

We could go on and on. The point we wish to make is that we are filling up the earth and for the first time making it dangerously dirty.

As our culture and technology roar ahead, we fill it faster and get it dirtier. But we do not solve our problems any faster. As they pile up we plaster them over and create new ones. Instead of cleaning up wastes, we make more by producing nuclear warheads, thousands more than are needed to incinerate the planet.

What is wrong with us that we continue to do such idiotic and lethal things when we should know better? We propose an evolutionary answer to that question. At heart we are still primitive people. We evolve far more slowly than our culture. Thus we are a species of an old-fashioned design trying to cope with a new-fashioned world that we have built ourselves. We are discovering that it is easier to invent things than to know what to do with them, or even to predict what they will do to us down the road. We have created a technology of appalling potency, and it is beginning to show signs that it may be out of control. It is running away from our emotions faster than from our wits. Unfortunately we make our most critical decisions out of passion, not out of reason, because in our guts we are passionate stone-age people.

Obviously we cannot prove this statement, nor do we wish to try. All we can offer in evidence is the history of the twentieth century, which may well turn out to have been one of the poorer ones in recorded history, despite all its technological achievements. Culture may dress up the appearance of people, but it seems to be too shallow to affect their nature deeply. In crises our nature reverts. We have always looked good in good times and absolutely awful in bad times. The ancient Greeks were at least as civilized as we are in many ways, though nastier in others. Go back from them for another 2,500 years, and we are deep in the stone age, with people essentially no different in nature from ourselves. For all the tinsel, we change slowly.

For a long time the growing gap between static, primitive emotions and a galloping culture made no real difference. What if a prairie was burned off, a mammoth driven to extinction, or a hundred cities sacked? The world was wide and, barring the mammoth, could recover. During this century it has survived the two most destructive wars in human history and some of the most dreadful massacres of people. The next war, if it goes nuclear, will be different. It could well end it for all higher life. We know that, and yet we are doing almost nothing to prevent it. We have not exorcised fear, wrath, jealousy, or revenge. Those ancient agitators are very strong in us. In crises they take over. They may have been useful to us once as part of our survival kit, but they no longer are and just might provoke an uncontrollable twitch of the finger on the wrong button.

There is a way out of this. It is not more weapons, more treaties, more garbage, more chemicals, or more smog. It is better people. Perhaps the next step in our evolution as a species will be for us to recognize that natural selection of our emotions has been too slow, and that we must speed things up, to keep pace with our culture, through applied genetics. For a species that, as presently constructed, has an abysmal record at solving the simplest social and political problems without violence, to suggest that it can ever get together on self-improvement standards acceptable to everybody may seem silly, an utterly improbable hope. It would seem so for Israeli and Arab, for Irish Protestant and Catholic, for Cambodian and Vietnamese, for black and white South African — actually for all of us when hard choices that touch our deepest emotions have to be made.

Improbable or not, this is certainly a matter that humans will find themselves thinking about more and more as the choices become harder and the decisions wilder. For we are now on the verge of having the scientific skills to do something about it.

If we should succeed in helping ourselves through applied genetics before vengefully or accidentally exterminating ourselves, then there will have to be a new definition of evolution, one that recognizes a process no longer directed by blind selection but by choice.

If not, then the old definition will still do. If we blunder along, getting ever more numerous, dirtier, and madder, we probably will not survive very long. Whether, to paraphrase T. S. Eliot, we go with a nuclear bang or a poisoned whimper will not matter. Either way, an observer from another planet would have to concede — along with Darwin, were he sitting here too — that intelligence was just another specialty that ran out of control. Like every other highly specialized organism, man carried his specialization too far. In the end his brain, which at the time seemed like such a safe and liberating development, did him in. Intelligence, one small blip among millions of other specializations that have wrinkled the line of life, turned out to be just that: one small blip. In its day it may have shaken the earth, as dinosaurs did. But when the big challenge came — change or go — it went. Darwin will have been right after all.

Bibliography

Ardrey, Robert. *African Genesis.* New York: Atheneum, 1961.

Avery, Oswald T., Colin M. MacLeod, and Maclyn McCarty. "Studies on the Chemical Nature of the Substance Inducing Transformation of Pneumococcal Types: Induction of Transformation by a Deoxyribonucleic Acid Fraction Isolated from Pneumococcus Type III," *Rockefeller Institute for Medical Research,* November 1943.

Barlow, Norah, ed. *The Autobiography of Charles Darwin.* London: Collins, 1958.

———, ed. *Charles Darwin and the Voyage of the "Beagle."* London: Collins, 1958.

Beadle, George, and Muriel Beadle. *The Language of Life.* New York: Doubleday, 1966.

Blunt, W. *The Compleat Naturalist: A Life of Linnaeus.* London: Collins, 1971.

Brent, Peter. *Charles Darwin: A Man of Enlarged Curiosity.* New York: Harper & Row, 1981.

Broom, Robert. *Finding the Missing Link.* London: Watts & Co., 1950.

Cairns-Smith, A. G. *The Life Puzzle.* Toronto: University of Toronto Press, 1971.

———. "The First Organisms," *Scientific American,* June 1985.

Cavalli-Sforza, L. L. *Elements of Human Genetics,* 2d ed. Menlo Park, Calif.: W. A. Benjamin, 1977.

Chambers, Robert. *Vestiges of the Natural History of Creation.* London, 1844.

Cherfas, Jeremy. *Man-Made Life.* New York: Pantheon, 1982.

Cherfas, Jeremy, and John Gribbin. *The Monkey Puzzle.* New York: Pantheon, 1982.

Crick, F. H. C., and J. D. Watson. "Molecular Structure of Nucleic Acids: A Structure for Deoxyribose Nucleic Acid," *Nature* 171: 737–738, 1953.

Dart, Raymond. *Adventures with the Missing Link.* Philadelphia: The Institute Press, 1967.

———. "*Australopithecus africanus*: The Man-Ape of South Africa," *Nature* 115: 195–199, February 1925.

Darwin, Charles. *On the Origin of Species.* New York: Atheneum, 1964.

Darwin, Francis, ed. *The Life and Letters of Charles Darwin.* New York: Basic Books, 1959.

Dawkins, R. *The Selfish Gene.* New York: Oxford University Press, 1976.

De Beer, Sir Gavin. *Charles Darwin.* New York: Doubleday, 1964.

Dobzhansky, Theodosius. *Genetics & The Origin of Species.* New York: Columbia University Press, 1977, reprinted 1982.

———. *Mankind Evolving.* New Haven: Yale University Press, 1962.

Dubos, René. *The Professor, the Institute and DNA.* New York: Rockefeller University Press, 1976.

Edey, Maitland. *The Missing Link,* rev. ed. New York: Time-Life Books, 1977.

Eigen, Manfred, William Gardiner, Peter Schuster, and Ruthild Winkler-Oswatisch. "The Origin of Genetic Information," *Scientific American,* April 1981.

Eisley, Loren. *Darwin's Century.* New York: Doubleday, 1958.

Eldredge, Niles. *Time Frames.* New York: Simon & Schuster, 1985.

Eldredge, Niles, and S. J. Gould. "Punctuated Equilibria: An Alternative to Phyletic Gradualism." In *Modern Paleobiology,* T. J. M. Schopf, ed. San Francisco: Freeman & Co., 1972.

Fox, S., ed. *The Origin of Prebiological Systems.* New York: Academic Press, 1965.

Futuyma, Douglas. *Evolutionary Biology.* Sunderland, Mass.: Sinauer Associates, 1979.

Goodman, Morris, and J. E. Cronin. "Molecular Anthropology: Its Development and Current Directions." In *A History of American Physical An-*

thropology, 1930–1980, Frank Spencer, ed. New York: Academic Press, 1982.

Gruber, Howard E. *Darwin on Man.* Chicago: University of Chicago Press, 1974.

Hoagland, Mahlon. *Discovery — The Search for DNA's Secrets.* Boston: Houghton Mifflin, 1981.

Huxley, Sir Julian. *Evolution — The Modern Synthesis.* New York: Harper & Row, 1942.

Iltis, Hugo. *Life of Mendel.* New York: W. W. Norton, 1932.

Irvine, William. *Apes, Angels and Victorians.* New York: McGraw-Hill, 1955.

Johanson, D. C. "On the Status of *Australopithecus afarensis,*" *Science* 207: 1104–1105, March 1980.

Johanson, D. C., and Maitland Edey. *Lucy, the Beginnings of Humankind.* New York: Simon & Schuster, 1981.

Johanson, D. C., and M. Taieb. "Plio-Pleistocene Discoveries in Hadar, Ethiopia," *Nature* 260: 293–297, March 1976.

Johanson, D. C., and T. D. White. "A Systematic Assessment of Early African Hominids," *Science* 203: 321–330, January 1979.

Judson, Horace Freeling. *The Eighth Day of Creation.* New York: Simon & Schuster, 1979.

Leakey, R. E. "Skull 1470," *National Geographic,* June 1973.

Leakey, R. E., and Roger Lewin. *Origins.* New York: Dutton, 1977.

Lyell, Sir Charles. *Principles of Geology,* 3d ed. London, 1834.

Malthus, Thomas Robert. *An Essay on the Principles of Population,* facsimile of 1st ed. London: Macmillan, 1926.

Marchant, James. *Alfred Russel Wallace.* New York: Harper, 1916.

Mayr, Ernst. *The Growth of Biological Thought.* Cambridge: Harvard University Press, 1982.

———. "Isolation as an Evolutionary Factor." Proceedings of the American Philosophical Society. New York: Columbia University Press, 1942, reprinted 1982.

Moorehead, Alan. *Darwin and the Beagle.* New York: Harper & Row, 1969.

Peterson, R. T. *A Field Guide to Birds East of the Rockies,* 4th ed. Boston: Houghton Mifflin, 1980.

Pilbeam, David. "Miocene Hominoids and Hominid Origins," *American Journal of Physical Anthropology* 52: 268, February 1980.

Robbins, Chandler S., Bertel Bruun, and Herbert S. Zim. *Birds of North America.* Racine: Western Publishing Company, 1966.

Sarich, Vincent. "A Molecular Approach to the Question of Human Origins." In *Background of Man,* V. Sarich and P. Dolinhow, eds. Boston: Little, Brown, 1971.

Sarich, V. M. "Molecular Clocks and Hominoid Evolution After Twelve Years," *American Journal of Physical Anthropology* 52: 275–276, February 1980.

Sarich, V. M., and J. E. Cronin. "Molecular Systematics of the Primates." In *Molecular Anthropology,* M. Goodman and R. E. Tashian, eds. New York: Plenum Press, 1976.

Sayre, A. *Rosalind Franklin and DNA.* New York: W. W. Norton, 1975.

Shine, Ian, and Sylvia Wrobel. *Thomas Hunt Morgan.* Lexington: The University of Kentucky Press, 1976.

Simpson, George Gaylord. *The Meaning of Evolution.* New Haven: Yale University Press, 1967.

Stebbins, G. Ledyard. *Darwin and DNA—Molecules to Humanity.* San Francisco: W. H. Freeman, 1982.

Stebbins, G. Ledyard, and Francisco Ayala. "The Evolution of Darwinism," *Scientific American,* July 1985.

Sturtevant, A. H. "Genetic Studies on *Drosophila simulaus,*" *Genetics* 5: 488–500, 6: 179–207, 1920–1921.

Walker, A., and R. E. Leakey. "The Hominids of East Turkana," *Scientific American,* August 1978.

Wallace, Alfred Russel. *My Life.* New York: Dodd Mead, 1905.

———. "On the Law that has Regulated the Introduction of New Species," *Annals and Magazine of Natural History,* 1955.

Watson, James D. *The Double Helix.* New York: Atheneum, 1968.

Weinberg, Robert A. "The Molecules of Life," *Scientific American,* October 1985.

White, T. D. "Evolutionary Implications of Pliocene Hominid Footprints," *Science* 208: 175–176, April 1980.

Wilson, Allan C. "The Molecular Basis of Evolution," *Scientific American,* October 1985.

Woese, Carl R. "Bacterial Evolution," *Microbiological Reviews,* June 1987.

Illustration Credits

Charles Darwin—Neg. No. 326672. Courtesy Department of Library Services, American Museum of Natural History.

Carolus Linnaeus—Neg. No. 334420. Courtesy Department of Library Services, American Museum of Natural History.

Comte de Buffon—Neg. No. 326703. Courtesy Department of Library Services, American Museum of Natural History.

James Hutton—Neg. No. 124665. Courtesy Department of Library Services, American Museum of Natural History.

Jean Baptiste Lamarck—Neg. No. 124768. Courtesy Department of Library Services, American Museum of Natural History.

Thomas Malthus—Neg. No. 326701. Courtesy Department of Library Services, American Museum of Natural History.

Robert Fitzroy—Neg. No. 326705. Courtesy Department of Library Services, American Museum of Natural History.

T. H. Huxley—Neg. No. 326882. Courtesy Department of Library Services, American Museum of Natural History.

Charles Darwin (cartoon)—Mary Evans Picture Library.

Alfred Russel Wallace—Library of Congress.

Charles Darwin—Neg. No. 326698, Photograph by H. P. Robinson. Courtesy Department of Library Services, American Museum of Natural History.

Gregor Mendel—Neg. No. 238141, Photograph by Kirschner. Courtesy Department of Library Services, American Museum of Natural History.

Hugo De Vries—Neg. No. 259823, Photograph by Kirschner. Courtesy Department of Library Services, American Museum of Natural History.

August Weismann—Neg. No. 19723, Photograph by Kirschner. Courtesy Department of Library Services, American Museum of Natural History.

Oswald Avery—Courtesy of the Rockefeller Archive Center.

T. H. Morgan—Columbiana Collection, Rare Book and Manuscript Library, Columbia University.

Calvin Bridges—National Library of Medicine.

James Watson—UPI/Bettmann Newsphotos.

Stanley Miller—UPI/Bettmann Newsphotos.

Carl Woese—Photograph by Alice Prickett.

Robert Broom—Photograph by Phillip Tobias.

Louis Leakey—UPI/Bettmann Newsphotos.

Taung Child skull—Institute of Human Origins.

Skull 1470—Photograph by Bob Campbell, copyright © National Geographic Society, 1973.

Zinj—Institute of Human Origins.

Australopithecus afarensis skull—Institute of Human Origins.

Lucy—The Cleveland Museum of Natural History.

Tim White—Photograph by D. Johanson.

Vincent Sarich—Photograph by Sylvia Hixson, Institute of Human Origins.

Index